Building Europe on Expertise

Making Europe: Technology and Transformations, 1850–2000

Series editors: Johan Schot (Eindhoven University of Technology, the Netherlands) and Phil Scranton (Rutgers University, USA)

Book series overview:

Consumers, Tinkerers, Rebels: The People Who Shaped Europe
by Ruth Oldenziel (Eindhoven University of Technology, the Netherlands) and Mikael Hård (Darmstadt University of Technology, Germany)

Building Europe on Expertise: Innovators, Organizers, Networkers
by Martin Kohlrausch (KU Leuven, Belgium) and Helmuth Trischler (Deutsches Museum, Germany)

Europe's Infrastructure Transition: Economy, War, Nature
by Per Högselius (Royal Institute of Technology (KTH), Sweden), Arne Kaijser (Royal Institute of Technology (KTH), Sweden) and Erik van der Vleuten (Eindhoven University of Technology, the Netherlands)

Writing the Rules for Europe: Experts, Cartels, and International Organizations
by Wolfram Kaiser (University of Portsmouth, United Kingdom) and Johan Schot (Eindhoven University of Technology, the Netherlands)

Communicating Europe: Technologies, Information, Events
by Andreas Fickers (University of Luxembourg, Luxembourg) and Pascal Griset (Paris-Sorbonne University, France)

Europeans Globalizing: Mapping, Exploiting, Exchanging
by Maria Paula Diogo (New University of Lisbon, Portugal) and Dirk van Laak (University of Giessen, Germany)

Initiator: Foundation for the History of Technology (Eindhoven University of Technology, the Netherlands)

The Foundation for the History of Technology (SHT) seeks to develop and communicate knowledge that increases our understanding of the critical role that technology plays in the history of the modern world. Established in 1988 in the Netherlands, SHT initiates and supports scholarly research in the history of technology. This includes large-scale national and international research programs, as well as numerous individual projects, many of which are in collaboration with Eindhoven University of Technology. SHT also coordinates Tensions of Europe (TOE), an international research network of more than 250 scholars from across Europe and beyond who are studying the role of technology as an agent of change in European history. For more information visit www.histech.nl.

Building Europe on Expertise

Innovators, Organizers, Networkers

Martin Kohlrausch
and
Helmuth Trischler

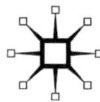

© Martin Kohlrausch, Helmuth Trischler and The Foundation for the History of Technology 2014

All rights reserved. No reproduction, copy or transmission of this publication may be made without written permission.

No portion of this publication may be reproduced, copied or transmitted save with written permission or in accordance with the provisions of the Copyright, Designs and Patents Act 1988, or under the terms of any licence permitting limited copying issued by the Copyright Licensing Agency, Saffron House, 6–10 Kirby Street, London EC1N 8TS.

Any person who does any unauthorized act in relation to this publication may be liable to criminal prosecution and civil claims for damages.

The authors have asserted their rights to be identified as the authors of this work in accordance with the Copyright, Designs and Patents Act 1988.

First published 2014 by
PALGRAVE MACMILLAN

Palgrave Macmillan in the UK is an imprint of Macmillan Publishers Limited, registered in England, company number 785998, of Houndmills, Basingstoke, Hampshire RG21 6XS.

Palgrave Macmillan in the US is a division of St Martin's Press LLC, 175 Fifth Avenue, New York, NY 10010.

Palgrave Macmillan is the global academic imprint of the above companies and has companies and representatives throughout the world.

Palgrave® and Macmillan® are registered trademarks in the United States, the United Kingdom, Europe and other countries

ISBN: 978–0–230–30805–3

This book is printed on paper suitable for recycling and made from fully managed and sustained forest sources. Logging, pulping and manufacturing processes are expected to conform to the environmental regulations of the country of origin.

A catalogue record for this book is available from the British Library.

A catalog record for this book is available from the Library of Congress.

Contents

Making Europe: An Introduction to the Series   ix
Acknowledgements   xvii

Introduction   1

Part I   Cultivating Experts, Ordering Knowledge

1. **Educating Experts**   21
   Technical Education as a Strategic Issue   28
   Copying & Circulating Educational Models   32
   The Quest for Respect   39
   Engineers as Experts   50

2. **Technical Experts as New National Elites**   55
   Old Regimes & New Elites   59
   Heralds of the Nation   65
   Reinventing the Nation in War & Crisis   70

3. **Architectures of Knowledge**   79
   Orders of Knowledge: Mundaneum   87
   Visual Education: Isotype   97

Standardizing Knowledge: Wilhelm Ostwald ... 102
The Social Promise of Technology ... 107

Part II Endangered Experts, New Social Orders

4. Expertise with a Cause ... 115
Rationalizing the Factory & Beyond ... 125
Transatlantic Fordism ... 132
Self-Empowerment & Social Engineering ... 139

5. Faustian Bargains in Totalitarian Europe ... 143
Hitler's Mastermind of Annihilation: Konrad Meyer ... 147
Best Man: Guglielmo Marconi & Benito Mussolini ... 154
Stalin's Rainmaker: Trofim Lysenko ... 161
Experts in the Age of Extremes ... 169

6. Experts in Exile ... 177
Stalin's Operation Osoaviakhim ... 177
Intellectual Reparations ... 179
Émigré Experts as Agents of Modernity ... 187
Émigré Experts as an Endangered Species ... 195
Forced Migration & New Knowledge Orders ... 197

Part III Cooperating Experts, Building Institutions

7. Geographies of Cooperation in Nuclear Europe ... 205
The European Super Machine ... 205
Europe's Model of Technoscientific Collaboration ... 208
Ways of Collaborating in Nuclear Europe ... 216
Breeding & Fusing Europe ... 229
The Multiple Geographies of Europe ... 238

8. Contesting Europe in Space ... 243
Space Europe Eastern Mode: Interkosmos ... 243
A Fragmented Europe: ELDO ... 247
Space Europe Western Mode: ESA ... 260
Towards European Integration in Space ... 264

9. Experts' Europe from a Bird's-Eye View ... 277
Lisbon 2000: Building Europe on Knowledge ... 277
Flying High: Is there a European Knowledge Society? ... 280
Flying Low: Are there European Expert Communities? ... 287

Conclusion 299

Endnotes 313
Bibliography 335
Illustration Credits 369
Making Europe: Series Acknowledgements 375
Index 377

Making Europe:
An Introduction to the Series

In a typical conversation about twentieth-century European history, the subject of war will almost certainly arise—whether it is the Great War, the Second World War, or the Cold War. Similarly, historians who write about contemporary European history often view war as the twentieth century's iconic event. In fact, many scholars rely on Europe's political history, rife as it is with military conflict, to set the timeframe for their work. The influential historian Eric Hobsbawn, for example, defined the twentieth century as beginning with the First World War and ending with the collapse of the Soviet Union; Hobsbawn named this period—1914 to 1991—The Short Twentieth Century. Indeed, the topic of war and rupture has dominated the discourse on Europe in the twentieth century—and understandably so.

We, the editors and authors of the *Making Europe* series, however, have taken an alternative approach to our subject. We offer a European history viewed through the lens of technology rather than war. We believe that a European history with technology at its core can help to understand the continuities that have endured despite the rupture of wars. *Making Europe* places continuities—from the rise of institutions like CERN to the evolution of hacker networks—in a longer-term perspective. The *Making Europe* narrative suggests that recent European history is as much about building connections across national borders as it is about playing out conflicts between nation states. This view of technology from a transnational perspective has proven to be felicitous. As a phenomenon, technology has always been particularly mobile; this mobility has allowed new technologies to help shape international relations between countries, companies, organizations, and people.

To understand the role of technology in this history, we required ourselves to rethink the very meaning of technology: referencing far more than machines alone, technology also embraces people and values; ideas, skills, and knowledge. Technological change, in our view, is a deeply human process. Technology was—and still is—central to the creation of Europe. And given its centrality, technology has been hotly contested—politically, economically, and culturally—in the making of Europe.

Technology's role in shaping Europe coalesced around 1850, when a new era began, an era from 1850 to 2000 that we refer to as The Long Twentieth Century. It was during the mid-nineteenth century that a newly globalizing world began to emerge. This was a world in which the many new transportation and communication technologies played a decisive part. At this time, technology became a reference point for European superiority—both within and beyond Europe. Cross-border connections and institutions thrived; the knowledge-sharing practices that fostered these connections were widely circulated and adopted. This circulation of knowledge led to a worldwide imagining, negotiating, and experiencing of Europe that exists today. This was also the foundation for the formal process of European integration that gained traction in the 1950s. Our perspective simultaneously decenters the European Union and its direct predecessors—which, after all, comprised only one force of Europeanization—and places the process of European integration in long-term historical context. Acknowledging that this dynamic of integration continues today, *Making Europe* presents and interprets a history that is still in the making.

That said, it is clear to us as historians that the decade 1990–2000 marks another watershed: it was in this period that the digital revolution gained new momentum, as did shifting power relationships at the global level. This spurred the European Union to become a hegemonic force of Europeanization, and it helped globalization to enter a new phase. Simultaneously, however, the processes of integration and globalization in this apparent new phase have proven to be fragile: in light of the global economic crisis, Europe's future, called into doubt, has become a pressing issue, and one with a sharp political edge. Accordingly, Europe's past has also come under fresh scrutiny. We contend that technology will continue to play a central role in defining Europe; that the politics of Europe is the politics of technology as much as anything else; and that now is the opportune time to explore technology's historical role in the creation of Europe.

Making Europe provides a perspective on European history that transcends borders. The volumes in the series examine the linking—and, in some cases, the disruption—of infrastructures and knowledge networks that operate beyond nations and states. Also mapped here is the transnational circulation—and appropriation—of people, products, and ideas. The people and organizations featured in this series employed particular notions of Europe in building their cross-border connections. Indeed, they imagined and invented new Europes, often making clear distinctions between which people and places belonged and which were alien to the concept and the reality of Europe. *Making Europe* asks: Who projected their ideas of Europe? When did these projections take place—how, and why? The series looks at the people and the organizations that perceived themselves as central—and peripheral—to Europe, its colonies, and the transatlantic crossings that were part of the European imagination. Examined here are migrants and experts, foods and inventions, markets and regulations—virtually everything that was identified, experienced, and communicated as "European." This Europeanization, we find, had significant—and sometimes unintended—consequences: some connections between people and institutions were lasting, others broken, these continuities and ruptures shaping Europe as both an imagined place and a living community. *Making Europe* explores the stability and fragility of these European connections, communities, and institutions.

The majority of existing studies of Europe have been based on either of two approaches. First is the, often massive, single-author narrative. Second is the essay collection, which presents many voices, in some cases edited to align the authors' themes. In the field of European history, single-author volumes have tended to be broad-ranging and to address different timeframes and regions.

Often, single-author volumes are a compilation of national stories; at their best, compilations transcend their individual stories to posit a complete European picture. Essay collections, for their part, have generally assumed a sharper focus—on particular communities, ethnicities, and empires, for example. These usual approaches point to a distinctive feature of *Making Europe*: in this series, five of the six volumes have two authors; one book has three writers. These voices, thirteen in all, create multiple narratives. The six sets of *Making Europe*'s co-authors have worked as a team to draft a series of volumes with coordinated yet individual themes (see www.makingeurope.eu). These six volumes contain six distinct points of view; as editors, we have imposed neither uniformity nor the pressure to harmonize narratives. In our opinion, the most informative new contributions to European history embrace diverse actors and diverse meanings, a range of purposes and understandings. *Making Europe* captures this diversity, reflecting a dynamic European history that continues to unfold.

All of the authors in the series have drawn on the European Science Foundation's "Inventing Europe" collaborative research initiatives as well as the Foundation for the History of Technology's "Tensions of Europe" project, begun in 1998 (see www.tensionsofeurope.eu). They have profited from an intensive period of discussion and joint research and writing at the Netherlands Institute for Advanced Study in the Wassenaar dunes in 2010–11. The fruits of these initiatives include the *Making Europe* book series as well as a web-based exhibit "Inventing Europe, European Digital Museum for Science and Technology" that encompasses a dozen of Europe's technology and science museums (see www.inventingeurope.eu) and scores of scholarly publications. All aim to promote creativity in fostering a more inclusive understanding of technology's role in refashioning Europe—an ongoing process that is as fascinating as it is contentious. The authors of *Making Europe* have asked themselves what shape an open-ended European history of technology would take. They provide their answers in the form of this book series.

The first volume in the *Making Europe* series, entitled *Consumers, Tinkerers, Rebels: The People who Shaped Europe*, is written by Ruth Oldenziel and Mikael Hård. This volume spotlights the people who "made" Europe by appropriating and consuming a wide range of technologies—from the sewing machine to the bicycle, the Barbie doll to the personal computer. What emerges is a fascinating portrait of how Europeans lived during The Long Twentieth Century. Explored here are the questions of who, exactly, decided how Europeans dressed and dwelled? Traveled and dined? Worked and played? Who, in fact, can be credited with shaping the daily lives of Europeans? The authors argue that, while inventors, engineers, and politicians played their parts, it was consumers, tinkerers, and rebels who have been the unrecognized force in the making of Europe.

The second volume in the series, entitled *Building Europe on Expertise: Innovators, Organizers, Networkers*, is written by Martin Kohlrausch and Helmuth Trischler. Here the focus shifts from consumers of technology to a new breed of professionals: the technical and scientific experts whose influence soared from around 1850 onward. The authors show how these experts created, organized, and spread knowledge—enabling them to shape societies, create cross-border connections, and set political agendas. During Europe's Long Twentieth Century, technoscientific experts became a strategic resource for serving national, international, and transnational interests, the authors argue. They revisit experts' visions of Europe, showing how these visions manifested in the dictatorships of Nazi Germany and Stalinist Russia—as well as helping to build Europe's vast research networks during the Cold War. *Building Europe on Expertise* ends with today's efforts to reinvent the European Union—as a knowledge-based society defined by experts.

The third volume in the series, *Europe's Infrastructure Transition: Economy, War, Nature*, is written by Per Högselius, Arne Kaijser, and Erik van der Vleuten. This book elaborates on the first two volumes by introducing a new cast of historical actors: system-builders. These individuals and organizations helped to transform Europe by envisioning, constructing, and manipulating large-scale transport, communications, and energy systems. Their efforts reshaped Europe as a geographical entity by forming massive new material interconnections—and divisions—between places. This had far-reaching implications for European integration; for peaceful economic exchange; for military planning and logistics. System-builders challenged Europe's natural barriers, from the Alps to northern Europe's forests and the vast marshlands to the east. But

Europe's water, air, and land were not only connected, they were transformed radically, sometimes destroyed. In response, system-builders eventually turned much of Europe's environment itself into infrastructure, interlinking isolated ecosystems via human-made corridors and networks.

The fourth volume, *Making the Rules for Europe: Experts, Cartels, International Organizations* is written by Wolfram Kaiser and Johan Schot. Here, the focus becomes the norms and standards of technological innovation—discussed in depth for transport and heavy industry. Featured are the people and organizations that debated, negotiated, and regulated the cross-border issues raised by innovation. Presented here are individuals with special—and often interdisciplinary—expertise in technology, business, and law. Often, these experts sought to de-politicize issues by deeming them technical; this yielded workable solutions to shared problems. It also paved experts' way in rule-making for multiple, distinct yet overlapping, and frequently competing "Europes." In the pursuit of finding technological solutions, many institutions' transnational practices survived ruptures, including the two World Wars. After the Second World War, the European Union was obliged to accommodate—and to compete with—other institutions' established practices in order for the EU to gain greater influence in shaping Europe.

The fifth volume, *Communicating Europe: Technologies, Information, Events,* analyzes Europe's information and communication systems from roughly 1850 onward. Authors Andreas Fickers and Pascal Griset place these technologies at the very heart of European society. Presented here is a global vision of media, telecommunications, and computers that reveals the tensions inherent in designing and appropriating electrical and electronic devices. The authors argue that the control in the material realm by research and entrepreneurship and the emergence of new forms of creativity and new ways of life are two sides of the same coin, mostly driven by political and cultural forces. Examined in this volume are the political, economic, and cultural realities and meanings of information and communication technologies on a European level. This perspective, which extends over the long term, provides the tools for a new critical understanding of the digital revolution.

How did today's globalized, thoroughly mapped-out world emerge? What part did technology play in Europe's international encounters, colonial and otherwise? *Europeans Globalizing*, written by Maria Paula Diogo and Dirk van Laak, concludes the *Making Europe* series with a study of how Europe interacted with the rest of the world from 1850 until the close of the twentieth century. The volume details how technologies were applied and creatively adopted–from India to Argentina, South Africa to the Arctic. From the turn of the twentieth century onwards, we witness assumptions about Europe's technologically-based superiority being continuously challenged. And we discover that globalized Europe in its present form looks quite different from what Europeans once imagined.

Consumers and tinkerers; engineers and scientists; system-builders and inventors. Experts in technology, law, and business; communicators and entrepreneurs; politicians and ambassadors. This is a cross-section of the actors represented on *Making Europe*'s pages. These actors, through the institutions and organizations they cultivated, the connections they created, the rules and practices they fostered, co-created Europe. Narrated from contrasting as well as complementary viewpoints, the six volumes in the series create a collage of co-existent portraits that depict Europe's Long Twentieth Century; its technologies; and its meanings. Together, these histories form the view of modern Europe that we and the authors wish to contribute to the historical record at this time.

Johan Schot & Philip Scranton
Making Europe Series Editors
Amsterdam, the Netherlands & Camden, New Jersey, USA
July 2013

Acknowledgements

This book covers a wide range of topics—chronologically, geographically, and thematically. Different fields of history—technology, knowledge, politics, culture, social change—inform our argument. Covering such a wide array was possible only because we have been able to rely on the outstanding research of other scholars and the manifold support of our colleagues.

In the first place we ought to pay tribute on the inspiration we received within the "Tensions of Europe" network, a network of historians interested in placing the history of technology in a wider context. In particular we would like to mention those members of the group who, as authors in this series, engaged in numerous discussions with us, read the chapters we produced at an earlier stage, and jointly contributed to forming an idea behind the series which again guided our book: Maria Paula Diogo, Andreas Fickers, Pascal Griset, Mikael Hård, Per Høgselius, Arne Kaijser, Wolfram Kaiser, Dirk van Laak, Ruth Oldenziel, Johan Schot, and Erik van der Vleuten.

Exchange was particularly fruitful during five intense and rich months in the seclusion offered by the Netherlands Institute for Advanced Study in the Humanities and Social Sciences (NIAS) at Wassenaar. Before, during, and after our stay in the Dutch Dunes, numerous conferences and workshops involved many other scholars in the project, ensuing additional constructive criticism, support, and further ideas. Several foundations and institutions have generously contributed to financing the organization of and the trips to such events. In addition, they have supported parts of the research on which this book is based. Below, we list the collective bodies and individuals without whom this book could not have been written. We hope the list is complete; in case the authors have forgotten anyone, please accept their sincere apologies.

We received conceptual advice, references, and insight on numerous topics relevant for our book from our colleagues, both within and outside the "Tensions of Europe" network: Gerard Alberts, Alec W. Badenoch, Stefan Beck, Katja Bruisch, Robert Bud, Paul Erker, Rüdiger Graf, Dagmara Jajeśniak-Quast, John Krige, Veronika Lipphardt, Karl-Erik Michelsen, Matthias Middell, Thomas J. Misa, Gijs Mom, Kiran K. Patel, Pieter Raymaekers, Malte Rolf, Pierre-Yves Saunier, Mathieu Segers, Kees Somer, Katrin Steffen, Kilian J.L. Steiner, Aristotle Tympas, Antonio Varsori, Hans Weinberger, Stefan Wiederkehr, and all members of the MOSA-research unit at KU Leuven. David Freis and Dana Mechelmans provided research assistance.

Brenda Black, James Morrison and, towards the end of the process, Lisa Friedman, gave us valuable assistance in creating a smooth text. Philip Hillyer did the final editing of the text. Susan Boobis compiled the index.

Five anonymous reviewers commented on the entire manuscript. The authors are extremely grateful for their input. Numerous workshops within the Tensions of Europe framework gave us ample opportunity to discuss our work. We would also like to thank departments outside the framework for their critical feedback: the Utrecht Seminar Political History (Liesbeth van de Grift), the Center for Comparative European Studies at the University of Cologne (Jakob Vogel), the Munich Center for the History of Science and Technology (Ulf Hashagen, Ulrich Wengenroth, and Karin Zachmann), and the University of Vienna's program on the Sciences in Historical, Philosophical and Cultural Contexts (Mitchell G. Ash).

Of the archivists and librarians who assisted the authors, special thanks go to Deutsches Museum (Munich), CIAM-archive (Zurich), and NAI-archive (Rotterdam).

We believe this book would not work without the images, illustrating and explaining its arguments. In retrieving these pictures we received invaluable support from the image assistants Katherine Kay-Mouat, Jan Korsten, and Slawomir Lotysz, as well as the research team of the virtual exhibit http://www.inventingeurope.eu/.

We express special thanks to the contributors and co-editors of special journal issues and edited books of direct importance to the volume. Previous special issues and essay collections which helped to build our argument were:

- "Tensions of Europe: The Role of Technology in the Making of Europe." Special issue of History and Technology 21, no. 1 (2005).

- "Wissensgeschichte als Gesellschaftsgeschichte." Special issue of Geschichte und Gesellschaft 34, no. 4 (2008).
- "Technological Innovation and Transnational Networks: Europe between the Wars." Special Issue of the Journal of Modern European History 6 no. 2 (2008).
- Martin Kohlrausch, Katrin Steffen, and Stefan Wiederkehr, eds., Expert Cultures in Central Eastern Europe: The Internationalization of Knowledge and the Transformation of Nation States since World War I (Osnabrück: fibre, 2010).

We also wish to express our gratitude to the organizations that contributed financially to making this book possible: the European Science Foundation (ESF), the Foundation for the History of Technology (SHT), the Netherlands Institute of Advanced Studies (NIAS), the German Research Foundation (DFG), the German Federal Ministry of Research and Education (BMBF), and the Volkswagen Foundation. Our time at the NIAS was certainly the most productive phase of writing the book. Only because of the selfless support of the institutions we work at, did it become possible to stay for a semester in Wassenaar. Ulf Hashagen filled in for Helmuth Trischler at the Deutsches Museum, Wolfgang M. Heckl, General Director of the Deutsches Museum, granted a leave of absence, and Christof Mauch took the burden of heading the Rachel Carson Center for Environment and Society alone during this period. Martin Kohlrausch would like to thank Andreas Fickers for the inspiring rides back and forth to Wassenaar.

Finally, two people deserve a special mention: Johan Schot—without whose original approach and enthusiasm this book, and indeed the whole Making Europe series, would never have come about—and Phil Scranton, whose experience helped enormously to separate our more promising ideas from the less convincing ones and to eventually produce a readable manuscript. We can hardly think of more inspiring and supportive editors for such a venture. The very abilities of the two to pull us to distant places, to have us inquire into new directions, and to make us stay on track once promising paths had been discovered made for an intense and highly inspiring collective work experience, which in this form the two of us had not experienced until now. For our families, however, it also meant that we were less present than one would wish. Not only for this reason do we dedicate this volume to our wives and children: Anna, Artur, and Edgar, Andrea, Yannick, and Nicola.

Martin Kohlrausch and Helmuth Trischler
Leuven and Munich, August 2013

Introduction

In the beginning the scene was set in London. It was 1851 in Britain's booming capital, epicenter of the industrializing world. Technologies abounded, from elaborate railroad systems to the telegraph and steamship. Proud of its "progress," Britain took the lead in showcasing the latest technical achievements: a spectacular public exposition was planned, international in scope—the first World's Fair. Innumerable experts would contribute to creating both the fairgrounds and the thousands of individual displays within the show's massive main structure, the Crystal Palace. The event's official name was the Great Exhibition of the Works of Industry of All Nations. Ultimately, 6 million people would tour the exhibition. And millions more soon participated in the politics of comparison that this iconic event ushered in: the Great Exhibition provided the ideal forum for societies and nations to examine and compare their technologies and scientific expertise, measuring their power, their material wealth, and their lifestyles in the process. The event was a forum for nationalistic displays of technology, as well as for international technical cooperation.

In the decades preceding the Great Exhibition, poverty and social upheaval had rocked Europe. Salvation from these crises was promised by technical inventions and economic innovations,

Fig. 0.1 Europe and the World in One Showcase: *The globe depicted as a small spot at the heart of the British Empire. In London 1851 not only were global attractions exhibited, but also the people of the world, or at least of Europe, gathered in their millions, experts and lay visitors alike, to see what was on display.*

both of which were to be offered in abundance at the fair. When it opened in Hyde Park in May 1851, the Exhibition was an immediate sensation. The main building, comprised of cast iron and plate glass, represented a monument to technical expertise. Constructing the Crystal Palace had employed the ingenuity of architects, designers, and many variously trained engineers. Within the building, attentive visitors were drawn to the engineering displays: here were 100,000 objects—chiefly artifacts of technology and science—arrayed in exhibits organized along national lines. The stunning French section, for example, featured sumptuous tapestries and porcelain, fine silks from Lyon, and glossy Limoges enamels—along with the sophisticated machines used to produce them. The Russian exhibit impressed with luxury products including silver and ornate furniture. (The exhibition's complex logistics nearly overwhelmed the Russians, who arrived late due to ice blocking the Baltic Sea.) Experts themselves also turned out in force: many innovators demonstrated their inventions to colleagues and an eager public. British physicist Frederick C. Bakewell, for example, displayed his image telegraph, a precursor of the fax machine. American firearms manufacturer Samuel Colt, creator of the six-shooter, was on hand to demonstrate his brand new Navy revolver.

Despite being dismantled after the fair, the Crystal Palace had lasting impact: the location continued as a sustainable exposition site of international repute. And the exhibition's enormous profits funded London's Victoria and Albert Museum, as well as the Science Museum and the Natural History Museum that later evolved from the South Kensington Museum. Significantly, the remaining funds were channeled into fellowships for conducting industrial research.[1] The Great Exhibition succeeded in portraying technology as a key to a better future, inscribed in Europe's collective memory as a cosmos of technical creativity.

Historically, the Great Exhibition stands out as the nineteenth century's first large-scale experiment in knowledge exchange among experts.[2] At the Crystal Palace, professionals from every conceivable technical field could study state-of-the-art machinery—and take their new discoveries back home to fertilize their nations' efforts at industrialization. With the Crystal Palace the long and intricate process of sharing knowledge began in earnest, on an international scale. The 1851 London exhibition served as a welcome prototype: a forum in which experts could meet and discuss current issues of transnational significance.

Many succeeding expositions advanced the process of knowledge exchange. For example, the Paris *Exposition Universelle* of 1889, symbolized by the Eiffel Tower, provided the framework for as many as 87 of the 97 international scientific and technical congresses held that same year. In fact, more international organizations were formed in the single decade after 1850 than in the entire half-century before.[3]

During the second half of the nineteenth century, from the era of the London fair to the Paris exhibition and beyond, an internationalist dynamic was strongly in play. Specifically, experts worked to create universal standards of weights and measures, money and time. Their efforts to standardize telegraph systems and railroad gauges aimed at overcoming local, regional, and national barriers. Tellingly, the metric system began to receive broader acceptance only in the wake of the Great Exhibition, when visitors from the industrial and commercial sectors became sensitized to the metric system's technical and economic advantages. The drivers of standardization were mainly European technical specialists: they set up international regimes to supervise the new standards. These organizations included the *Conférence Générale des Poids et Mesures* (General Conference of Measures and Weights) and the *Bureau International des Poids et Mesures* (International Office of Weights and Measures), in Sèvres, France.[4]

Internationalism—in this case, the push to standardize systems—was a predominantly European project, which can only be understood against the background of rising nationalism. The existing so-called "standards" were strongly imbued with national pride. The meter, for example, was proudly seen as a French product—each European region tried to promote its own "standard" of measurement. To enable infrastructures to work, however—for international trains to run smoothly and on time, for example—references to everything from time to railroad-track gauges had to be unified, standardized. And London's 1851 event offered an arena for the politics of comparison that necessarily accompanied this standardization. At the Crystal Palace, numerous expert commissions studied the proposed best practices on display and communicated their assessments back home. How to determine technical advances and deficits—this question naturally fell to experts to answer. Similarly, catching up, and keeping up, technologically with other countries and regions became an important element in creating political legitimacy. Those experts who grasped the rules of the political game were those who triumphed.

International exhibitions were also an inherently European undertaking (although the U.S. quickly took a key role in the business of exhibiting). Europeans initiated the politics of comparison; it was a meaningful, valuable endeavor, as long as both nations under scrutiny were viable competitors. Being rivals and exchanging expertise were not mutually exclusive; they were, in fact, intertwined. Moreover, the "world's" fairs reflected a still predominantly European—albeit rapidly globalizing—world. They testified to the hierarchical organization of expert exchange, with Europe, the U.S., and, to a lesser extent, British colonies at its center. The rest of the world, with the exception of Japan, remained largely excluded from the expert networks that participated in these expositions.

How technoscientific experts shaped Europe—and how Europe in turn shaped these experts—is the subject of *Building Europe on Expertise*. During The Long Twentieth Century (1850–2000), experts

Fig. 0.2 **Knowledge Exchange versus Protection:** *Not every exhibit at the London World's Fair was as splendid as the most adored French one and not every pavilion openly showcased recent innovations based on the latest engineering knowledge. This lithograph (original in color) depicts the rather meager North German exhibit with its miscellaneous assortment of raw materials, manufactured goods, and works of art. When Queen Victoria visited the exhibit on May 19, 1851, she attributed the relatively poor display to "that foolish Protectionist feeling, which will do no one but themselves harm."*

became a huge force in fashioning the European landscape, both physically and intellectually. How did experts acquire that power? And what, exactly, did they do with their authority? In this volume, we the authors—historians who share an interest in the formative role of knowledge in the modern world—show how experts in technology and science catalyzed European development after 1850. We call the main actors in this narrative "technoscientific" experts, referencing the inextricable tie between science and technology in the modern age. They became a key resource for nation-states, for business and industry, and for societies. Experts fostered industrialization; supported urbanization; and played a key role in democratization—both for and against it. We focus on experts' role in both the physical building of European nation states as well as Europe's knowledge building: creating the cross-national networks, organizations, and institutions that structured Europe's knowledge base. We explore when, why, and how experts became aligned with emergent nation states. We also examine how experts simultaneously worked for their individual nations *and* maintained a decidedly European outlook. We argue that experts' nationalism and their international mindsets can be fully understood only in reference to one another.

Building Europe on Expertise documents the vital role of technoscientific experts in Europe's Long Twentieth Century; we portray experts as innovators, as organizers, and as networkers. We examine the ways in which nation states sought to harness experts' knowledge for the purpose of serving particular national interests—of extending the nation's "place in the sun," in the words of German chancellor Bernhard von Bülow. We also analyze the counterpoint to this nationalism: the *trans* nationalism which linked European countries to one another—and connected Europe to the rest of the world. This transnationalism was at times voluntary and deliberate, but it was often a mere byproduct of experts' various projects. This theme of inextricable nationalism and transnationalism pervades every chapter of this book. Our claim is that the building of Europe's individual nation-states and the creation of cross-national networks, organizations, and institutions—also referred to as Europeanization and European integration—are interconnected processes, both featuring technoscientific experts as the main actors. We challenge the view that nationalism and transnationalism were opposing forces during The Long Twentieth Century.

Predictably for a volume in the "Making Europe" series, *Building Europe on Expertise* focuses on the European continent. Perhaps surprisingly, though, we do not confine the concept of Europe to the atlases and maps that demarcate the "European" nations of the Eurasian continent, nor do we limit Europe to the Western part of the Continent. As in the other volumes in this series, our concept of Europe transcends traditional geography in order to map the multiple—and often overlapping—contours of expertise. In certain instances, our narrative of European history encompasses the United States and (Soviet) Russia, Turkey and Israel, for example—the many places in the world that offered platforms for European expertise in action. In this narrative, the term "European" not only stretches the usual geography but also includes social meanings, cultural connotations, and sociopolitical identities.

We apply "European" to *individuals* like the renowned Italian inventor Guglielmo Marconi as well as the unknown engineer who helped to build the gigantic Dnieper Dam in Stalinist Russia. "European" pertains to an *actor category*: those individuals in Europe who understood themselves to be members of "homo Europaeus"—the imagined European.[5] With this broader interpretation of

Fig. 0.3 The Rules of the Game called Politics of Comparison: *The Eiffel Tower testifies both to the iconographic clout of the world exhibitions—in this case Paris 1889—and the new standing of engineers. Gustave Eiffel's famous tower also serves as a perfect illustration for the politics of comparison. Nothing seemed more appropriate to demonstrate Bohemia's progress in an exhibition of 1891 in Prague than to build a—reduced—copy of the Paris original.*

"European" in mind, we speak of European experts as historical actors. "European" is also used to describe *organizations* and *institutions* such as CERN (*Conseil Européenne pour la recherche nucléaire*), the cross-national laboratory for particle physics. Various *places* are also subject to the term "European"—places like the European equatorial spaceport near Kourou, French Guiana. Yet another category deemed "European" highlights *artifacts* of technology and science. Among these are mundane objects, such as the thousands of index cards used by the *Office International de Bibliographie* (International Institute of Bibliography) in an attempt by Belgian Paul Otlet to organize the world's knowledge. In this category we also find objects that are highly symbolic, such as the space launcher Europa 1. In some contexts, places and artifacts fuse into what we call European technoscientific sites of memory. One example is the place called Peenemünde, the German army's research center where Wernher von Braun and his team designed the deadly V-2 rocket for Hitler.

What, exactly, is meant by "technoscientific experts"? This group includes all manner of scientists, from chemists to nuclear engineers to space scientists, for example. Experts also include trained professionals who, while not scientists per se, drew on scientific principles in their work. This encompasses everyone from engineers to architects, agriculturalists to urban planners. During The Long Twentieth Century, science and technology grew ever more entwined, yielding "technoscience"; the experts in question operated from knowledge that was mainly technoscientific. Often, these professionals were able to convert their expertise into political and societal influence; their specialized knowledge was accepted as relevant expertise. Yet despite this, the outcome of their efforts was often beyond their control. Experts were not entirely self-motivated and self-created; rather, they were shaped by their interactions with political and social actors who required experts' specific knowledge. The ever-more-important public, too, shaped experts. The London exhibition, with its millions of visitors, illustrated the power of the masses to confer status and influence on experts. By contrast with "specialists" and "professionals," experts were seen to possess capabilities reaching beyond their formal credentials. Equally important were personal character traits and individual reputations. For this reason, we emphasize the inner workings of expert groups, focusing on their internal knowledge bases, including their specialized—and in some cases unique—expertise. Examined here are the questions of how experts were educated; how they used their expertise; and how they built institutions. Also revealed is how they communicated with technoscientific communities—and with those in the political, economic, and social realms. All of this tells us as much about the historical trajectory of expertise as about the changing European societies in which experts operated.

We use the term "expert" primarily because it embraces various kinds of people having a breadth of knowledge—people well-integrated in European society. But although knowledge was diverse, experts were recruited from a narrow, elite group during most of The Long Twentieth Century. Access to this elite was long governed by legal and political circumstances, social and economic constraints, as well as by gender barriers. For example, until well into the 1900s, women were denied the right to vote and largely banned from receiving higher education. Meanwhile, the international exchange and communication required of experts called for costly travel, which only a small group of Europeans could afford. Large gaps in economic development between European nations underscored the problem of limited geographical access to experts; the knowledge community was concentrated around the Continent's centers rather than scattered across its peripheries.

The decision to start this narrative at roughly 1850 yields the question: is the rise of experts an essentially modern phenomenon in European history? If so, what distinguished the era of London's Great Exhibition as a watershed for knowledge exchange? In fact, during the early modern era (circa 1500 to 1800), Europe was already connected by a well-developed web of technical knowledge. A small elite with special knowledge traditionally traveled across Europe and disseminated technical innovations with remarkable speed, given the communication and transportation technologies of the time. Gutenberg's invention of printing, for example, spread throughout Europe, from Italy to Sweden and from Spain to Poland—within only a few decades. By the start of the sixteenth century, some two hundred print shops had been established, the high investment costs for printing technologies notwithstanding. Innovations in military and civil engineering spread with equal speed. This led to technical and economic problem-solving that set Europe apart from other regions of the world. Europe learned much from other cultures, of course, particularly by transferring and adapting knowledge from advanced scientific and technical practices in China and the Arab world. It was Europe's rapidly growing inventory of written technical treatises, manuals, and books on mechanical engineering that fueled development. This documentation provided emerging experts with a codified knowledge base. Also in process at the time was a distinctive, early-modern European culture of improvement.[6] During the eighteenth century, the rise of universities and scientific academies—and eventually the foundation of institutions for higher technical education—contributed to Europe becoming a place for producing, appropriating, and communicating technical knowledge. European sovereigns competed for the services of technical experts. This catalyzed a European identity and made technical progress more socially sustainable, given its implementation by expert elites within gradually emerging bourgeois societies.[7] Europe's culture of extensive technological capture culminated in the refined taxonomies of the Enlightenment. Carl Linnaeus established the nomenclature of botany; Antoine Laurent de Lavoisier revolutionized the understanding of chemistry; and Denis Diderot, with Jean-Baptiste le Rond d'Alembert, created the definitive *Encyclopédie, ou dictionnaire raisonné des sciences, des arts et des métiers* (Encyclopedia or a Systematic Dictionary of the Sciences, Arts, and Crafts), a crowning intellectual achievement. At the time, however, expertise in Europe was still supported by patronage rather than by open public funding.

Europe's rise of expert cultures and knowledge societies requires its own periodization. In this chronology, the mid-nineteenth century represents a watershed. Science and technology gained new ground after 1850.[8] Scientists' mission to improve society had, of course, existed long before the period we have chosen to explore; French Enlightenment projects like Diderot's *Encyclopédie* confirm this. Yet the advance of technology from the mid-nineteenth century had a new force, a new traction. The scope of experts' ideas expanded, and their concepts became more credible as they became tangible to people of all classes. Daily life was now shaped by technology and science, a fact to which anyone who had ever stepped onto a railroad car could attest. This was also the period when most European countries undertook rapid industrialization, establishing knowledge-based economies in the process.[9] Simultaneously, European experts from practically every academic discipline began to advocate knowledge exchange via international organizations. Associations with members from various countries all aimed to stimulate international exchange in both established and newly created fields of scientific inquiry and technical development, from meteorology to eugenics, from railroads to steel production. These emerging, cosmopolitan knowledge societies originated in Europe, for it was chiefly European experts who

Fig. 0.4 Starting Anew as Tradition: *This image depicts the representatives of the city of Düsseldorf on the occasion of the Neue Messe (new fair) opening in 1969. The painting in the back shows the negotiations on the great GeSoLei exhibition on health and other social issues at the same place in 1926.*

formed the cross-national organizations linked to nearly each and every discipline. Tellingly as well, the title "expert" began to gain social currency during the second half of the nineteenth century. Only then did it become widespread practice for experts to train professionally in public institutions and earn formal qualifications. Only then did Europe's increasingly complex technoscientific knowledge give rise to distinct groups of experts with bargaining positions vis-à-vis state and society. And only then did innovation become planned, professional, and systematic, securely embedded in European life. The acceleration of technological progress invoked the need to re-evaluate the role of experts; those who commanded the relevant knowledge soon found a ready place in society and politics. Lasting until the First World War, this period of intense knowledge exchange was shaped largely by expert communities which were often transnational in character.

Within The Long Twentieth Century, expertise's next developmental period spanned the First and Second World Wars and the early Cold War, all of which exemplified the era's extreme societal breaks and upheavals. Our research revealed that experts continued to grow their organizations through periods of revolution, war, and shifting political systems. Typically, ruptures like the First World

War have been assumed to foster long-term developments and basic processes in technoscience; and indeed, Europe's wartime leaders often provided new resources that yielded new opportunities for some experts. But other assumptions about experts during periods of strife proved false—for example, the notion that experts steadily accumulated political influence and social status. The reality was that, during times of upheaval, large numbers of experts were often disqualified from practicing, and the expertise available in many regions of Europe declined. Moreover, the development of expertise was anything but linear. Ruptures such as the World Wars also motivated experts to question the ways in which the knowledge that they themselves advanced was being used. In what follows, we pay particular attention to the otherwise under-narrated stories of expertise: how wars and similar ruptures affected the individual trajectories of experts, generations of experts, and entire expert communities.

When referring to the "politics of expertise," we invoke the intricate relationship between experts and socio-political objectives. Experts operated in areas between nationalism and universalism; between geographic centers and peripheries; between the state and society, and between special interests and the interests of the many. All these relationships were European. Accordingly, we do not treat cross-border collaboration in, for example, nuclear energy and space research in Europe as merely a regional manifestation of a global phenomenon. Rather, we analyze what these patterns of collaboration in these technical fields reveal about European history.

During this period, modern mass media, from magazines to radio, gave technoscientific experts prominence, conferring new meaning on their causes and contributing to their growing independence vis-à-vis state and society. The fierce politics of contrasts among European nations turned experts into leading actors, sometimes initiating contrasts and formulating problems that defined experts as part of the solution. We argue that, during The Long Twentieth Century, experts amplified their bargaining power, even in the face of nations' increased attempts to control "their own" experts. When seeking this control, nation states subjected experts to various forms of pressure, even tyranny. Many suffered from Europe's ill-fated totalitarianism, particularly numerous experts with a Jewish background, people who, after 1933, were forced to flee Nazi Germany or were killed throughout Nazi-occupied Europe during the Second World War. This also holds true for the countless members of the *technical intelligentsia* who lost their lives in Stalinist Russia during the Great Purges of the late 1930s. At the end of the Second World War, German specialists who had been captured by the Allies also suffered, albeit to a significantly lesser extent, when compelled to supply intellectual reparations, of a sort, to the Allied countries.

In selecting the technologies and technosciences to feature in this book, we "followed the actors," as sociologist Bruno Latour would say. We focus here on experts in the dynamic spheres of science and technology, particularly those fields at the center of public debates. We sought the professions that were firmly connected to political developments. This implies giving little or no attention to certain major developments that were relevant to technoscience in general but not meaningful to our inquiry. One example in this category is the rise of the laboratory as the principal place of scientific research. Thus the chemists whose innovations may have marked milestones are less pertinent for us than, for example, a certain breed of modernist architects adept at linking technological insights with social problems. Relevant here are the engineers who rethought social organization, as well as the scientists who left their ivory towers in order to build European research centers, thus becoming key players in European integration.

Setting priorities for any narrative means bypassing whole groups of actors. In the case of *Building Europe on Expertise*, this translated to focusing on individuals of influence; absent here are the large groups, the rank-and-file of technical expertise, many of whom worked in industry. We also chose to narrate the development of science-based technical knowledge, hence of science-based experts. Readers will find that the closer the book moves towards the second half of The Long Twentieth Century, the more technoscientific expertise emerges, culminating in the race for technical innovation that marked the Cold War.

Building Europe on Expertise is organized chronologically in chapters structured around case studies. Each chapter focuses on a set of interconnected issues relevant to linking society and experts in Europe after 1850. Each case study demonstrates—if not typifies— the dynamics of expertise of the time. The cases are examples only, however; they do not imply that the events discussed are significant only to the timeframe in question. Specifically, the case study that we believe reveals the complex relationship between experts and the state takes place in the period around 1900; yet the same problem between experts and the state recurred—in the postwar period, in fact. Consider also the relationships of experts with large-scale research. This phenomenon was not unique to the period after 1945, the Cold War era; rather, we simply present a prototypical case. Similarly, we have chosen to discuss the education of experts in the nineteenth century because this was the time in which particularly significant processes took place in education. Yet technical education issues pervade the twentieth century as well.

The book's three main parts represent three phases and time periods within The Long Twentieth Century. Each part establishes the trends and developments of the era. Starting with the mid-nineteenth century, Part I, *Cultivating Experts, Ordering Knowledge* explores the education of experts, specifically engineers. We trace how these professionals attained a new position in Europe's emergent knowledge societies. We see how educating these experts became imperative for nation states—and how the relationship between the state and experts became one of mutual dependence. This conferred certain freedoms on experts, but the interdependency also put pressure on these critical actors. Indeed, experts answered the call of the state; at the same time, however, they blurred national boundaries by cooperating on a transnational level. Consequently, experts expanded their influence beyond their individual fields. Part I ends by outlining how European experts tried to reshape knowledge in the 1920s and 1930s in order to establish new European—and even universal— orders of peace. We describe, for example, Paul Otlet's bold attempt to gather and classify all of the world's knowledge in a single place called "Mundaneum." Equally instructive is Otto Neurath's no less audacious effort to create a universal system of representing quantitative information with graphic symbols called "Isotypes." At first glance, these may seem to be extreme cases of Europeans trying to structure the expanding cosmos of knowledge. Deeper consideration shows, however, that Otlet and Neurath lived the experience of certain prototypic European experts: those possessing the idealistic if not missionary conviction that the act of making knowledge universal could restore peace to the fragmented European continent—if not the world. Their fate, as well as the fates of many other prominent techno-intellectuals and experts, illuminates how expertise became increasingly politicized. Politics influenced whether experts were, for example, sought out or exiled by the state—regardless of whether the government or regime was National-Socialist, in the Fascist and Communist camp, or part of the ever-shrinking Democratic contingent.

Part II of the book, *Endangered Experts, New Social Orders*, illustrates how, in the Europe of the interwar period, technology was translated into new concepts for ordering society. Technicians, for example, believed that they could solve even the gravest of social problems. Architects and urbanists reinvented themselves as engineers with a social calling. With new technologies transforming the practice of architecture, they drew on the enormous symbolic meaning that technology had acquired: technology now represented power and wealth, progress and social welfare. That said, experts' self-empowerment as designers of the social realm, of course, did not lead directly to authoritarian regimes. But experts' growing scepticism toward democracies—that is, the belief that the contentious nature of democracy impeded experts' plans for fighting social crisis—did help pave the way for the totalitarian Europe of the 1930s. The authoritarian regimes of the "Dark Continent"— the moniker one historian has given Europe in this period—were based largely on the skills of technoscientific experts.[10] Quite a few such experts became the "architects of annihilation."[11] Indeed, technoscientific expertise enabled the devastating power games played by Europe's totalitarian regimes. Without experts' collaboration, Hitler could not have annihilated millions of Jews in the Holocaust, nor could Mussolini have waged a one-sided war in Ethiopia. Stalin could not have controlled a mass society through oppression and terror without expert assistance. Willingly, for the most part, experts entered into a Faustian bargain with totalitarian regimes and pursued their now-infamous goals. We ask the question: why were experts on the "Dark Continent" so susceptible to striking these bargains with the state? We show that the goals of totalitarian regimes often harmonized with experts' ambitious schemes, although the consequences of the deals struck were often unintended. The most dramatic unintended consequence was the rise in the global circulation of exiled experts. This worldwide knowledge exchange caused sharp frictions and ruptures, but that self-same knowledge-sharing allowed for technoscientific and political integration, as well. Nation states at Europe's peripheries—Spain and Portugal, for example—were not spared the tragedies that resulted from experts' widespread collaboration with regimes. The main stage, however, was East-Central Europe, which saw the most severe social and political upheavals during and after the World Wars.[12]

Proceeding from the 1950s, Part III, *Cooperating Experts, Building Institutions*, charts how and why powerful cooperative European institutions of technoscience were built. In postwar Europe, the relationship between the expert and the state was renegotiated. Generally, transitioning from war to peace meant cutbacks on states using expertise for military purposes (although the Cold War renewed demand for military innovations). In this period, the two superpowers, the U.S. and Soviet Russia, harnessed expertise for the purposes of nation-building and preparing for what they perceived as the imminent Third World War. In contrast, European nations responded to peacetime by fostering transnational collaborations. These became most visible in nuclear development and aerospace, for example. There, experts could access technologies like particle accelerators, designed to explore the hidden principles of cosmology; nuclear reactors to provide purportedly unlimited energy resources for Europe's economic reconstruction; and rockets to launch Europe into space. Thereby, collaborating experts were able to colonize new sites for the technosciences, from the Kourou spaceport to CERN's expansive laboratories in Meyrin, near Geneva. But the politics of expertise in postwar Europe also reveal a countercurrent to this collaboration—trends consistent with the previous period, in fact. Hence, we narrate various political attempts on the part of Western Europe to leverage the work of experts who had obviously struck Faustian bargains with past authoritarian regimes. We also show how, beneath the surface of these new, Europe-wide institutions that brought experts together, national competition persisted.

In addition to its three-part structure, this book is also organized around various levels of expertise. The narrative moves from the level of individual experts to the level of organizations and the communication of expertise. Ultimately, we broaden the discussion, exploring the institutionalized expertise eventually embedded in the European governance framework. These three levels correspond with the three discrete but overlapping historical phases of European expert knowledge that we have outlined above.

Building Europe on Expertise employs a conceptual framework to enhance the reader's appreciation for the era's historical complexity. These concepts are outlined in the relevant chapters. One is so pervasive, however, that it bears mentioning here. "Hidden integration" is the idea that invisible, taken-for-granted connections exist among individuals, organizations, regions, and nations. This "hidden integration" arises from the transnational circulation of technoscientific knowledge. As an unintended consequence of expertise, "hidden integration" constitutes a platform for continuously creating and sharing knowledge, a process with substantial social and political implications. Recognizing this allows us to examine closely how actors in the realms of politics and expertise have designed and used technology to both unify and fragment Europe during The Long Twentieth Century.[13]

By definition, any book that addresses such an array of problems must leave out other, related problems. Most of these are set aside deliberately, as they are central to other volumes of the "Making Europe" series. Volume 1, *Consumers, Tinkerers, Rebels,* focuses on users as non-professional experts who made their mark on Europe. Volume 3, *Europe's Infrastructure Transition*, looks at technical infrastructures, which appear in this book only in the course of making the case for experts. Volume 4, *Writing the Rules for Europe*, addresses the international organizations and cartels (including experts) who shared in governing Europe.

The process of building Europe on expertise is ongoing and perennially controversial. In fact, at the start of the current millennium, when the European Union released the so-called Lisbon Agenda, the process of building Europe could be seen as having taken on a fresh intensity; European leaders agreed on a new vision for the European Union. The stakes were high: Within ten years, Europe was to become "the most dynamic and competitive knowledge-based economy" in the world. Europe would be transformed into a fully integrated space of knowledge creation. More than ever, knowledge-building was at the core of Europe's identity.

Today, we are well aware that the March 2000 Lisbon Agenda was far too ambitious. By 2010, when the ten-year time frame had expired, the European Union was mired in the most severe economic crisis of its existence. Despite those challenging circumstances, the original vision—that Europe needs experts' knowledge to overcome crisis and ensure a dynamic future—has become clearer still. Leaders expect to restore Europe's once-flourishing economies with the help of techno-scientific experts who openly cooperate across national boundaries.

Whether or not this expectation can be fulfilled remains to be seen. Some historians believe that those who best understand the past are best equipped to visualize the future. We believe, in contrast, that historians are the least reliable of professionals in predicting the future, bound, as we are, to the present. And so we will conclude this volume by questioning critically the current discourse of a newly-emerging, fully-integrated European knowledge society by placing that debate in its broader historical context. To provide a deeper understanding of experts' roles in building, molding, and rebuilding Europe during the troubled yet dynamic Long Twentieth Century: that is what this book is all about.

I
Cultivating Experts, Ordering Knowledge

1
Educating Experts

The Inauguration of the *Politechnika Warszawska* (Warsaw University of Technology, WUT) in the autumn of 1915 was one of the stranger acts in the Eastern Theater of the Great War. On November 15, three months after German troops had conquered Warsaw, the former Polish capital, from the Russian occupation power, German military personnel set out on a seemingly peaceful task. Zygmunt Kamiński, soon to become a WUT professor, recorded the events as follows:

> German General-Governor [Hans] von Beseler, the famous conqueror of Antwerp, came to take part in the ceremony in a huge landau drawn by four white Arabian horses. The horses were booty from the stables of the Belgian King. Von Beseler was accompanied by officers from the Guard, among which the tall Count Hutten-Czapski—impressive in his richly decorated uniform of the Skull Hussars [Totenkopf-Husaren]—aroused particular attention. The latter was the official director of Warsaw University and Warsaw University of Technology, appointed by the German administration.[1]

After a number of speeches from both the new occupiers and the newly occupied, the WUT opened. The new institution did not emerge out of the blue, however. Like the entire eastern sector of the former state of Poland, Warsaw was then part of the

Russian Empire. When the WUT commenced operations in 1915, it moved into buildings of an earlier institution erected between 1899 and 1901 from funds that, as one chronicler proudly recorded, "the Polish society collected on its own initiative."[2] Yet these buildings had not been used exactly as the donors (namely, the technical section of the Society for the Advancement of Commerce and Industry) had hoped. In 1898, Russian Czar Nicolas II had agreed to fulfill a longstanding demand from Poland's elite for an institution of higher technical education; it was to be named after him. The *Instytut Politechniczny im. Cara Mikołaja* II (Czar Nicolas II Polytechnic Institute, PI) was established that autumn in Warsaw, albeit with Russian instead of Polish as the language of instruction. Clearly, the Czar and his on-site governor saw the Institute as part of the "Russification" that would secure the dynastic loyalties of this troubled western province. When Nicolas II made plans to visit Warsaw in 1897, interested circles collected funds to help erect the needed buildings. In many respects, the Czar's visit stirred the hopes of both "positivists," that is, those interested in organic change through education and social improvements, and Polish elites for the modernization of the city.[3] Warsaw's spectacular electric illumination on this occasion underscored the symbolic merger of past and future, with the Czar as a decisive factor.[4]

While the Russian imperial administration and Polish local elites could hardly have been more different, their interests did converge, albeit unintentionally, around educating technical experts. This story reveals much about the central role that technical education had acquired in the preceding century. Although the Russian imperial administration was hardly concerned with economic problems in Poland, it had an increasing need for technical experts which traditional institutions for educating the nobility were no longer able to meet. In order to fill this gap, the empire established a whole range of new institutions across its territory, from Tomsk to Kiev to Warsaw. If the Warsaw Polytechnic Institute gained particular importance as a training ground for a national technical elite, this was accepted but certainly not intended. The Russian government in St. Petersburg explicitly defined the role of the new institutions as producing agents of modernity—technical experts for the whole empire.[5] On the other hand, the local Russian authorities understood quite well that the constant brain-drain of capable—and embittered—technical elites presented a long-standing political

Fig. 1.1 Technical Education & the Nation: *The opening of institutions of higher technical education saw multifaceted interests at play. The case of Warsaw's University of Technology (WUT) captures this nicely. Here Count Bogdan Hutten-Czapski is honored in 1931 with a doctor honoris causa for his instrumental role in the re-opening of the institution in 1915. Next to him we see WUT's rector Andrzej Pszenicki. Mind the inner courtyard of the WUT, which is strongly reminiscent of the TH Charlottenburg-Berlin. The impressive architecture of buildings such as these reflects the ambition of the new institutions of technical education.*

problem that had to be tackled. After the 1830 Polish nationalist uprising, it was long forbidden for Poles to study outside Russia, and within Russia their access to education was restricted. One study counted only 300 graduates in all disciplines in Russian Poland, a province of 4.5 million people in the 1830s. Even if this number is underestimated, it indicates the dramatic effect that politics had on how knowledge was acquired and circulated.[6]

The Polish elites, meanwhile, saw the Czar's Institute as a means of keeping aspiring youth in the country, which would sustain a technical knowledge base for potential nation-building. Though the learning environment at the Institute remained restricted, it did produce experts who played decisive roles in the new Polish nation state established in November 1918.

The construction of the building which the PI occupied is characteristic of the interests and desires that shaped the institution's foundation. Looking for a location that was suitable both practically and symbolically, the initiators and the city of Warsaw agreed on an area at the city center's southern edge, which had served as the grounds for a large international hygiene exposition in 1896. Much like the newly founded Institute, this exposition had been conceived as a way to advance the nation by improving strategic knowledge. However, the Russian administration did not want to involve local architects. It proposed using building plans that had been placed second in an architectural competition for the Polytechnic Institute in Kiev, while the Poles insisted on something more original. The two architects eventually chosen completed their plans in only seventy-two days. As they had no experience with such buildings, they embarked on a European trip before starting work in order to familiarize themselves with the state-of-the-art in building edifices for technological institutes. Among the polytechnic institutions they visited were those in Vienna, Graz, Milan, Zurich, Strasbourg, Darmstadt, Paris, London, Berlin-Charlottenburg, and Leipzig. As a result, they saw almost all the centers of technical education before starting their own project, which was conceived as a synthesis of these buildings.[7]

The same was basically true for the institutions the building contained, not only for the PI, but also for its successor the WUT. WUT professors had, necessarily, been educated mostly outside Poland, often at the above-mentioned institutions. Therefore, the new curriculum carried a rather European blend of different traditions, particularly the French *École Polytechnique* and the German *Technische Hochschule*.[8] Because the new institution had little previous tradition to govern the form and content of its teaching, its organization was also noticeably modern. For example, the Department of Architecture, founded in 1915, was one of the first in Europe to feature a chair for urban planning and to include the social dimension of architecture that was so important for the profession's role in a prospective Polish state.[9]

Another aspect worth noting is the strong role of civil society in the WUT's organization. After the Polytechnic Institute was essentially closed to Polish students following student strikes over the language of instruction (insisting on Polish instead of Russian) in the context of the 1905 Russian revolution, self-help organizations like the *Towarzystwo Kursów Naukowich* (Society for Academic Courses) and its polytechnic committee effectively filled the gap and then played a key role in the 1914/15 opening. The fact that the Institute had been founded using comparatively large funds provided by Polish civil society highlights its importance for nation-building.

Even today, the WUT looks modern and has stood the test of time. Yet, it was certainly not a direct outcome of any science policy on the part of the German occupiers.[10] Their motives were hardly more philanthropic than those of their Russian predecessors. Rather, the Germans' weaker position as newcomers to Warsaw forced them to make substantial concessions, particularly concerning the use of the Polish language, and to have the institution—transformed into a proper polytechnic university—run mainly by Polish personnel. While approximately 70 percent of the students at the earlier institute had been of Polish origin, Poland supplied only 20 percent of the staff. Against the background of a highly-politicized higher education, the WUT had become an important piece on the Germans' Eastern Front chessboard and in its attempt to create Poland as an ally or at least as buffer state.[11] Moreover, it seemed likely that the university could profit from the new networks established during the war, for example, when German advisers evaluated an urban master plan for Warsaw conceived by the WUT professor Tadeusz Tołwiński.

Zygmunt Kamiński, the professor at the WUT who reported the Warsaw events, was very much part of the story himself and was well aware of the ambiguities and pitfalls of his situation. From the day the University opened, the Polish professors were accused of collaborating with the occupiers. Indeed, at the same time as the Germans were exploiting Poland materially, they were also presenting themselves internationally as bringers of culture to an allegedly deserted area of Europe. Nevertheless, Kamiński believed that there was no alternative that would enable Polish technical students to stay and form the elite that the country so desperately needed, especially with national independence becoming a much more plausible option late in the war. WUT students were taught in four departments (architecture, engineering and electrical engineering, chemistry, and civil engineering), and trained by a heterogeneous corps of professors who had necessarily gained their knowledge in practice rather than in academia. Thus, when Poland achieved independence in November 1918, WUT graduates formed the basis for a new elite with immense professional prospects and influence.

Although Warsaw was undoubtedly unique, its experience is relevant for understanding what was happening in many European countries. In Finland, then still part of the Russian Empire, the Helsinki *Teknillinen Korkeakoulu* (University of

Technology) was founded in 1908. Norway, which gained full independence from Sweden only a few years before Poland was re-created, also followed a pattern similar to Poland's. By 1900, five years before separation, Norway's regional parliament, the *Storting*, had already passed a manifesto demanding the foundation of an institution of higher technical education. In 1910, five years after independence, that technical university opened in Trondheim.[12]

Not only due to its resonance with examples in Finland or Norway, the Warsaw episode represents much more than just a curious example from Europe's imagined educational periphery. It exemplifies a number of broader themes which we will elaborate in this chapter.

First, the Warsaw case shows how much the education of technical elites had become a strategic issue of utmost importance, both to aspiring nation states and struggling empires. The rapid expansion of technical infrastructures and industrial development meant that no state could do without those people who controlled strategic knowledge that went far beyond a narrow understanding of technical problems. States might acquire technical information, but they certainly could not implement it without these experts. As the institutes founded by the "almighty" Russian Czar showed, the lack of technical knowledge put even the most authoritarian state structures at risk.

Second, the WUT example highlights how strongly various national paths to technical education were intertwined. The WUT was international in character in at least two respects. New professors (few of whom had formal qualifications to hold such positions) had trained all over Europe, displaying a complex, but still systematic geography of education.[13] Moreover, the curriculum merged a variety of national traditions, reflecting not only the professors' international training, but also a much wider process taking place all over Europe during the nineteenth century. This Europeanization of technical education was epitomized by the itinerary traced by the Polytechnic Institute's architects in 1898. While French institutions long served as an important example, Poland never directly adopted France's model of technical education.

Third, we also see the lasting relevance of the prior non-technical elites for the establishment of new knowledge-based formations. This process can hardly be adequately described as a fierce opposition of the old and the new, however, as we will discuss in the next chapter.

Fourth, technical experts were, on the one hand, seen as apolitical specialists, who regardless of their national backgrounds could work in diverse places—in this case the Russian empire in the 1830s and German-occupied Poland in 1915—where they were needed. They were at the same time highly nationalized figures, as conflict about the power of the Polish versus the Russian language shows. The strategic power of education applied to all sides involved. Thus engineers could rise to national symbols by building on the very real need for, and relevance of, their profession. This also helps to explain the extremely high expectations for engineers and technical elites in general.

Fifth, the WUT's establishment demonstrates the importance of local initiative and commercial dynamics for professional development. Its foundation—setting aside how it was instrumentalized for national and imperial purposes—was based on a local effort in which business interests dominated. This, as well as the general dynamic of a fledgling market economy, helps to explain why the status of formal education was still very tenuous even as late as 1900. Most WUT professors were employed due to their practical accomplishments and skills. In these decades, most engineers worked without formal qualifications.

There is, however, an additional aspect worth mentioning; it will be taken up in more detail later on. Periods of upheaval—wars and revolutions—played a significant role in generating positions of status and importance for experts. Challenges for the state generally enhanced the standing of those potentially able to provide solutions. After all, such upheavals demanded strategic decisions and arrangements from those. This is most evident in the First World War's role as catalyst.

Finally, the WUT's startup was significant, because by 1915 technical education in Europe had largely consolidated into a model that remained intact throughout the century. Thus, it is worth looking closer at the roots of the situation. Technological progress and the speed of innovation posed obvious difficulties for most of Europe by the mid-nineteenth century. As a result, it became imperative to educate experts and create institutions for technical training. The question was how. Using the example of engineers to show the place of experts vis-à-vis state and society, we will now explore how technical education both generated novel tensions and confronted existing ones.

Technical Education as a Strategic Issue

Engineers were essential for spreading the industrial revolution and, at the same time, benefited immensely from technical change. These "foot soldiers of industrialization" dealt with problems that arose with the diffusion of increasingly complex technical devices, while advancing this process at the same time.[14] New technologies, such as railways, automobiles, radio transmission, and electricity, all demanded increasingly specialized educational programs, including new institutes and new kinds of laboratories. The growing costs of such infrastructures increased dependence on the funding-providers, whether states or companies. However, economic flows only partly explain why engineers became key figures in modern states and societies. State-like organizations long depended on specially-qualified personnel who exerted influence or extended their scope. This was true to some degree for medieval political entities, and indeed, this very mechanism anchored what has been described as absolutism in early modern Europe.[15] However, there are good reasons why the importance of these personnel rose in the process of industrialization.

Technology's increasing significance was most evident for military decision-makers. From a technological perspective, the Napoleonic wars had been fought in a rather traditional way, whereas the Crimean war (1853–56) and the three military conflicts preceding German unification in 1871 could not have been won without useful knowledge about the latest transport and communication technologies. Even the most narrow-minded conservatives learned to accept that courage and military virtue alone would no longer suffice. This understanding pertained to the entire organization and administration of the state, now challenged to supervise technology-driven production sites, to build technological infrastructures, and to educate the very technical elites needed. These elites joined with legal professionals in running modern states and societies.

Fig. 1.2 Watching the New Times Taking off: *Captain Paul Engelhard, apprentice of the Wright brothers, flying over Johannisthal near Berlin, a pioneer ground for early aviation in Germany (ca. 1910). The cavalry officer evaluates the potential of the new branch of the service (and the decline of his own position). Still, it was often the old elites deciding which representative of the new technical elites would get chance.*

This link was generally in place already by the eighteenth century. In France, the first engineers were high-ranking civil servants and usually served as officers in the army at the same time. Thus, the term "engineer" referred to a person's status within the state administration.[16] Engineers worked as generalists who organized and managed approaches to the complex problems critical to the state's technical infrastructures. Indeed, "The birth of the modern French engineer coincided with the establishment of the French modern state."[17] In the Netherlands, as well, the connection of hydraulic engineering and state formation was already obvious before 1800.[18]

These engineers' early activities focused on construction and were strongly connected to military needs. In Britain, France, and other European countries, the first structured training in specialized schools available to engineers took place as preparation for the army.[19] Early military engineers, the best known of whom was Sébastien Le Prestre de Vauban (1633–1707), were officers who served the state in peacetime civil functions.[20] This helps explain the close ties between the developing engineering profession and the state, and also the profession's strong *esprit de corps*. The *corps* (as it was soon to be called) of engineers experienced an increasingly standardized education.

By the 1740s, the first schools of this type had opened in France, but they were not unique. In Prague, for example, a school for engineers commenced in 1707. The Ottoman Empire supported from 1773 the Imperial School of Naval Engineering (*Mühendishane-i Bahr-i Hümayun*) in Istanbul. Still, the French institutions, particularly the *École des Ponts et Chaussées* (founded in 1747), achieved what was labeled *l'invention de l'ingénieur moderne*.[21] A key to its success was the establishment of highly theoretical general entry exams, along a pattern Vauban developed, thus enhancing the *corps*' external prestige.[22] Innovative as it was, the school was soon supplemented by other more specialized entities, such as the *École des Mines*.[23]

Tellingly, the system underwent another change in the Revolutionary era that had long-lasting effects. The relevance of the new comprehensive institution, the *École Polytechnique*, had a lot to do with the circumstances of its establishment.[24] As a result of the Revolution, France lacked the technical elites that older institutions like the *École des Mines* or *École du Génie* had produced. The recruitment procedures of these institutions were highly biased towards the upper bourgeoisie and the aristocracy, with the *École du Génie* even refusing ennobled aristocrats.[25] Because the old technical elites had been so close to the *ancien régime*, the new state managers regarded them as potentially disloyal to the new system—if they had not been exiled, imprisoned, or beheaded anyway.

In this critical situation, a council of scientists proposed creating a single educational institution that would prepare students exclusively for state service and that would integrate the different specializations that had previously been run separately. Originally called *École central des Travaux publics*, the institution was rebranded as *École Polytechnique* in 1795; the neologism proved so catchy that it was taken up in a number of languages and other countries later on.[26] Significantly, and in line with the traditional perception of technical training, Napoleon integrated the *École Polytechnique* into the War Ministry only ten years after its foundation.[27]

The state's quasi-monopoly in providing high-quality academic training for prospective engineers further enhanced and secured graduates' critical position in the era of modernization. Of course, their improved education mattered, but equally weighted culturally were their prestigious diplomas and their strong links with the administrative hierarchies. The education of engineers reflects not only the joint rise of the modern state and the engineering profession, but also their increasing interdependency. As many as 85 percent of French graduates, called *polytechniciens*, entered the armed services between 1806 and 1813.[28] This extreme identification of the state and its qualified elite is best expressed in the apparently heroic fight of the young students of the *École Polytechnique* in the 1814 battle of Paris, where they were among the last continuing to identify with the Bonapartist regime.

The flip side was that the restored Bourbon regime regarded the school as a stronghold of Bonapartism and consequently closed it in 1816. However, royal administrators could not ignore the the fact that it created technical capabilities that were essential for the new state. Thus it had to reverse its decision a year later.

The unparalleled success of engineers in France derived largely from their character as a strictly and rigidly defined elite group. Alexis de Toqueville described the *Corps des Ponts et Chaussées*, one of the precursors of the *École Polytechnique*, as "the creature of enlightened despotism," the "first professionalized French elite of public engineers [...] dressed in iron-gray uniforms with silver buttons displaying the Bourbon fleur-de-lys in gold as they faced provincial and parochial opponents of royal authority."[29] In this, the upper echelons of the French technicians may well be seen as technocrats *avant la lettre*, a phenomenon, for better and for worse, familiar to Europeans across the two centuries to come.

This narrowly defined group sought to undertake a broad range of tasks that stretched far beyond the technical.[30] Engineers served on local committees formed to improve general health, sanitation, and housing. They identified with progress as much as with the nation. More than their counterparts in Germany, Britain, and the United States, French state engineers proved remarkably successful in promoting a common ethos of national service.[31] Moreover, the central authorities in Paris saw to it that *polytechniciens* commanded the skills and standing (in short, the ethos) they needed to successfully socialize within local frameworks. Marriages between the *polytechniciens* and the daughters of local elite families were systematically encouraged and supervised by Paris. The intention was to shape a new composite elite of meritocrats from the nation's center linked to local money and influence. Local elites seemed to happily accept this arrangement, which further enhanced the

CENTENAIRE DE L'ECOLE POLYTECHNIQUE
1794 - 1894
Le Comité remet au Président Carnot l'Histoire de l'École depuis cent ans.

polytechniciens' exceptional standing.³² Against this background, engineers unsurprisingly saw themselves as representatives of the state, *per se*, and many *citoyens* shared this perception. In France, the process of turning "peasants into Frenchman"³³ was influenced significantly by a corps of engineers who extended their range of activity well beyond bridges and roads.

Copying & Circulating Educational Models

This system, and the *École Polytechnique*'s strong mathematical grounding in particular, became a point of reference throughout Europe in the nineteenth century, from the *Escola Politécnica* in Lisbon (founded in 1837) to the Saint Petersburg State Polytechnical

Fig. 1.3 Polytechniciens in Battle: *The education of technical experts served the needs of industrialization. Yet, the graduates of the early schools also fulfilled the role of a social and political elite, visible—until today—in the uniforms of the French Polytechniciens. The latters' close attachment to Napoleon, who gave the school its primary role, came dramatically to the fore in the last stand of Polytechniciens in the battle of Paris in March 1814. This drawing on the occasion of the school's hundreds anniversary commemorates this event so central for the institution with the motto, "Pour la patrie, les sciences et la gloire."*

Institute (Санкт-Петербургский Государственный Политехнический Университет, 1899). Spain even imported curricula and textbooks directly from France.[34] France's pre-eminence, in the eyes of the many aspiring nations in and beyond Europe, derived in fair measure from the strong interconnection between nation-building and fabricating training institutions.[35] Yet, while the French educational model could be emulated, adapted, and transformed, it was far more difficult to transfer the engineers' function as a social elite, even considering that the *polytechniciens* represented a small proportion of all French engineers.

Of course, the direct institutional transfer of models may not have been the most influential form of circulation.[36] The transnational movement of students, often within empires, steadily developed. Certain centers of learning—primarily Paris, but also Vienna and, later in the century, Berlin-Charlottenburg and smaller places like Karlsruhe—dominated the scene. Moreover, the imperial structures in Eastern Central Europe allowed for an astonishing degree of mobility, both for students and faculty.[37] In fact, this transnational movement was often involuntary; for example, for Polish students who had to study in a Russian, Austro-Hungarian, or German institution operating within the borders of their respective political entities. The same was true in principle for many of the smaller nations. Even more striking, but beyond our scope, is the huge influx of American students to European universities, which mounted to many thousands across all disciplines.[38]

While the French tradition of technical education became an attractive model for countries in Europe and beyond, the close alliance between nation, state, and the engineering corps was not replicated elsewhere. The sense of the expression *pantouflage dans la privé*, referring to the allegedly easy life in the private sector as compared to the state sector, was hard to grasp in Britain or the German states, even at a time when French was still a universal language in Europe.[39] In France, after all, experienced government engineers moved regularly into work for private companies, a "retirement" that was actually a reorientation. French firms profited from capturing skills that state institutions had taught and that state service had developed. In Britain, with shop-floor-based technical knowledge creation dominant well into the twentieth century and with no institution of comparable rank to the French *Grandes Écoles*, engineers could achieve impressive social standing, but they rarely gained close ties to state structures.[40] With its rapid industrial development, Britain's private sector had far greater importance. Thus, higher technical education was of minor relevance in Britain until well into the nineteenth century.[41] There, with a merchant tradition and a history of strong private involvement in infrastructure construction, the interrelation between state, society, and engineers was less pronounced. Britain's position as an industrial forerunner meant that foreign models did not initially exert an influence comparable to continental countries. However, over time anxiety arose that Britain would lose the economic race of high imperialism, and technical experts became more valued, as can be seen in the establishment of the National Association for the Promotion of Technical Education in 1887.[42]

German institutions certainly reflect the idea that the "ambiguities of the *École Polytechnique* helped make it the world's most influential school."[43] The *Polytechnikum* in Karlsruhe, capital of the former *Rheinbund*-state Grand Duchy of Baden, was founded in 1825 with direct reference to the French model. At the same time, its system of technical education also included practical training—something handled by a different set of institutions in France. Moreover, the *Technische Hochschulen* (THs, as they were usually called after 1871) were much closer to the existing "classical" universities.

The major difference was that the German *Technische Hochschulen* retained or developed a considerable degree of self-administration, whereas French *Écoles* responded to central government controls. One manifestation of the THs' autonomy was their gigantic new buildings. The edifice of the TH Berlin–Charlottenburg, inaugurated in 1884, was after Cologne's cathedral the largest building in Prussia, and served as a design base for the Mining School in Madrid.[44] In 1879, the TH Hannover moved into the huge palace of the deposed Guelph dynasty. These examples speak to the high significance technical education held in the German lands, also carrying the hopes of many for Germany's political transformation.

With a curriculum not unlike the THs and an equally practical approach, a specifically-Swiss and highly-successful model was the *Eidgenössische Polytechnikum*, later *Eidgenössische Technische Hochschule* (ETH), founded in Zurich in 1855. The general objective was to promote national renewal and modernization, this time in the framework of Switzerland's modernizing constitution of 1848,

Fig. 1.4 Demonstrating the Principles of Technology: *This picture of 1887 shows the cantilever principle used by engineer Benjamin Baker for the building of the Forth Railway Bridge in Scotland. The design was theoretically already well established, but apparently still in need of practical demonstration. In the center we see Japanese engineer Kaichi Watanabe. Watanabe had been trained in Scotland in the preceding years, attesting to the global reach of engineering education.*

which explicitly mentioned the federal state's right to erect such a school. Therefore, as a "machine for the future," ETH may be seen both as an "effect and a realization" of the recently confirmed federation. Training technical experts was the central task, but the ETH was also entrusted with furthering a "civic identity supportive of the state."⁴⁵ As the designation *eidgenössisch* (federal) indicates, the institution—unlike the German THs—had support from the central state, which ensured its funding and raised its status considerably, despite initially substantial opposition. Interestingly, when the ETH started, explicit references to the French *École Polytechnique* not only acknowledged a strong tradition, but also addressed fears of Germanization which stemmed from its location in a German-speaking canton. The ETH's inaugural celebration was intended to demonstrate its legitimacy; it also stressed how the technical curriculum would meet the particular challenges posed by the Alpine federation's topography.

Country	City	Institution	Founded
Czech Republic	Ostrava	Bergschule	1716
Germany	Braunschweig	Collegium Carolinum	1745
France	Paris	École royale des ponts et chaussées	1747
France	Mézières	École du Génie Militaire	1748
Slovakia	Banska Stiavnica	Bergakademie	1762
Germany	Freiberg	Kurfürstlich-Sächsische Bergakademie zu Freiberg	1765
Germany	Berlin	Bergakademie (1770) / Bauakademie (1799) / Gewerbe Institut (1829)	1770
Turkey	Istanbul	*Imperial School of Naval Engineering*	1773
Germany	Clausthal	Clausthaler montanistische Lehrstätte	1775
Hungary	Budapest	Institutum Geometrico-Hydrotechnicum	1782
France	Paris	École des Mines	1782
France	Paris	École centrale des travaux publics	1794
Germany	Berlin	Bauakademie	1799
Spain	Madrid	Escuela de Caminos y Canales	1802
Czech Republic	Prague	Polytechnisches Landesinstitut	1806
Germany	Schweinfurt	Balthasar-Neumann Polytechnikum	1807
Russia	St. Petersburg	*Institute of Engineers of Ways of Communication*	1809
Russia	St. Petersburg	*Saint Petersburg military engineering school*	1810
Austria	Graz	Joanneum	1811
Austria	Vienna	k. k. Polytechnisches Institut	1815
Germany	Berlin	Vereinigte Artillerie- und Ingenieurschule	1816
Germany	Bochum	Bergbauschule	1816
Ukraine	Lviv	Lviv Polytechnic	1817
Sweden	Marieberg	Artilleriläroverket och Artilleri- och Ing. Högskolan	1818
Germany	Nuremberg	Polytechnische Schule	1823
Germany	Berlin	Lehr- und Forschungsanstalt für Gartenbau	1823
United Kingdom	Liverpool	Liverpool Mechanics' School of Arts	1823
United Kingdom	Leeds	Leeds Mechanics Institute	1824
Germany	Karlsruhe	Polytechnische Schule	1825
United Kingdom	Huddersfield	Huddersfield Science and Mechanics Institute	1825
Poland	Warsaw	*Preparatory School for the Institute of Technology (1826) / Emperor Nicolas II University of Technology (1899)*	1826
Germany	Munich	Polytechnische Schule	1827
Sweden	Stockholm	Teknologiska Institutet	1827
Germany	Dresden	Kgl.-Technische Bildungsanstalt Sachsen	1828
United Kingdom	Preston	Institution For The Diffusion Of Useful Knowledge	1828
Denmark	Copenhagen	Den Polytekniske Læreanstalt	1829
France	Paris	École centrale des Arts et Manufactures	1829
Germany	Stuttgart	Vereinigte Real- und Gewerbe-Schule	1829
Sweden	Göteborg	Chalmers tekniska högskola	1829
Russia	Moscow	*Imperial Vocational School*	1830
Germany	Hannover	Höhere Gewerbeschule	1831
Belgium	Brussels	École royale militaire	1834
United Kingdom	Wolverhampton	Wolverhampton Mechanics' Institute	1835
Belgium	Mons	École Provinciale des Mines du Hainaut / École des Mines (Guagnini)	1836
Germany	Darmstadt	Gewerbeschule	1836
Germany	Chemnitz	Gewerbeschule	1836
Belgium	Liège	École Spéciale des Mines	1837
Greece	Athens	*Royal School of Arts*	1837
Belgium	Gent	École du Génie Civil	1838
Belgium	Gent	École des Art et Manufactures	1838
Belgium	Liège	École des Art et Manufactures	1838
United Kingdom	London	Royal Polytechnic	1838

Continued

Country	City	Institution	Founded
Austria	Leoben	Steiermärkisch-Ständische Montanlehranstalt	1840
Netherlands	Delft	Royal Academy for the education of civilian engineers, for serving both nation and industry, and of apprentices for trade	1842
United Kingdom	Sheffield	Sheffield School of Design	1843
United Kingdom	London	Royal College of Chemistry	1845
Czech Republic	Brno	k. k. Technische Lehranstalt - c. k. technické učilište	1847
Finland	Helsinki	*Aalto University*	1849
Czech Republic	Příbram	*Mining high school*	1851
United Kingdom	London	Royal School of Mines	1851
Switzerland	Lausanne	École Spéciale de Lausanne	1853
France	Lille	École des arts industriels et des mines	1854
Switzerland	Zurich	Eidgenössisches Polytechnikum	1855
Spain	Valencia	Escuala Superiore de Ingeniores Industriales	1855
Spain	Gijón	Escuala Superiore de Ingeniores Industriales	1855
Spain	Barcelona	Escuala Superiore de Ingeniores Industriales	1855
Spain	Sevilla	Escuala Superiore de Ingeniores Industriales	1855
Spain	Vergara	Escuala Superiore de Ingeniores Industriales	1855
France	Lyon	École Centrale lyonnaise pour l'Industrie et le Commerce	1857
Italy	Turin	Scuola di Applicazione per gli Ingegneri	1859
United Kingdom	Plymouth	School of Navigation	1862
Czech Republic	Prague	Polytechnisches Institut	1863
Italy	Milan	Istituto Tecnico Superiore	1863
Romania	Bucharest	*Institute of Civil Engineering*	1864
United Kingdom	Portsmouth	Portsmouth and Gosport School of Science and Arts	1869
Germany	Aachen	Polytechnikum	1870
United Kingdom	London	Royal Engineering College	1871
Austria	Vienna	Hochschule für Bodenkultur	1872
France	Lille	Institut Industriel du Nord	1872
France	Paris	École Municipale de Physique et de Chimie Industrielles	1882
United Kingdom	London	Central Institution / Central Technical College	1884
Ukraine	Kharkiv	*Practical Technological Institute*	1885
France	Nancy	École Nationale Supérieure des Industries Chimiques	1887
United Kingdom	Dundee	Dundee Institute of Technology	1888
United Kingdom	London	Woolwich Polytechnic	1890
France	Marseille	École d'ingénieurs	1891
France	Grenoble	Institut d'électrotechnique	1892
United Kingdom	London	West Ham Technical Institute	1892
United Kingdom	London	Borough Polytechnic Institute	1892
Germany	Illmenau	Thüringisches Teknikum Illmenau	1894
United Kingdom	Newcastle upon Tyne	Rutherford College of Technology	1894
Russia	Moscow	*Moscow State University of Railway Engineering*	1896
Russia	Tomsk	*Tomsk Technological Institute*	1896
United Kingdom	London	Northern Polytechnic Institute	1896
Ukraine	Kyiv	*Kyiv Polytechnic Institute*	1898
Russia	St. Petersburg	*Saint-Petersburg Polytechnical Institute*	1899
United Kingdom	Kingston upon Thames	Kingston Technical Institute	1899
Spain	Bilbao	Escuela de Ingenieros Industriales	1899
United Kingdom	London	Ponders End Technical Institute (1901, among many other schools merged into Middlesex University)	1901
United Kingdom	Sunderland	Sunderland Technical College	1901

Continued

Table 1.1 Building Technical Education: *The question of what exactly an institution of higher technical training was—and still is—as difficult to answer as the question of when these institutions actually commenced their work. No doubt, such schools existed by the eighteenth century and flourished in the following century. This table, without claiming to cover all cases, gives an idea about the dynamics at play as well as the geographic diffusion of technical training institutions throughout Europe.*

Country	City	Institution	Founded
Poland	Gdańsk	Königliche Technische Hochschule zu Danzig	1904
France	Toulouse	Institut d'électrotechnique et de mécanique appliquée de l'université de Toulouse	1905
Czech Republic	Prague	Czech University of Life Sciences Prague	1906
France	Toulouse	École Nationale Supérieure de Chimie de Toulouse	1906
Russia	Nowotscherkassk	*South Russia State Technical University Nowotscherkassk*	1907
France	Toulouse	École Nationale Supérieure Agronomique de Toulouse	1909
Norway	Trondheim	Norges tekniske høgskole	1910
Poland	Wrocław	Technische Hochschule Breslau	1910
Turkey	Istanbul	*Kondüktör Mekteb-i Âlisi/ Conductors School of Higher Education*	1911
United Kingdom	Trefforest	South Wales and Monmouthshire School of Mines	1913
United Kingdom	Staffordshire	Central School of Science and Technology	1914
Estonia	Tallinn	Ingenieurschule der Estnischen Ingenieurgesellschaft	1918
Czech Republic	Brno	*University of Agriculture*	1919
France	Nantes	Institut Polytechnique de l'Ouest	1919
France	Nancy	École nationale supérieure des mines de Nancy	1919
Poland	Poznań	*Higher State School of Machinery*	1919
Poland	Kraków	*University of Mining and Metallurgy*	1919

Main source: Anna Guagnini, "Technik." In *Geschichte der Universität in Europa*, III, edited by Walter Rüegg, 487–514. Munich: C.H. Beck, 2004.

Note: For the cities the present names are used; for non-Western languages names are translated into English and italicized.

The British naval engineer John Scott Russell highlighted ETH's role in modernizing the Swiss nation, as well as the European dimension of that endeavor. He saw its founders as trawling "the annals of pure philosophy and applied science, for the names of those men best known for science, skill and love of teaching, and these men from every country, they selected and entreated to come and teach their children."[46] With four faculties (civil engineering, mechanical engineering, chemistry, and architecture), ETH very much reflected what would become the standard in the "TH" sphere of Europe: indeed this division later served as a template for many institutions, including the WUT. Owing to its high reputation, the ETH attracted very large numbers of students from throughout Europe. In Greece ETH graduates returning home exerted a particularly pronounced influence and became known as the Zurich Circle.[47]

What is European in this story is not only the circulation of models of technical education, but also the pattern behind it. Looking beyond the borders of one's own state was driven by a genuine interest in identifying best practices for effective teaching and technical innovation. Yet, it was also driven by politics of comparison, which often gained momentum from political motives or projections, from fierce competition and outright nationalism. Periods of particular tension, like wars, led to the questioning of national models of European technical education (or what were perceived as such).[48] In any case, these so-called models were always the product of fusing different traditions and influences and only gained (or appeared to gain) a particularly French, German, or Swiss character when viewed from outside. Moreover, their adoption always resulted in new entities, which local conditions shaped more strongly than any foreign examples could.[49] Finally, the circulation of ideas was never restricted just to Europe, but also included its colonies and, beyond them, Japan and the U.S., the latter attaining the status of a technical education center by the late nineteenth century.[50] In a sense, national frameworks of knowledge had a paradoxical effect: the nation state was a major dynamic force in creating academic infrastructures, which, in turn, developed in extremely international ways.[51]

The Quest for Respect

Despite engineers' impressive rise to new social status, cultural significance, and even direct political influence, institutionalized training was only part of a much larger trajectory. Most engineers had worked before and around 1900 without formal qualifications. This was particularly obvious in Britain, where academic engineering education commenced later than on most of the Continent. In Britain, the early and revolutionary development of industrialization essentially occurred without academically trained engineers. Furthermore, many British engineers worked on the Continent as expatriates sharing critical practical experience, thereby achieving an essential knowledge transfer largely outside academic institutions. Though hard to trace due to their informal practices, these engineers' influence can hardly be overestimated. Know-how and commerce went hand in hand. Establishing successful businesses or aligning themselves with them, British engineers had key functions in the modernization of European infrastructures and economies and spread the seeds of industrialization. Particularly striking was William Lindley (1808–1900), who started with successfully counseling the city of Hamburg on railway construction in the 1830s. He progressed into constructing sewer systems, actually introducing these to continental Europe, and later into equipping European cities with gasworks. With his design for Warsaw's huge sewer system in the 1870s, Lindley made his name as an engineer across the whole Continent. In the second generation, around 1900, Lindley's company took on projects in places as remote as Baku, Bucharest, Iaşy in Moldavia, and Lithuania's Kaunas, thus impressively highlighting this knowledge entrepreneur's transnational reach.[52]

Indeed, emphasizing the significance of technical education, schools, diplomas, and titles should not overshadow the fact that other forms of technical training remained important until well into the twentieth century. Indeed, many individuals who had been unsuccessful in academic settings were still able to rise to great heights based on their practical skills. The Prussian "*Lokomotivkönig*" August Borsig, a pioneer of engine construction, had had to withdraw from Berlin's *Gewerbe-Institut* because he was "technically untalented." Werner Siemens, founder of the technology-based enterprise, could not afford to study at the *Berliner Bauakademie,* which meant his only training was with the artillery.[53] The educational record of the luminaries of modern interwar architecture, who were so proud of their technical versatility, was strikingly poor. Walter Gropius withdrew from his studies without a degree, Le Corbusier only attended a vocational school for watch-engraving, while the Dutch pioneer of modernist architecture J.P. Oud studied at a range of vocational handicraft schools. Ludwig Mies van der Rohe made his way into the profession as the son of a building contractor without special training. The architect of the century, as some have judged, Mies never studied architecture and was, aged 72, asked by the U.S. authorities to mend his lack of a formal qualifications when finishing the famous Seagram Building in New York.[54] Gerrit Rietveld was schooled by his father in furniture making, and Theo van Doesburg attended theater classes.[55] These examples do not undercut the rising professionalization of engineering and other technical professions. Yet, they highlight how, well into the twentieth century, informal education and practical experience remained important resources. Moreover, these examples show how other factors such as symbolic capital or visionary designs helped in attaining expert status.

While the *École Polytechnique* was certainly the most visible example of educational institutions, it by no means dominated the scene even in France. For example, the much bigger *École Centrale*, founded in Paris in 1829 in part as a successor to the original, pre-Restoration

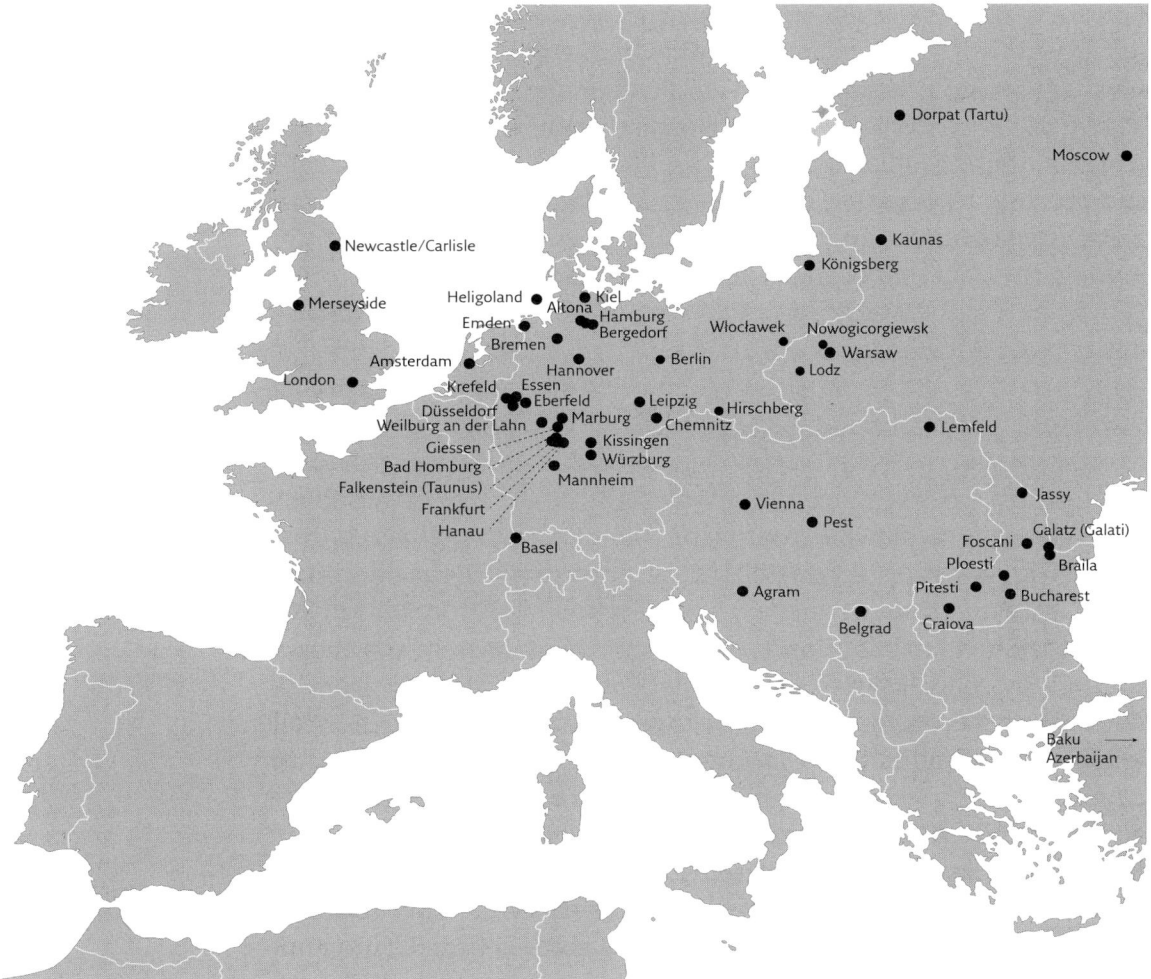

Fig. 1.5 **Know-how for the Continent:** *The Lindley brothers were one of the countless British engineering companies setting the pace of progress all over the Continent. They stood for the many less visible British engineers selling their know-how, mostly acquired on the job. The Lindleys excelled in a number of fields from railway construction to urban planning, but became particularly famous for their expertise in sewage systems which the firm pioneered in a number of places. This map captures the astonishing geographic reach of Lindley & Sons.*

Polytechnique, offered a private alternative to the state-run institutes.[56] Even the prestigious *polytechniciens*, so strongly tied to the state, used their expertise widely in private undertakings. Thus, technical and managerial expertise percolated through the state to the private sector.[57] While the state was an important actor, its interests were not always the only factor governing its actions. Instead, technological change and innovation, as well as the needs of the ever-increasing private sector, dictated the direction and pace of establishing educational institutions. Hence, the state funded schools (such as the *École Supérieure d'Electricité* in Paris, 1893) that reflected the specialization of knowledge and contributed to extending it. In Britain, decision makers began to orient themselves more closely towards continental institutions of technical training, and important regional centers grew in Glasgow and Manchester. At the same time, continental observers were trying to emulate British and American shop culture with its emphasis on developing practical skills. Officials in France and Germany believed, for different reasons, that supplementing the existing curriculum with practical training would better meet industrial requirements in the second half of the nineteenth century.[58] Newly conceived institutions of applied learning in Germany were, in turn, highly valued and studied by the U.S. in the 1920s.[59]

As the WUT example showed, the call for effective technical institutions often came from influential circles in prosperous regional centers. Its foundation began as a influential initiative in which business interests reinforced more general ideas about the value of education and national progress. This, as well as the general dynamic of the evolving market economy, helps to explain the fragility of formal technical education even as late as 1900. Most WUT professors had secured employment on the basis of their practical merits. Even in centralized France, many institutions, such as the *École Supérieure d'Electrotechnique* at the University of Grenoble, reflected local initiatives.[60] In Germany, the comparatively strong role of the *Mittelstand* (locally-rooted mid-size companies) strengthened the importance of applied technical education. Though technical education was largely standardized before 1900, the engineering profession remained extremely heterogeneous.[61] This makes it difficult to estimate the number of nineteenth-century European engineers. The term "engineer" generally referred to anyone who worked in an engineering position, with or without formal education. Thus, the number of graduates from technical institutions does not offer a complete picture. Approximate figures can be secured through the increasingly large and specialized professional organizations. The growing membership of Germany's prestigious *Verein deutscher Ingenieure* (VDI, Association of German Engineers) illustrates the trend. Having 2,000 members in 1871, the VDI reached 4,000 within ten years and an astonishing 22,000 by 1908. However, even this growth meant that the VDI had failed in its original goal of enlisting the bulk of German engineers, as Germany held an estimated 120,000–180,000 engineers of various kinds around 1900, including 30,000 with higher education. Throughout the nineteenth century, the supply of engineers in Germany outstripped demand.

Although it certainly helped industrial development, the "overproduction" of engineers led to occupational social stratification and out-migration. In most European countries, on the other hand, the education of engineers was much more closely linked with market demand.[62] In Great Britain the availability of German engineers even reinforced an argument against building more British technical institutions.[63]

The extent to which this process can be understood as professionalization remains a matter of dispute. Certain characteristics can be seen across Europe.[64] Professional associations were an important element, pioneered by Britain. The Association of Civil Engineers was the first, founded in 1771, with more specialized engineering associations following from the mid-nineteenth century. In France,

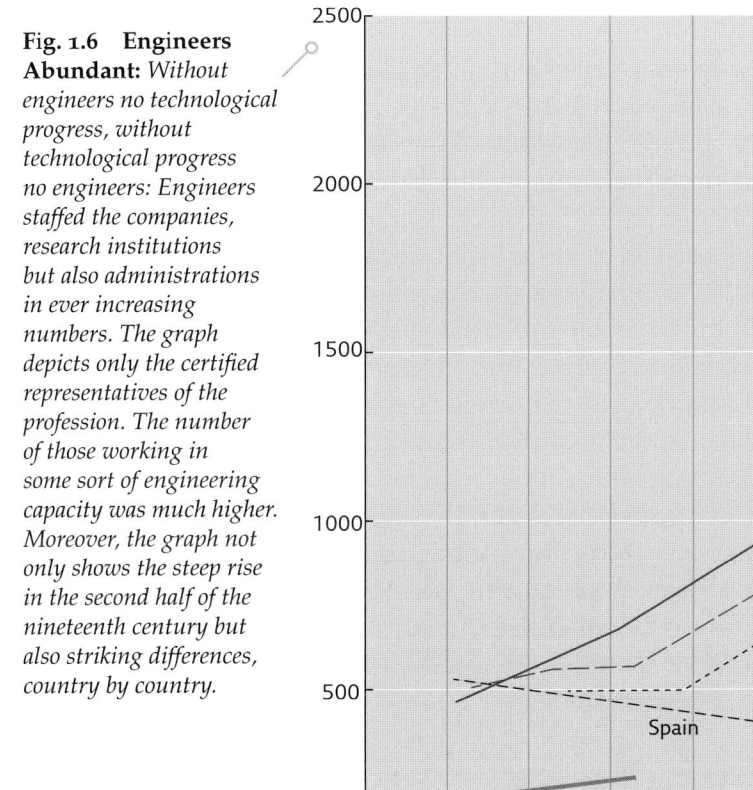

Fig. 1.6 **Engineers Abundant:** *Without engineers no technological progress, without technological progress no engineers: Engineers staffed the companies, research institutions but also administrations in ever increasing numbers. The graph depicts only the certified representatives of the profession. The number of those working in some sort of engineering capacity was much higher. Moreover, the graph not only shows the steep rise in the second half of the nineteenth century but also striking differences, country by country.*

the *Société des Ingénieurs Civils* started work in the context of the 1848 revolution, the German VDI in 1856.[65] Even in partitioned Poland, at least in the Austrian territory, such associations surfaced in *Lwów* (Lviv) and Krakow in 1862 and 1877 respectively. Tellingly, the Polish Polytechnic Society (*Towarzystwa Politechnicznego Polskiego*, 1835) originated in Paris, representing the technical expertise of Polish emigrants. The significance of these associations lay not least in their publication of technical journals that communicated the most up-to-date knowledge and ideas.[66] As has been shown for the Netherlands, engineering associations also had a key role in developing professional ideologies.[67]

However, the process had important national differences which derived mainly from the institutional structures in which professionalization developed. Whereas the British organization was rather elitist, the VDI in Germany was more inclusive, in spite of the strong hierarchies prevailing in the system of technical education. Most important was the role of the state, with a durable tradition of centralization and civil servant engineer positions in France, Prussia, Austria-Hungary, and Russia contrasting with the free "gentlemanly" occupations in Britain and Switzerland.[68]

Degree-holding engineers were a minority those in Germany. Far more numerous were those who attended a TH but did not graduate, attended one of the many technical schools of varied specialization and academic standing, or had only been trained on the job. This process has been described as "simultaneous professionalization and deprofessionalization."[69] Indeed, even while one group—the *polytechniciens* in France and graduates of other prestigious schools across the continent—devised increasingly sophisticated ways to bolster their status, the population of engineers outside these structures continued to grow. German engineers, for example, tried to emulate their French colleagues' status by stressing the importance of theory, particularly mathematical knowledge.[70] As in France and other European countries, constructing technical training institutions also always followed social considerations, with more or less explicit assumptions about which social classes were expected to attend the new schools.[71]

Germany's late, but very dynamic industrialization provides some important insights into the interconnection between technical education and social standing.[72] For rapidly-growing German industries, the availability of immense numbers of engineers without formal academic training offered huge opportunities: the massive oversupply kept wages low, and companies could hire as they pleased and promote rigorously on the basis of knowledge and proven ability. All the titles that individuals had fought for at technical institutes counted for very little, except in government and administrative positions.[73] This also implied that it was not clear whether a young engineering student's diploma would pay off, both economically and in career terms. In fact, educational institutions' elitist tendency to create standards to distinguish themselves, rather than to meet practical demands, presented a severe obstacle to success after graduation.[74]

This development helps explain why graduates who had invested funds and energy to secure these titles felt increasingly uneasy. For them, professional courts of honor were an essential part of the modernization of professions, rather than a "holdover" from the past. Professionalization helped groups like engineers to advance and, at the same time, to maintain a certain status.[75] In general terms, degrees attained the function in Germany that the prestige of the school had in France and that professional success had in Britain.[76] Titles held extreme importance in an environment of infighting and external positioning, a game that was never played only for salaries. The *Verband Deutscher Diplom-Ingenieure* (VDDI), founded a decade after the *Diplom* appeared in Germany, heightened expectations. A defender of formalities, this organization could be ignored had it not reflected a much broader trend. While credentialed engineers in Germany sought to establish more regulated working conditions in the private sector, echoing the civil service along French lines, practitioners saw such rules as potentially anti-business and technocratic. This was a discussion about know-how (favored by industry) versus credentialed knowledge (favored by degree-holders). However, the question of who might be called an engineer unsettled a group that, until only a few years earlier, had established itself as a profession with social standing. Thus those titled "engineers" soon found themselves in a war on two fronts: firstly, against the old elites' intent on retaining their positions in the administrative hierarchies; and secondly, against a huge group (a "mass," in the language of the time) of newcomers who challenged their status, influence, and professional standing. Meanwhile, professional power became concentrated in the hands of a small elite of managerial and entrepreneurial engineers.[77]

In most countries, engineering was a typical profession for achieving upward social mobility, although patterns differed considerably.[78] Its social composition depended on location; in Germany the nobility played almost no role at all, while in Poland it had a considerable share.[79] Social pre-selection was generally stronger in peripheral states where there were few specialized schools or where these were hard to access for certain ethnic groups. Being able to study abroad depended on raising the necessary funds.

Because one reason for the success of the THs was that their academic rigor imitated the older universities, new, more open, and basic schools became necessary in order to meet the massive demand for engineers. However, this development endangered the social integration of engineers at the very moment they thought they had achieved it, for example, in the Netherlands, Austria, or Switzerland.[80] This status-anxiety of academic newcomers, particularly profound in Germany, certainly spilled over into the twentieth century. The price for the immense formal and symbolic successes that engineers had made was the subsequent fear of losing ground again. By the First World War, this had already led to radical, mainly right-wing, political visions becoming popular within the VDDI.[81] Indeed, the 1920s' economic crisis intensified attempts to bring about professional exclusiveness for engineers.[82] These debates became fierce because the question of educating engineers had always touched on the self-description and self-perception of societies. In the Netherlands, for example, a state-sponsored commission inquired in the 1890s about the future of higher technical education and the question of how to relate its theoretical and practical aspects.[83]

This exploration of European engineers' status has not yet touched upon the equally severe struggles concerning what engineers actually dealt with at work or were trained to do. Generally speaking, the establishment of specialized educational institutions often lagged far behind technological development. For a long time, the traditional fields of expertise, such as mining, road construction, and bridge building, shaped the engineering of new technologies. Indeed, trained mining engineers proved decisive in the early phase of railroad construction.[84] Only in 1867, however, did the *Polytechnico* in Milan establish a special curriculum for railroad engineering.[85]

The complex nineteenth-century struggle between architects and "engineers proper" serves as a particularly-illustrative example here. As hinted earlier, the craft of building in many respects provided the ground upon which modern engineering developed. The growing scientification and specialization of engineering separated engineering, construction, and architecture into distinct fields, albeit along different national and regional paths. In some countries, like France, this process was institutionalized with special *Beaux Arts* schools for architects. This "sibling rivalry" between architects and engineers extended far beyond the question of which profession should be in charge of designing and constructing bridges or streets, into the field of the social influence.[86] With emerging urbanism and technology's immense role in conceptualizing modernist architecture, this distinction became much more than simply a problem of education.[87] On the other hand, the rivalry indicates the importance of cultures of construction. Technical drawing long remained a key skill in both fields, partly derived from certain models of education and partly shaping them.[88]

Against this background of struggles, both inside the profession and externally, the prospect of women becoming engineers did not appear particularly compelling, even to those who did not represent then-prevalent misogynist modes. While the entry of women into academia and engineering was generally a painful process, it was not the same in all European countries. Decisions about if and when women would be admitted to academic technical education were not only important to aspiring candidates; they also revealed a great deal about educational systems and societies. France, a leader in technical education in so many ways, was one of the last countries to admit women to its prestigious institutions. Not until the 1960s could female engineers enter the structure at its highest strata, the *Grandes Écoles*.[89] Russia, on the other hand, which had a number of high-level institutions for technical training, but no reputation for social progress, introduced a special Women's Polytechnical Institute in St. Petersburg following the 1905 Revolution (and opened completely new paths for women after the Bolshevik revolution).

It is not far-fetched to conclude that the very prestige of the French schools and their system's general inflexibility hindered the entry of women, even though this was not the case for traditional universities there. The character of the institutions in place played a decisive role. In the Swedish model, the rigidly demanding *Kungliga Tekniska högskolan* (Royal Institute of Technology) had almost a monopoly on the official higher education of engineers. Apparently, the function of this instituion, which was to train a technical elite to fill rare leaderhip positions in the economy and administration, led it to exclude women before the 1920s. This was the case despite Sweden's active feminist movement, strong progressivist tradition, and universities open to women in the 1870s.[90] In Germany, by contrast, women were, in practice, unable to enter any universities before the turn of the century and were first admitted to the THs not long before the First World War. The situation was similar in Austria, but not in more progressive Hungary. Germany's first female engineer graduated from the TH Darmstadt only in 1913. The cover of the popular *Berliner Illustrierte Zeitung* presented her surrounded by male peers, indicating this event's sensational impact.

The question of if, how, and where to train female engineers was thus closely related both to the profession's general struggle for status, indexing the place of engineers in society, and, of course, to larger social norms about gender equality. Threatening the new positions their male colleagues had fought so hard to acquire, emancipated women and the elevation of engineers were, in part, contradictory. Because the process of professionalization was well advanced before women's entry, the new instruments of titles, honor courts, and professional organizations were tools to preserve engineering's predominantly male character. The French case, meanwhile, had less to do with general discrimination than with the enormous importance of the *Grandes Écoles* in the reproduction of the French elite.

Yet, entering the profession via formal education was only one of many paths. In Britain, with its less-developed formal academic training and greater flexibility, women became engineers earlier.[91] This said, the growing stress on practical on-the-job training, in work environments that were highly coded and almost exclusively male, made occupational access difficult for women throughout Europe.[92]

Provided that women seeking to study engineering had the necessary funds, they could circumvent institutional constraints by traveling abroad. The large number of foreign students at technical training institutions in France, and also in Weimar Germany,

Fig. 1.7 Making the News: *Female engineering students were featured in the illustrated press as late as 1913. The widely-read* Berliner Illustrirte Zeitung *pictured Serbian-born Jovanka Bontschits as Germany's first "Miss Dipl.-Ing.," combining respect for the title with a somewhat patronizing view on women and technology. Bontschits had studied architecture at the Technical University Darmstadt.*

once these became accessible for women, suggests that Eastern European women in particular grasped chances that French or German women did not perceive or could not secure for social and cultural reasons.[93] The geographic movements of female engineering students reflected their enormous mobility.[94] Switzerland was particularly important in this respect. Suzanne de Dietrich, regarded as the first female French engineer, graduated in Swiss Lausanne.[95] The politics of comparison also helped the introduction of female engineers. Those who applied pressure to admit women to the *École Centrale des Arts et Manufactures*, a school below the level of the *Grandes Écoles*, referred to the example of Russia and Hungary. When the *École Centrale* finally admitted women in 1917, about half of its female students soon came from outside France.[96]

Moreover, certain engineering fields were seen as being more suitable to female talents and thus more accessible for women.[97] The emerging world of the laboratory also became an accepted female workplace.[98] On the other hand, these new opportunities had much to do with the First World War, the ensuing lack of engineers, and the rise of democratic systems in Central Europe after the War.

Engineers as Experts

In focusing on the relation of technical experts with state and society, this chapter has only explored a segment of engineers' increasing role in nineteenth-century Europe. However, it is essential to understand engineers in a broader sense; that is, to also see their stake in fields beyond technology narrowly construed. Only a small number of engineers went directly into planning and fields detached from technology. Still, the notion that an academically trained engineer needed to be able to think about the broader picture was an essential part of the French system, particularly general education at the *École Polytechique*. It also lay behind the emphasis on the *Diplom* as being distinct from less rigorous degrees in Germany and countries with similar systems. The concepts behind this education always reflected commitments about where expertise should be located vis-à-vis state and society. In this, they add depth to the emergence of technoscientific experts, whose scope was much wider than just solving technical problems. Unsurprisingly, both engineers and many social thinkers concluded that engineers should also play an active political role, that the whole society should be arranged by the new masterminds of industrial organization. The French thinker Henri de Saint-Simon (1760–1825) had already expressed such ideas by the beginning of the nineteenth century.[99] Hardly surprisingly, Saint-Simon became and remained the foremost intellectual reference point for the *polytechniciens*.

From Zurich to Warsaw, from Lisbon to London's Imperial College, the decisions to erect institutions of technical education always also reflected ideas about new technical elites and their potential to organize better societies and more competitive nations. It is hardly accurate to describe these foundations as models directly transferred from one place to another. Learning was mixed with rejection, admiration with competition. The purpose of this chapter has been, through reviewing European technical education, to understand how institutions functioned as incubators for experts, as places where states and societies, economies and scientific communities negotiated both using expertise and positioning technoscientific experts among state, society, and economy. The arguments around who might call themselves an engineer, who could enter which organization, or whether women could join the profession anticipated problems that have remained characteristic for technoscientific experts till today. In this sense, and in line with the definition of experts given in the Introduction, one could regard a certain stratum of the engineers, albeit a small elitist group, as prototypical experts. This was true when they took up expert functions in organization, risk management, assessment of the future, and linking economy and politics—increasingly important tasks for complex state agencies and technology-based companies. In many fields engineers were necessarily in much closer contact with the political sphere than, say, laboratory scientists. They were engaged in the great projects of their time, often with a Europe-wide dimension, and thus were involved in issues with strong political and social reverberations.

Engineers as technoscientific experts perceived themselves as agents of technological change, and as unbiased forces for progress outside the political sphere.[100] However, this was, at best, only part

Fig. 1.8 An Acropolis of Technology: *One obvious representation for the growing role of technological education was the structure housing the new institutions. These schools not only competed architectonically with the most characteristic buildings states had on offer, but also took prominent locations. The picture illustrates the first Portuguese school fully dedicated to engineering* (Instituto Superior Técnico) *at the top of a hill above Lisbon, following the Greek concept of the acropolis.*

of the truth. The focus on the education of engineers as technical experts in this chapter has emphasized that their rise was impossible without catalysts from the state. The constant development and growth of critical knowledge that states and economies required certainly contributed to a growing autonomy and authority of those commanding this knowledge, as the opening WUT example demonstrated. On the other hand, the ever more complex structures upon which these new technical experts relied increased their dependence either on large-scale commercial interests, company laboratories or administrative units or on quickly growing and steadily differentiating state bureaucracies, educational systems, and research organizations. Obviously, the role of these experts also changed in the wake of economic progress and rationalization. While nineteenth-century engineers pulled the strings, managers gradually took over in the twentieth.[101]

Did these early experts also think of themselves as Europeans and as solving European problems? Probably most did not, at least not in a direct and reflective way. They were, however, part of international organizations, effective agents of international exchange that mainly functioned through European networks.[102] The following chapter, which focuses on the early twentieth century, will look at the changing degrees of autonomy, authority, and coercion that technical experts faced vis-à-vis state, nation, and society. The chapter will explore the positions technoscientific experts took in nation states while highlighting how they inscribed a European pattern. The third chapter, concluding Part I, will then analyze the emergent dynamics of organizing knowledge and expertise beyond the nation state after the First World War.

5 luty 38
XVIII BAL

Warszaw[a]
swej
Politechnic[e]

2
Technical Experts as New National Elites

On May 15, 1891, the General Land Centennial Exhibition opened its gates in Prague. It bore the title "centennial" in reference to Habsburg Emperor Leopold II's coronation as King of Bohemia a hundred years earlier, and specifically to an exhibition—one of the first major industrial exhibitions in Europe—that had taken place on the occasion of that coronation. However, despite this evocation of the past and the aristocracy's key role, the Centennial Exhibition looked to the future. By this time, Bohemia had become one of the technologically most advanced regions in Europe, with prosperous and modern knowledge-based industry and high-level technical institutions in Prague and *Brünn* (Brno). However, the exhibition was not just a place to display new technologies; it was also the site of a struggle between the rising Czech nationalist movement and the Austro-Hungarian imperial rulers, both wishing to take credit for Bohemia's impressive accomplishments. Similar political tensions had shaped the Hungarian exhibition in Budapest in 1885, which served as an example for the Czech organizers.[1] Czech nationalists hoped to gain greater political autonomy like the Hungarians, or even, in the long run, to form an independent Czech or Czechoslovak state. The German population, which was well represented in Bohemian industry and technology, was reluctant

Fig. 2.1 **Exhibiting national progress:** *Technical Exhibitions reflected the complex motivations of making oneself internationally visible but insisting on national achievements at the same time. In Empires such as the Austro-Hungarian this was even more complicating. The* Kaiserpavillion *("Imperial Pavilion") of the General Land Centennial Exhibition in Prague 1891 showcases the modernity of the cast iron construction of the surrounding industrial palace, modeled after the construction of the "Galerie des Machines" in Paris. At the same time the Pavilion inside the palace called on the highest support of the established powers, here Emperor Franz Joseph.*

to accept this identification of technology and nationalism and boycotted the Prague event. The imperial government in Vienna and Emperor Franz Joseph I himself avoided taking a clear position. While imperial rulers often gave their names to technological and scientific institutions—strengthening the link between the empire and technological progress—the politically charged atmosphere of the Czech exhibition required them to act with caution. These tensions were played out in one of the main displays. The imperial government sponsored a *Kaiserpavillion* ("Imperial Pavilion") prominently positioned in the industrial palace. However, the exhibition journal of the Bohemian Architects and Engineers' Association referred to the *Kaiserpavillion* only as the *Königspavillion* ("King's Pavilion")—that is, stressing the national Czech dimension and displacing the imperial dimension. At the same time, it used the pavilion on its title page in order to profit from its official status.[2] Dynastic symbolism and state-of-the-art technology, represented by Prague's cast-iron industrial palace, merged in manifold ways.

What had started as a regional exhibition was transformed into a "national pilgrimage of the Czechs."[3] Technology and industry, both of which, in principle, followed universalist standards, became highly charged with nationalism. Economic pressure groups tried to improve the region's industrial level and communicate to a wider world what had been achieved. More important, exhibiting technology seemed an ideal means of demonstrating the aspiring nation's cause. This was particularly visible in all matters related to education, particularly technical education. The exhibition featured kindergartens, schools and, not least, the project to build a polytechnic institute in Prague.

In total, 2.5 million spectators visited this "technologically perfected city of the future."[4] As spectacular as many of the individual displays may have been, the exhibition as a whole did not live up to the standard of other world's fairs, such as that in Paris in 1889. Instead, Prague was part of a significant trend to merge technology and nation-building, as were the Swiss *Landesausstellung* or *Exposition nationale* ("National Exposition") at Zurich in 1883 and at Geneva in 1896, and the German *Jahrhundertausstellung* ("Centenary Exposition") in Breslau (1913), with its stunning reinforced-concrete hall commemorating the anti-Napoleonic Wars of Liberation. These exhibitions moved beyond a purely commercial focus and attempted to merge the nation and technology.

Exhibitions were not the only way technology could demonstrate national accomplishments. Museums of technology, which often emerged out of industrial celebrations or world's fairs, flourished in the early twentieth century, and oscillated between cherishing national cultures of technology and displaying the internationalism of technoscientific knowledge. One of the first of these museums was the *Deutsches Museum von Meisterwerken der Naturwissenschaft und Technik* (German Museum of Masterpieces of Science and Technology) in Munich. Founded in 1903 by the internationally renowned electrical engineer Oskar von Miller, it soon developed into an "Olympic Stadium of Technology" and became a model for similar institutions worldwide.[5] The technical museums in Vienna and Prague, both established in 1908, followed Munich in celebrating the ingenuity of engineering through modern cathedrals of technology, as did the London Science Museum in the same year.

One important function of these demonstrations was to measure national standing through international comparisons. The Czech knowledge displayed in Prague was compared with that of France, and in particular the Paris world exhibition two years earlier. "We even possess a small Eiffel Tower," explained one Czech commentator, referring to the 60-meter *Petřín* Lookout Tower, a replica of the original.[6] Those who organized the exhibition used a reference system that was at once national, international, and dynastic-imperial. In the center, however, stood the self-assured representation of the progressive forces, the technical experts. When the Czech architects' and engineers' journal enthusiastically exclaimed that the exhibition would bring the Czech nation to the cusp of the future, all while evoking the national flag as a symbol of progress, the editors also made it clear which group would play a decisive role in this progress. A representative article was headlined by a quotation from the famed British chemist Sir Humphry Davy: "What I am I made myself." This reflected the nation, but even more the self-understanding of the engineers who organized the exhibition and whose works it featured.[7] The General Land Centennial Exhibition documents two important characteristics of technological development in Europe before the First World War: the lasting influence of the old non-technical elites and the growing influence of the nation state on the development of technoscientific experts. The identification of technology and engineers with national progress was not unique to the Bohemian lands, but can be observed across Europe. It was also a European phenomenon in another sense: the national struggles for recognition and status in a highly contested arena of technological change always used other European nations and not the rest of the world as a point of comparison. This chapter starts by discussing the enduring presence of old-regime structures, in which technical experts rose to new significance. Their influence was, particularly in comparison with the U.S., a marked characteristic of Europe. We then take a closer look at technical experts as national personae and the price paid for shifting away from the more universalist developments described in the preceding chapter. Finally, this discussion touches on some of the changes that the First World War brought for the status of technoscientific experts.

Old Regimes & New Elites

Although the German sociologist Max Weber famously warned around 1900 that technical progress would form an "iron cage" in which modern society would be imprisoned, his Austrian colleague Joseph Schumpeter stressed that modern technology was established under conditions persisting from the *ancien régime*:

> It is with the *state* that the bourgeoisie with its interests seeks refuge, protection against external and even domestic enemies. The bourgeoisie strives to win over the state for itself, and in return serves the state and state interests that are different from its own. Imbued with the spirit of the old autocracy, trained by it, the bourgeoisie often takes over its ideology, even where, as in France, the sovereign is eliminated and the official power of the nobility has been broken.[8]

Schumpeter's evaluation of the bourgeoisie may be somewhat harsh, but his observation is highly relevant to the development of the new European elites based on technoscientific knowledge. Their rise can best be understood by considering local traditions and the general importance of the old regime, particularly the established elites. This influence was much greater than just the symbolic Bourbon *fleur-de-lys* in gold on the *Corps des Ponts et Chaussées* or the Grand-Duke of Baden opening a *Polytechnikum*. The new technical elites and the political and social forces that held power were mutually dependent. In ever-changing constellations, this interdependence formed a decisive structure for Europe into the twentieth century. As recent experiences had shown, there were manifold opportunities to establish and express the alliance between the still-potent powers of the past and the modern representatives of technology and science.

Germany embarked on industrialization later then Britain and France, but developed rapidly towards the end of the nineteenth century, while keeping its traditional authoritarian political system in place. The alliance between a new technical class and the old elites is particularly apparent there. As Berlin's *B.Z. am Mittag* (the self-proclaimed "fastest newspaper in the world") commented on the occasion of Kaiser Wilhelm II's fiftieth birthday, "having an Emperor for whom it matters whether his fast train travels eighty or eighty-two kilometers per hour, creates a feeling of inner solidarity among men who themselves always calculate how soon they will be able to arrive and which route they want to take."[9] The representatives of tradition and those of technical innovation at least claimed to be pursuing common interests. In 1890, the *Verein deutscher Ingenieure* (Association of German Engineers, VDI) and Wilhelm II formed an unofficial alliance to improve the standing of the *Realschulen* (practical schools) in order to stress technical education as the basis of "Germany's position in the world in peace and war." From 1900 onwards, traditional humanistic education and more scientific-technical education were given equal value, at least in theory. The VDI did not forget to thank the Emperor, stressing the many technical achievements upon which the "safety of the fatherland rests as much as the thriving development of humankind."[10]

One strategy adopted as part of the quest for respect was to compete with mainstream universities by creating more systematic academic training for engineers, including a more theoretical approach than had previously been characteristic of the polytechnic institutes. Indeed, presenting engineering as a scientific field in its own right proved successful, combining empirical and practical findings and problems. Graduates who had received a diploma liked to stress that academic engineers were equal to general staff officers.[11] What would ultimately distinguish these *Technikwissenschaften* (technosciences) was not least their connection with economic power and national might, which already hinted at technocratic notions flourishing in the years to come.

The *Technische Hochschulen* (THs) prospered in part as a result of royal backing. As Crown Prince, Wilhelm II had attended the inauguration of the Charlottenburg TH's new buildings in 1884. In his years as Kaiser, he was far more likely to be seen there than at the university located much closer to his palace. While the Kaiser may not have understood the content of the scientific lectures organized especially for him, he was perfectly receptive to the THs'

Fig. 2.2 Exhibiting Progress: *In 1903, the internationally renowned electrical engineer Oskar von Miller founded the* Deutsches Museum *of Masterpieces of Science and Technology which quickly developed into the role model for museums of science and technology worldwide. The term* Deutsches *signalled that it was not just a Bavarian but rather a national undertaking, although the collection and exhibition policy aimed at displaying the international dimension of science and technology. The central part of a triptych by Georg Waltenberger from 1916 shows Emperor Wilhelm II laying the foundation stone on November 13, 1906.*

demands for equality with the universities, both in a formal and particularly in a symbolic sense. Ceremonial chains for the rector and robes for the faculty were far more than merely ornamental. Decorated with the monarch's emblem, they were seen as expressions of the new THs' significance.[12]

An even more powerful symbol of the new relevance of engineers was the so-called *Promotionsrecht*, the right to grant the title of doctor. This was a highly symbolic matter but also one of practical relevance in title-fixated imperial Germany. In 1899, on the occasion of the hundred anniversary celebration of the TH Charlottenburg, Wilhelm II, symbolically dressed in the uniform of the engineer corps, officially granted the right of THs to bestow the *Dr. Ing.* (DEng) and *Dipl. Ing* (MEng). titles. This measure was more than just an organizational change. Wilhelm II stressed in his speech that the THs had proven in the preceding years that they were on par with the universities. What engineers regarded as a "patent of nobility" was indeed an attempt to forge a new elite to solve new problems. The THs and the engineers, the Kaiser explained, were to undertake the great "social" tasks of the future, challenges going far beyond purely technical matters. In the years that followed, engineers frequently referred to this so-called "social mission."[13] When in the same speech Wilhelm II noted the immense need for technical expertise in the colonies and imperial infrastructure, he confirmed that this was in fact a very political mission.

While the increasing social status of engineers was certainly based on a real need for their skills in infrastructure and industry, public recognition of engineers' accomplishments was still often a matter of politics and having the right personal connections. This point is well illustrated by Adolf Slaby, pioneer electrical engineer, president of the VDI and the Association for Electrical Technologies (*Verband der Elektrotechnik*, VDE), and rector of TH Charlottenburg. Slaby's informal contacts with the Kaiser (he built an electrical illumination for the royal gardens) helped substantially to bring about the formal changes described above. Like the rectors of the other Prussian THs in Aachen and Hanover, in 1898 Slaby became a member of the Prussian *Herrenhaus*, the exclusive First Chamber of the Prussian Parliament, on the tenth anniversary of Wilhelm II's reign.[14] Public reception was apparently positive. As one newspaper remarked, "the position of the technician was thus rightly honored."[15] Arousing wide public attention, this change illustrates how closely intertwined politics was with the engineering professions, and how strongly this link was still based on personal connections. The new seats had no legal basis, resting instead on royal prerogative. In its proceedings, the VDI reprinted the telegram notifying TH rector Adolf Slaby of his entry in the *Herrenhaus* as an indication of the new social and cultural importance of the entire engineering profession.[16] Thanking the Kaiser, the VDI stressed that the engineers would return the honor by helping him gain new glory for the fatherland. In fact, to celebrate the twenty-fifth year of Wilhelm II's reign, all the German THs gathered to confer on Wilhelm II an honorary *Dr. Ing*. On the surface, these actions were part of a discussion about the organization and improvement of technical education. At the same time, however, it also concerned the social position of engineers. In an extremely short period, a group that many had once looked down upon came to be regarded as essential to the Empire's welfare.[17] As Wilhelm II explained, a new aristocracy for a new age could emerge only if the "best families would direct their sons towards technology."

At first glance, such power politics and obsessions with formalities seem to be a particularly Prusso-German phenomenon. Upon closer examination, however, these exchanges touch on general problems that were relevant for the whole Continent. In Belgium, a forerunner in nineteenth-century industrialization, more than two hundred members of the prestigious Order of Leopold were engineers, and thirty-seven Members of Parliament held an engineering degree.[18] By 1900 engineers had also been knighted in Britain.[19] In spite of the many regional and national variations, an overall European pattern took shape: The prestige of tradition was combined with the appeal of the modern, adapting old structures without challenging them directly. Referring to tradition offered a way to legitimate new technologies and help them gain social acceptance. Ship launchings or the completion of dams became powerful and highly emblematic events in which technology was affirmed as a national endeavor, in which the old elites also paid tribute to the new technical class.

A figure like Slaby, whose immense influence in science policy far outweighed his scientific accomplishments, also shows the ambivalence of this strategic alliance for both sides. Politics risked being identified with technical failures in new and highly complex fields of innovation. The controversy surrounding Slaby's experiments in radio communication was not serious enough to ruin either his or the Kaiser's reputation. Yet other technologies endorsed by the Kaiser, such as Zeppelins which often bore royal names, were riskier. A Zeppelin crash had the potential to undermine both the technology and its royal supporter. Negative results of a policy of actively advancing technology could not be deflected in into the anonymous realm of the bureaucracy.

The alliance between the state—originally the dynastic state, but more and more the nation state—and the experts became very important with the outbreak of the First World War. A striking example is offered by Germany's *Kaiser-Wilhelm-Gesellschaft* (KWG) (today *Max-Planck-Gesellschaft*), an organization set up before the First World War to conduct large-scale research. After the war started, Wilhelm II granted the KWG the right to fly the imperial flag "in orange-green-diagonal stripes with the imperial code-of-arms." Adolf von Harnack, an eminent theologian and the KWG's first president, reacted with an emphatic statement: "This will always remain unforgettable for our society, and will fill her with grateful pride to have received the flag from the headquarters during the period of the victorious World War, out of the hands of its ardently loved and admired emperor and protector."[20] Harnack's sentiments were widely shared, underscored by the fact that the KWG kept its name after the end of the monarchy in 1918 (thanks to a deliberate decision by its senate) in the face of severe attacks from republican politicians.

The closeness between the old regime and its technical elite was not simply a result of opportunism on the latter's part. Historically, technical experts have tended to show a strong sympathy

for the relatively fast decision-making processes of authoritarian structures and an equally strong irritation with the more complex procedures of democratic societies. The structure of research institutions in Germany reflected this; they were created according to Harnack's idea that supporting particularly exceptional scientists was the best way to make progress in research and technology. Research centers were therefore often built around a single prominent individual.[21] While *Grossforschung* (Big Science) such as pursued by the KGW was a result of huge advances in science and engineering, it also emerged from this constellation, which focused on the accomplishments of brilliant individuals. Harnack developed into a kind of meta-expert on the relation of society, state, and technoscience, skillfully using the most glossy catchwords of applied science in memos that he produced for the Kaiser and the public. In an equally skillful way, he appealed to Germany's national pride and fears of being overtaken by other countries, particularly the U.S. In Harnack's view, only radical

Fig. 2.3 Technical Icon: *The Portuguese Canal do Tejo was one out of numerous technical projects all over Europe which turned into national icons—due to their technological ambition or their importance as an infrastructure (or both). Those in charge of conceiving such projects could increase their standing considerably and sometimes turned into national icons themselves.*

technical advances could secure Germany's place in the race of nations. Wilhelm II was certainly not the sole promoter of large-scale research, but "using" the emperor made it much easier for Harnack and many others to establish a research structure that had not been part of the German academic system.[22]

Finally, the flag anecdote illustrates the price for this seemingly win-win arrangement between the powers of the past and those of the future. The KWG's technoscientific experts, such as Harnack and Slaby, identified to a high degree—indeed, much more than seemed necessary—with the monarch and the national project. The chemist Fritz Haber epitomizes how a scientist's allegiance to the ruling powers led him to develop technology which took an incredible human toll. Haber became the founding director of the Kaiser Wilhelm Institute of Physical Chemistry and Electrochemistry in 1912. Only two years later he started research on using poison gas for warfare.[23] Granted the rank of captain in the armed forces by Wilhelm II, Haber personally oversaw German gas warfare advances in 1915 and became one of the first examples of the moral failures of German scientists in the twentieth century.

As Haber's example shows, loyalties to the old regime were not the only ties that shaped experts' actions. The First World War proved the power of nationalism, which few experts in the participating countries resisted—though most did not compromise themselves to the degree that Haber did. The bond between nationalism and technical expertise was established much earlier, however.

Heralds of the Nation

Technical experts professed to be apolitical specialists who, regardless of their national background, could work in any place or for any political entity when required. At the same time, and partly for the same reasons, they identified strongly with the nations and states in which they were educated or raised and where their technical expertise could allow them to become national symbols. This also helps explain the high expectations for engineers and technical elites that far surpassed their essential function. "The engineer is the king of our epoch," said the 1873 edition of the *Larousse* encyclopedia in the article on this profession. In the same vein, the German industrial architect Peter Behrends stated in the early 1900s that "the engineer is the hero of our age."[24] Clearly, engineers personified a major transformation taking place in society, in which technology played a key role beyond its narrower technical functions.

Nevertheless, technical experts could not have acquired a prominent public position without the sounding board of nationalism, or at least an increased national consciousness articulated by the mass media. One reason was the simple fact that engineers built national infrastructures. Large-scale railroad and telegraph networks tied nations together, and created a symbolic connection between the new technologies and the state. This applied both to the forerunners of industrialization as well as to less-industrialized countries: indeed, engineers were particularly valued in countries that were eager to catch up with their more developed neighbors.

In Portugal, for example, technology rather than politics played a decisive role in the "regenerationism" —the attempt to lead Portugal back to its old glory—during the second half of the nineteenth century.[25] Landmark projects such as railroad construction and the newly conceived polytechnic school in Lisbon helped engineers rise to a new prominence. Portuguese engineers figured as national heroes, erecting technological monuments that used a symbolism that referred to European examples, but in fact contained a strongly cohesive national force. One Portuguese newspaper addressed the Alviela Canal, the aqueduct that catapulted Lisbon into urban modernity, as "an eternal monument reminding us of the audacity and intelligence of those who planned it and built it."[26] Spectacular opening ceremonies for such technological artifacts provided the pictures and stories that acquainted the nation's illiterate majority with its technological progress and with those who had made it happen.

In much the same way, Spanish engineers fashioned the technological transformation of Madrid's city space: specifically, the connection of the capital with the railroad network and a modern water supply via the Isabel II Canal. The official opening ceremonies in 1858 celebrated the technology and the engineers who had enabled such an undertaking. Directly inspired by these engineers, newspaper writers used biblical metaphors to convey the magnitude of the achievement. The "mass extravaganza" of the opening provided "a vote of thanks to those who have done it so well." Unsurprisingly, a civil engineer guided the Queen through the opening choreography.[27]

In Switzerland, the Gotthard Tunnel, inaugurated in 1882, turned into a powerful and lasting symbol of Swiss technical

Fig. 2.4 Investing in Expertise: *When states started to heavily invest in technical education in the nineteenth century, they hardly followed humanist ambitions. Rather, investments in education reflected the insight that technoscientific experts became essential for all activities a modern state relied on. This table shows how dramatically expenditures for engineering schools grew in the case of Prussia, one of the most dynamic states in nineteenth-century Europe.*

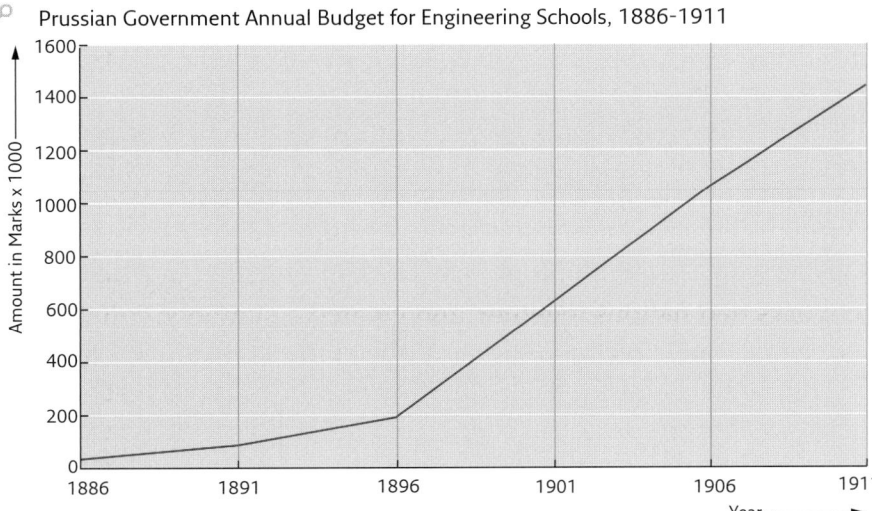

Prussian Government Annual Budget for Engineering Schools, 1886-1911

genius. Interestingly, the international composition of the building committee, including Italy and Germany, as well as the engineers erecting the tunnel, did not prevent the infrastructure from being identified with Switzerland. Rather, it helped turn the tunnel into a symbol for the Swiss nation's ability to combine the traditional and the modern, in affirming that the Swiss were on par with the most progressive developments elsewhere in Europe.[28]

Well into the nineteenth century, Germany was an industrial latecomer compared to France, Belgium, and Britain. Even so, German engineers successfully used technical infrastructure like dams to present themselves as benefactors of the nation.[29] Not only were engineers actively involved in bringing technological processes and amazing architectural accomplishments to Germany, they also presented themselves as taming nature for the sake of civilization. Initially, the inspiration came from outside Germany: civil engineer Otto Intze applied French technology he had admired while visiting the world exhibition in Paris in 1878 on behalf of the Prussian Ministry of Public Works, and learned from the "symbol of Belgium," the Gileppe Dam in the Ardennes. That modern German dam building started in annexed Alsace was no coincidence. Officials felt that the major civil engineering project would impress the residents of this region and overcome their reluctance to suddenly becoming German citizens. Intze also pioneered in propagandizing his endeavors in dam building, transforming

Fig. 2.5 **Visualizing the Modern Nation:** *The Stockholm Exhibition was held in 1930. Like many other exhibitions at that time, it served to demonstrate the modern character of its host country—in this case Sweden. More so than in previous exhibitions, the idea of modernism and innovation was understood broadly, encompassing potentially all areas of life. Tellingly, constructing a positive national self-image would only work when internationally reaffirmed.*

himself into a publicly-celebrated technical expert. Public trust in experts who oversaw critical and potentially dangerous technology became palpable. The identification of engineers with nationally significant technological projects enhancing national pride increased respect for engineers throughout Europe.[30] As French businessman Georges Besse later remarked on the polytechniciens: "they strove 'to build something for France.'"[31] French engineers employed by the government—such as those executing the great schemes of Baron Hausmann's Paris or the famous star-shaped railroad system—achieved the status of heroes.[32] Tellingly, Gustave Eiffel, one of the most famous engineers of the nineteenth century, not only constructed one of Paris's characteristic landmarks, but also provided its name.

In England, naval engineers constructed much more than ships, helping to construct a national identity.[33] Here, engineers and inventors had already started to rise to fame by the early nineteenth century, reflecting the impact of what they had created. Inventors like Robert Stevenson and James Watt prompted cults which always also reflected the self-image of the nation as ingenious and able to compete with others.[34] Widely known engineers like Intze, or

France's far more influential and glamorous Ferdinand de Lesseps, who organized building the Suez Canal, as well as the many other technological heroes in practically all European countries, speak to what historians have called the "myth of the engineer."[35]

Until he became entangled in the pitfalls of the turning political tide, prominent Russian railroad engineer Iurii Vladimirovich Lomonosov embodied that country's modernization.[36] Technical experts like Ferdinand Count Zeppelin in Germany[37] became national heroes celebrated for their scientific and technical achievements, while to an even greater extent representing the nation's genius and, therefore, its future.[38] In Italy, radio pioneer Guglielmo Marconi's fame extended far beyond his field of expertise.[39] The myth of the engineer lasted well into the twentieth century, and indeed, the most impressive example of the link between social expectations on technology and the rise of "technocelebrities" is early space flight in both the Soviet Union as well as in Central and Western Europe.[40]

The prominence and celebrity of experts was not limited to engineers, however: nineteenth-century scientists were breaking new ground in areas of great concern to almost everyone, particularly in medicine and hygiene, and they, too, gradually gained public acknowledgment. When German bacteriologist Robert Koch welcomed more than 5,000 participants to the tenth International Medical Congress held in Berlin's voluminous *Circus Renz* building, he presided over more than just a special interest event.[41] Louis Pasteur transformed the French political and social landscape through research that resulted in pasteurization and vaccination, thereby becoming a public icon.[42] Marie Skłodowska Curie's work in physics and her discovery of radioactivity had less immediate social impact, but her iconographic status in France was no less great.[43] Significantly, the 1858 unveiling of a statue for the British physician and pioneer of smallpox vaccination Edward Jenner on London's Trafalgar Square met with severe disapproval. Supporters of the military successfully argued that the vicinity of Lord Nelson's monument had to be reserved for generals, not scientists, and had Jenner's statue removed.[44] Although the episode was a setback for those wanting to give the nation's scientific heroes appropriate recognition, it still attests to the public standing eminent scientists were achieving—in some cases, even far beyond their home countries.

Largely thanks to their fame within the national sphere and a nationalist discourse, public scientists sometimes became important political players. A case in point is the chemist Ignacy

Mościcki, who rose to Poland's presidency in the 1920s.[45] Such technoscientific personae relied on a nationalized public sphere, which was again situated in a European framework of cross-references. To become proper heroes, scientists had to represent their country's accomplishments abroad. Therefore international recognition, such as the acclaim that came from receiving a Nobel Prize, was essential.[46] The rise of such figures also interacted with the emergence of mass communication, popular novels, and, later, films.[47] Against the backdrop of a growing fascination with technology and science, as popularized by Jules Verne and his many imitators, these individuals could claim increasing independence and autonomy from their homelands as a result of popular backing. Heinrich Seidel's "Engineer's Song" (*Ingenieurlied*), which boldly sketched the wide range of engineers' activities spanning the globe, came to the clumsy but telling conclusion: "Long live the engineers! They represent the true spirit of the most modern times—from country to country, from ocean to ocean—the engineer."[48]

Engineers throughout Europe would have subscribed to the emerging notion of heroically mastering nature "from country to country, from ocean to ocean." "There is a faith in us, this faith is our power, the future will belong to us," engineers of the Russian Agricultural Academy proclaimed even before the Revolution.[49] In this sense, engineers also symbolized the heights reached by a century of technical education. In almost all European countries, the public clout that technical experts commanded translated into new chances for influence. However, this was not by any means an automatic process.

Reinventing the Nation in War & Crisis

Given their swelling social value, engineers across Europe increasingly demanded a say in their nation's affairs. The expression of engineers' demands spread across Europe through lively interchanges and the inspiration of different national examples, which increasingly included the U.S. From at least 1918 onwards, U.S. technocratic models became influential, including the theories of the American economist and sociologist Thorstein Veblen on the desirability of technical experts having a stronger political voice.[50]

Even before the First World War, engineers throughout Europe had agreed upon a professional ideology which in many ways took up Veblen's points. Although the Netherlands hosted no autonomous technocratic movement, Dutch engineers developed homogeneous and far-reaching demands for greater participation by the technical elite in what could be described as a social-democratic-cum-technological transformation of society. Based on more effective professional organizations, and on engineers' rising presence in public administration, they successfully advocated technoscientific solutions to social problems in public health or urbanization. Immensely-influential engineer Cornelis Lely, head of the Dutch Ministry of Water Management, Trade, and Industry, claimed that "we [the engineers] should not remain technicians exclusively." Another prominent Dutch engineer demanded the training of "social engineers" able to assess the social impact of their work.[51] While such views were shared by forward-looking members of the profession, many others were also happy to stick to a merely technical interpretation of their work. Yet, all could agree on the need to improve their social position, in particular with regard to the frustrations of their day-to-day experiences in the lower or middling ranks of private business.

Engineers claimed not only to be generalists, able to meet all future challenges, but also asserted that they had an ethos which led them to solutions dictated only by objective demands and not by political or other criteria. Engineers, generally successfully, presented themselves as men in the middle, between theory and practice but also between national economic interests and those of the workforce.

Based on the core ingredients of purportedly neutral technical expertise and an elitist, potentially anti-democratic politics, the engineering ideology included the claim that only their small technical elite possessed the knowledge necessary to command nature and society. The First World War did not substantially alter this line of thought. Instead, the war and the ensuing discussions on technology's role in winning or losing it intensified the pre-war discourse.[52] The war economy had a catalytic effect on technology in all countries involved, although to different degrees. In Germany, where the need for resources was particularly pressing, Walther Rathenau developed the concept of a "New Economy," which was very widely discussed throughout Europe. Rathenau headed the Raw Materials Department. Using his experiences with reordering the German economy through enforced planning, this new economy would be built on state intervention and thus naturally give technocrats an advantage over classic enterprise managers.[53]

While modern states had for a long time increasingly relied on technical expertise, this became even more true after 1918. With the establishment of new states, or at least new political systems, as in Germany and Austria, and in some of the victorious countries, governments needed to prove their ability to solve social problems in order to demonstrate their political legitimacy.[54] Providing health services or adequate housing would demonstrate that new governments could surpass the administrations which had collapsed after the First World War. This was even more so where military defeat and social crisis coincided, or where the war had changed the political regime and brought about completely new political entities, regardless of whether this was seen as positive or negative.

In an extreme way, the changes of 1918 highlighted a much older development. Just before the outbreak of the First World War, Gyula Hevesi, the "mastermind" who mobilized Hungarian engineers during the war, had complained: "There is a much greater problem here to be solved, a problem of which my individual grievance is but a tiny, though inseparable, part. The task is to destroy the barricades that have been raised by today's society in the way of all who possess nothing but their talent and knowledge."[55] An ardent reader and follower of Schumpeter, Hevesi believed that know-how was to become the decisive factor of his time.

This was not only a strong symptom of a growing "engineer consciousness." It was also a stepping stone for a fervent anti-capitalist, state-oriented ideology that guided the majority of Hungarian engineers after the war that their empire had lost so dramatically. When engineers suffered a drop in salaries that was sharp and disproportionately dramatic compared to other professions once involved in the war effort, radical solutions became more attractive. Like their colleagues in other European countries, Hungarian engineers lamented about being sidelined in capitalist production. They also believed that a scientific approach to politics would harmonize with a scientifically-grounded state socialism. Supporters of such an "Engineers' Utopia" closely followed Russia's war socialism, as well as the U.S. public debate Veblen initiated.

What was distinctive about the Hungarian engineers' technocratic vision was how close they believed they were to the political center. After all, as these engineers informed "technical science workers" in and outside Hungary via telegraph, there was a chance "that the technical sciences in their purity will play the leading role in production."[56] In fact, the outcome was much more prosaic, with bureaucrats rather than engineers initially assuming political power.

Still, the challenges for the state posed by the First World War generally enhanced the technical elites' position. While those commanding critical knowledge often could stabilize their own positions, such upheaval still demanded strategic decisions and arrangements by technical elites. In Poland, soon after the war, engineering associations voiced demands for a stronger political role, albeit to a less radical degree than in Hungary. During the First World War, Polish engineers effectively aligned their cause with that of their hoped-for nation. They believed that their expertise would also enable them to deal with social problems whose solution was essential for the country's future. Hence, engineers should occupy key positions in the higher administration. The training of engineers should receive particular attention, not just in terms of technical knowledge, but also in the skills necessary for the formation of a democratic society.[57] Similarly, Czech engineer Albín Bašus proposed a more integral technical education which was to produce "organizers" and "leaders," the technical experts needed to shape a modern nation.[58]

Such demands—for example, attempts to ensure that no foreign specialists would take qualified positions in the administration that could be filled by Polish experts, or to prevent state meddling with what engineers perceived as their deserved rights—had a clear professional edge. However, the engineers' reasoning and political agitation went far beyond such issues. Continuing after the war, their professional organizations argued that only their expertise would make it possible to overcome social differences and, in particular, problems of party politics. In line with technocratic conceptions, engineers claimed to offer a neutral basis for establishing effective government. They also could to transfer the necessary technical and organizational knowledge into government policies. A prominent example was the electrification of the country, a project that extended far beyond simply providing energy for industry, also touching on such crucial matters as hygiene and public health, understood broadly.[59] Although such interventions did not yield direct political influence, technocratic models popular with Polish engineers had an extremely strong influence. This was true for the whole notion of the *Sanacja* ("healing") political regime, established in 1926, but also for the leading role of "state technicians," such as Eugeniusz Kwiatkowski, who oversaw huge planning schemes in the 1930s.[60]

However, technical experts were scarce and found mainly in Poland's former Austrian territory, where the relevant training had been easier to secure. A 1931 estimate counted only 25,000 technicians and engineers—only about 10,000 of whom had degrees— in a country of 32 million inhabitants, due to the restrictive policy of the partition powers that lasted until 1919.[61] Having spent almost the entire second industrial revolution, that is, the rise of science and technology production methods before the War, without relevant training facilities, the new Polish state relied mainly on two sources, in addition to the small technical elite trained before 1918. The government actively encouraged the return of Polish experts from Western Europe and the U.S. Secondly, it profited from the initially small group of experts who had trained under the auspices of the new republic.[62]

After the First World War, the Warsaw University of Technology (WUT) retained its substantial political relevance and its tight connection with the fate of the newly-founded Poland. This was true, coincidentally for the transformation of its main building into the headquarters of General Józef Haller's general staff during the Polish–Russian war of 1920. In a much more structural way, this was also true for the institution's standing as the country's major training ground for technical elites. The bare numbers (in 1918, some 2,500 students enrolled at the WUT, the total doubling by 1939) fail to provide an adequate measure of their significance.

The new elite, trained mainly at the WUT, had significantly more women and younger people entering high positions than in Western European states, often in newly founded planning bodies. For example, of the 813 interwar graduates from WUT's architecture department, ninety-six (12 percent) were women. This compares to some 300 female students (not graduates), many of whom were from Eastern Central Europe, matriculated at the eight (later nine) THs of the progressive Weimar Republic, or about 4 percent of Germany's 7,000 students of architecture.[63]

| Technical Experts as New National Elites | 75

Fig. 2.6 National Hero of Technology: *When Dutch engineer Cornelis Lely died in 1929 a whole nation mourned him. Unsurprising in a Dutch context, Lely bought his ticket to fame through his tremendous achievements in waterworks. His plans prepared the enclosing of the Zuiderzee. The city of Lelystad, built on the land resulting from Lely's efforts, commemorates one of the technical heroes found in all European nations.*

As in Greece or Portugal, Poland's technical experts depended heavily on the state. Nearly half of the Polish engineers in the interwar period served the state or state-dependent institutions.[64] In Greece, the ratio was even higher: the nineteenth-century state was nearly their sole employer. State dominance in the labor market contributed to this group's strong elitist self-conception and technocratic edge. Following the massive changes Greece underwent after the Balkan wars and the ensuing doubling of the nation's territory, Greek engineers claimed leadership in what they perceived as a necessary transformation of the state.[65]

When stressing the alignment of engineers with the national cause highlighted here, it is important not to lose sight of the European perspective. In 1920s and 1930s technical exhibitions below the level of world's fairs, the European dimension becomes very evident. One example, particularly telling, is the 1930 Stockholm exhibition.⁶⁶ Reacting more or less directly to the Paris Exposition of Industrial and Decorative Arts (1925), Sweden attempted to merge what was perceived as European or even global modernism with the essence of "Swedishness."⁶⁷

Czechoslovakia, the new state encompassing the former Bohemian lands, staged a national exhibition in 1928 that was even more ambitious than the one in 1891. In line with the new nation's desire to look towards the future, Brno was deliberately chosen this time, rather than Prague. Instead of the capital's historical eminence, the organizers opted for the country's dynamic second city because it "had no tradition and no past." They argued that the exposition would allow Brno to represent the new state (only ten years old at the time). The workshop and the laboratory represented the new model of the state.⁶⁸ This visual identification of engineering and scientific knowledge with a new nation, new technology, and social reform culminated in the "electrical farmhouse," one of the show's major attractions.⁶⁹

This second Czech exhibition encapsulated an important transformation of technical expertise between the late nineteenth century and the interwar years. Technical expertise now functioned, in a rigid and once unimaginable way, in the service of the nation. For technical experts, this suggested opportunities. Their new status was visible and plausible to the general population. Moreover, in line with the older visions of Henri de Saint Simon and Auguste Comte, such experts were offered a voice in political and social questions. This was particularly true in periods of transformation. Following postwar demobilization, states pursued new applications of engineering and technology.⁷⁰ Their increased need for legitimacy in fields such as public health and housing generated challenges and opportunities for experts. In some cases, experts were able to overcome traditional hierarchies. However, those who rose to new eminence through "national engineering" projects also had to realize that they needed to decide where their loyalties lay, beyond their mere professional identities. The flipside of this process brought increasing constraints and coercion for technical experts who aligned themselves more closely than ever with political regimes.⁷¹ This strain was only partially eased by participation in international networks and in a European space of communicating knowledge, which eroded during and immediately after the First World War but recovered to a surprising extent in the 1920s.⁷² The War obviously strangled the vibrant internationalist tendencies of

Fig. 2.7 Celebrating "Our" Experts: *A new elite for a new state—technical but more glamorous than might be expected. The poster advertises a ball-event dedicated by the city of Warsaw to "her University of Technology" in the 1930s. Such events and posters illuminate the important role such institutions played far beyond their narrower purpose of educating experts.*

the late nineteenth century. It brought the marriage of convenience between experts and nationalism to new heights. Yet the radicalization the war brought, and the new position achieved by experts in its wake, also formed the preconditions for more intense and visionary forms of exchanging, applying, and developing expertise. These developments will anchor the following chapter.

3
Architectures of Knowledge

In 1929, two of the most brilliant thinkers in modern architecture became involved in a heavyweight intellectual clash. Karel Teige, the Czech *enfant terrible* and an ardent proponent of scientifically grounded architecture, then only 29 years of age, published a fierce critique of the blueprints for a "Mundaneum" in Geneva conceived by Franco-Swiss architect Le Corbusier. In so doing, Teige started what became known as the Mundaneum Affair.[1] In order to understand why Teige believed this issue demanded a general debate, it is worth looking at his characterization of the project at stake, formulated and promoted by Belgian intellectual Paul Otlet. The Mundaneum, Teige explained, was to be built on international territory near Geneva, "a city of world culture." In a first stage, the new city would comprise "the five traditional institutions of intellectual creativity: Library, Museum, Scientific Societies, University and Institute." Moreover, the Mundaneum should serve as a center for "professional, scientific, philosophical and artistic unions, social and artistic movements and the headquarters for educational and hygiene groups, and archives." As Teige put it, in a manner that matched the ideas of Otlet and his followers, the Mundaneum was to become "a center of the modern world," a home for a "wider and more realistic League of Nations," thus developing a "great

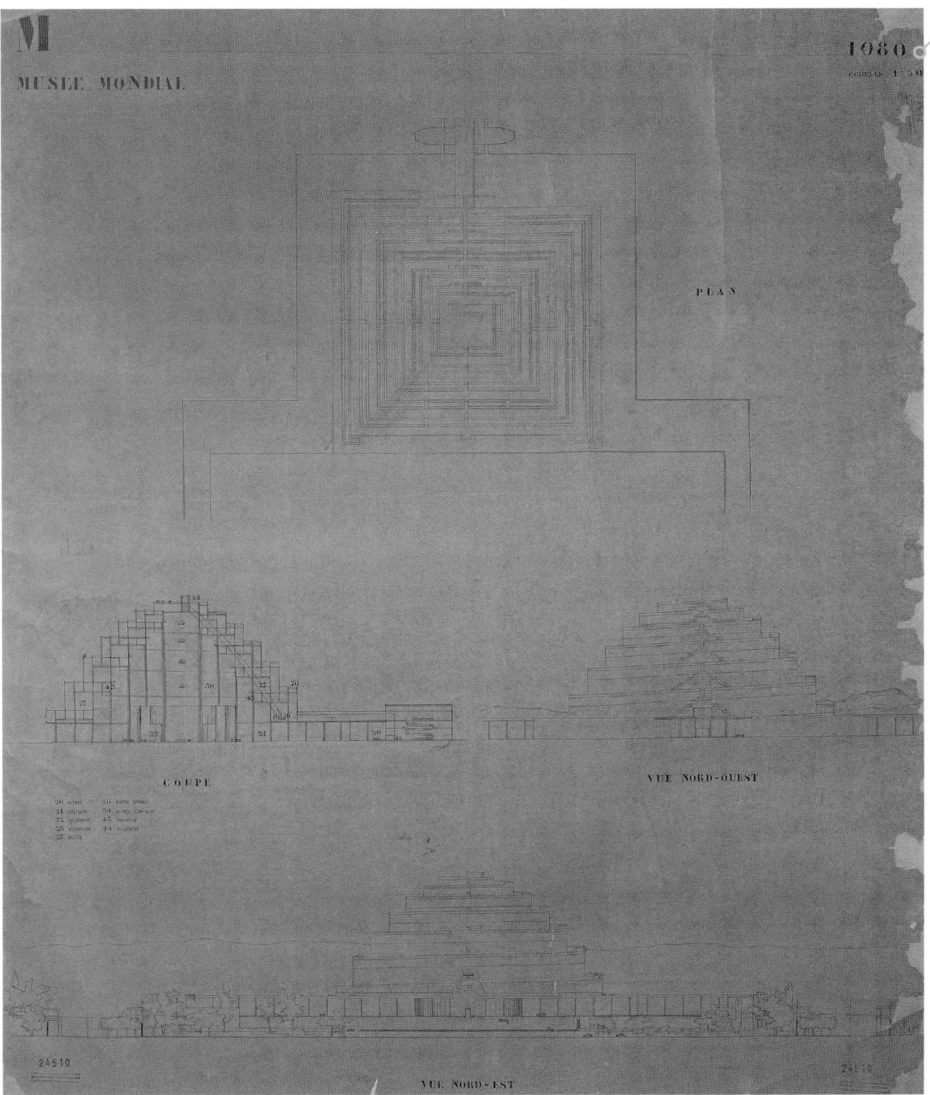

Fig. 3.1 Drawing the Dream: *Choosing Le Corbusier for concrete designs of Paul Otlet's longtime Mundaneum project was hardly a coincidence. Le Corbusier had made a reputation as a planner of large-scale structures and whole new cities, including a hyper-modernist entry for the competition for the Palace of the Nations in Geneva. In his Mundaneum design Le Corbusier displayed his visionary potential without going for the outright utopian.*

work of peace after war." In this way, it would express a new epoch characterized by "cosmopolitanism, internationalism, and mondialism, appropriate to a time when world-wide measures and opinions dominate the lives of nations and individuals more than provincial and personal ones do." The Mundaneum, in short, would both capture and propel the progress of mankind.

Teige subscribed to Otlet's criticism of the League of Nations. As a union of governments and treaties, rather than nations and cultures, the League would not be able to secure world peace and international cooperation; peace was a universal concern, not just a political one. Thus, in Teige's eyes, Otlet's plea for a "wider League of Nations" was justified. Incorporating hundreds of international associations, whether scientific or not, would allow the creation of a "center of world intelligentsia." This "monument to contemporary men" would be the "modern equivalent" of the library of Alexandria, the French academy, or other historical institutions of knowledge. Knowledge was to replace politics, in its narrowest sense, as the driving force of postwar recovery and international cooperation.

Donations from interested individuals and institutions would finance the Mundaneum, "in the same way as world exhibitions are created." Switzerland, on the other hand, was expected to provide the territory for free, as the Mundaneum would enhance Geneva's role as a hub of international organizations. Already, Teige stressed, 250,000 foreigners visited Geneva each year. It is not surprising, therefore, that transport and communication facilities were a large part of Le Corbusier's concept. The plan comprised a port for ferries, a harbor for motor boats, and a railway station, not to mention an airport and radio station and light towers that would illuminate the whole complex at night.[2]

Le Corbusier's architecture took up Otlet's ambitious projection. For example, the museum, an "idearum" picturing "the thoughts that are hidden under facts," was to be structured in three aisles featuring the categories of time, place, and type, unwinding in a spiral. Even more ambitious was the library, "cataloging all 'problems of ideas,'" which would become "a modern documentary encyclopedia." Le Corbusier placed particular emphasis on providing books that were censored in the states in which they had been published or where, due to war, they were inaccessible or threatened. The library would be linked to existing institutions, such as the International Book Exchange Service. Moreover, it should contain not just books, but also other media such as photographs, records, and films. Glass and steel pneumatic chutes and freight elevators would architecturally correspond to the modern artifacts housed in the structure.

In much the same vein, the architecture of the other buildings spoke to the global, modern, and grandiose reach of their purposes.[3] The Mundaneum incorporated a number of existing ideas and organizations, such as the 1920 plan for a Confederation for International Student Cooperation or the International Institute for Advanced Studies that already existed in Geneva. Of course, the Mundaneum could only be realized gradually, which the case for its construction acknowledged. Buildings for the library and museum were to be erected first; these would offer a provisional home for the international university and international unions. As the International Labour Organisation (ILO) already had its headquarters close to where the Mundaneum was to be sited, Le Corbusier's plans had the ILO forming an entrance to the Mundaneum. Exhibitions on various activities such as urban planning, already ubiquitous in Europe and America, would find a permanent place at the Mundaneum. The planned World Institute, on the other hand, would reshape the trajectory of technical laboratories and offices of social work, transforming them into "a center of composite knowledge" and a synthesis of the sciences and social reorganization. With such a holistic approach, unsurprisingly Le Corbusier added sporting facilities and planned to integrate the premises of the Olympic Committee into the project.

Teige's criticism of the project was not directed at what would seem to us today to be its obvious flaws. The hubris of the design was not his concern, nor did he initially reject the idea of the Mundaneum itself. Teige questioned Le Corbusier's architectural solutions, saying they were too traditional and not fit for the purpose they were to serve. For Teige, Le Corbusier went only halfway. Teige advocated a left-wing, radical strand of modern architecture holding that architecture and its purpose, form, and function could not be separated. Teige quoted the Swiss architect and Bauhaus director Hannes Meyer, saying that all things in this world are a product of the formula "function times economics," meaning that none of these things are works of art. Accordingly, it was not only Le Corbusier's architectural solutions that Teige felt were old-fashioned. The entire notion of international intellectual organization was, to Teige, a relic of the nineteenth century, at least if not inspired by the transformation from a bourgeois to a socialist society and the attempt to realize it. In his view, only a clear-cut political mission—that is, a socialist program—could guide a project like the Mundaneum. The latter had to be purely functional, serving clearly defined social goals, such as better health or housing. Collecting knowledge for the sake of collecting knowledge was a waste of energy. Therefore, Teige saw Le Corbusier's plans as not only inconsistent and contradictory, but also as symptomatic of the much larger problem of how to transform modern societies by using the knowledge produced decades before.

The importance of the issue at stake was confirmed by Le Corbusier's reaction in an article published in an avant-garde Czech journal, written "on the way to Moscow."[4] Le Corbusier

Fig. 3.2 Linking the World: *The Mundaneum was to be a complex of buildings replicating the established institution of the same name based in Brussels. In the first place, however, it was a way of organizing knowledge. Otlet believed that new technological solutions would deliver the key to dealing with the vast amounts of modern knowledge. Here we see his visionary ideas of how to communicate knowledge with new devices.*

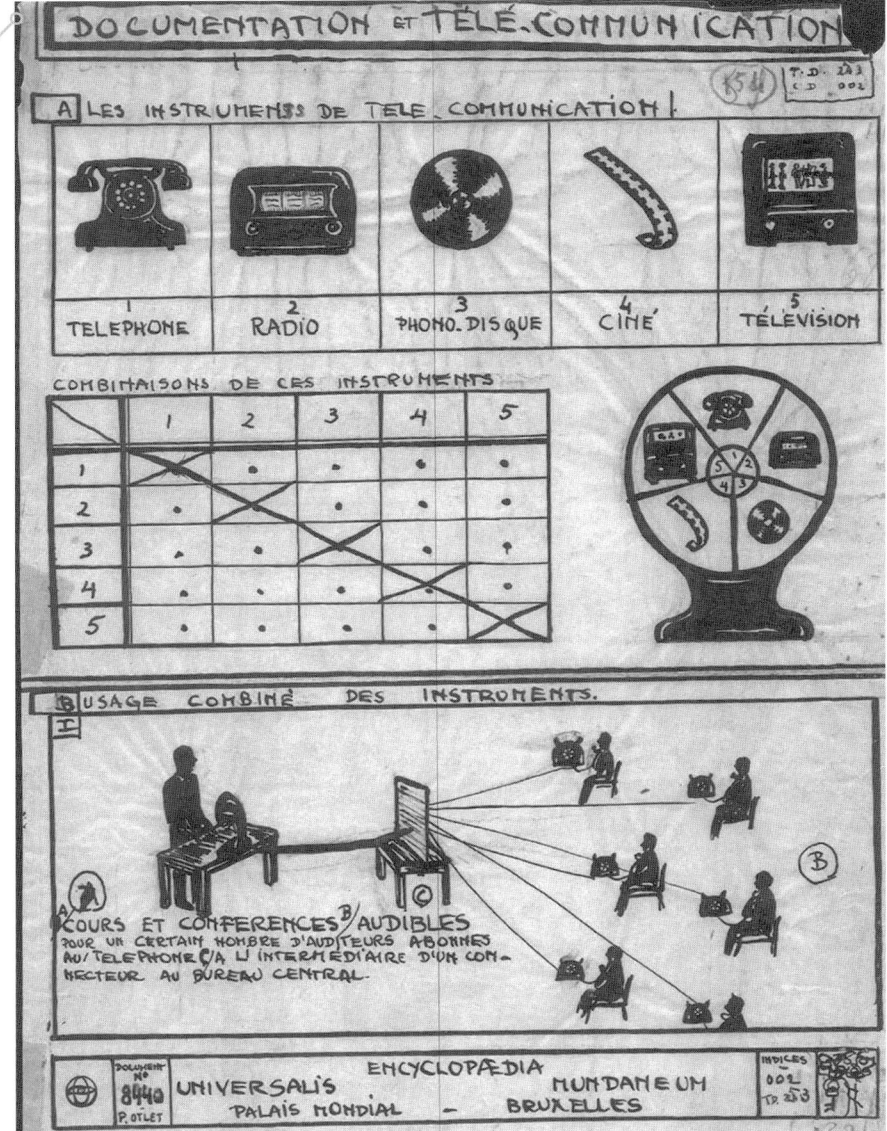

started off by declaring that this was the first time he had ever replied to criticism, a statement that was plausible to anyone who knew him.[5]

First, he confirmed the congruence of his interests with those of Teige, whom he held responsible for the "Czechs hav[ing] shone so brightly in the emerging sky of the new times." In Le Corbusier's opinion, their difference was one of terminology. He criticized Teige for his "romanticism of the machine." Tools of progress were not goals in themselves. Progress was a means rather than an end, and Meyer's formula of "function times economics" was far too simple. Le Corbusier defended himself against accusations of committing the crime of *lèse-Sachlichkeit*; that is, departing from purely functionalist solutions. He stressed the artistic value of architecture, but only on the basis of architectural laws that were "drawn from the very source of technology."[6]

Moreover, Le Corbusier defended his alliance with Otlet on the grounds that the latter represented the great progressive movement of the late nineteenth century, "the visionaries, the organizers of ideas, the generators of magnetic currents, the receivers, and emitters of waves." Le Corbusier stressed that Otlet had been instrumental in setting up the Union of International Associations, and drawing up the League of Nations' statutes. He also defended Otlet's concept of healing the world via knowledge, saying "it is indispensable to know the comparative states of nations, peoples, races, and cities" via "urbanism." Le Corbusier shared Otlet's desire to "activate this knowledge" (that is, what science and technology had achieved) and to communicate it via images and graphics; that is, "through iconography." Explaining his architectural solutions, Le Corbusier also made it clear that the design of the Mundaneum project reflected his conviction that architecture was the all-encompassing art, providing solutions to the most pressing tasks of the time: "And I promise you sincerely—a fact that reassures me —we are all, at this moment, at the foot of the same wall."[7]

Le Corbusier's design was so important and relevant to a wider public because the visualization of knowledge was a key to implementing the core of Otlet's idea. The plan was meant to be a climax to what had been achieved in the widespread movement toward international intellectual organization, educational reform, new systems of documentation and information, and, finally, the visualization of knowledge. This shared assumption, that it was possible and desirable to transform the world through a scientific approach to all aspects of life, made the Teige–Le Corbusier argument important for both sides. The idea of the Mundaneum embodied the integral and far-reaching character that Teige wanted architecture to exemplify. Teige, largely in reaction to what he perceived as the new postwar world took a radical stance. Building was to be grounded on a strictly statistical and scientific basis. Technological innovation should serve the education of the new man in a socialist society, with a scientifically grounded architecture playing a key role.[8] In this way, Teige saw Le Corbusier's ongoing emphasis on aesthetics as a bourgeois residue, which he contrasted with his understanding of architecture as being explained by "technology, economy, and sociology." Moreover, Teige criticized the design as an expression "of the cultural consciousness of the ruling classes."[9] In his rejoinder to Le Corbusier's reply, Teige attacked Otlet as a "locarnist" and an "ideologist and theoretician of imperialism and the world economy."[10] The common ground that Le Corbusier referred to no longer existed: "Those who belonged to one camp are so far split now, the ones go to the left, the others to the right." Teige felt there was no middle ground; and with the benefit of hindsight, his was a prophetic statement in central Europe.

The conflict between Teige and Le Corbusier marked an important threshold: challenging the vision of universal science and at that same time, signaling the rise of utopian notions of a new knowledge-based society. In many respects, the Mundaneum clash represented the climax of pre-war internationalism and scientific exchange. As Le Corbusier had stressed, the project stood in a long continuum. Teige questioned exactly this assumption. In his view, after the First World War and against the background of a deep political, social and economic crisis in Europe, it was time for a decisive break.

In Mundaneum practice, using universal forms of communication in order to reach a common understanding would thereby prevent new conflicts. This vision reflected high expectations about the exchange of knowledge. The very idea of the Mundaneum flowed from these expectations. For Teige and the strand of thought he represented, however, it was more important to draw lessons from the war. Primarily, this meant that pre-war solutions had lost their legitimacy. Scientific knowledge was to be used for an improved society. In addition, the Mundaneum conflict reflected the rise of new "techno-intellectuals" such as Teige and Le Corbusier, who were at the forefront of ubiquitous discussions about how to translate technological progress into a new social and political reality.

For Teige, Le Corbusier, and Otlet, as for many other scientists and intellectuals, the need for building a new society driven by science was obvious after the First World War. Equally evident, from their perspective, was that science had to transcend the boundaries of the

Fig. 3.3 **The World Centre of Communication:** *Plans for the Mundaneum and a city of knowledge reached back to the time before the First World War. Already then the organization of knowledge, new forms of communication and the building of new cities went hand in hand. The architectural solutions, however, still drew on traditional forms. American sculptor Hendrik Christian Andersen and French architect Ernest Hébrard conceived their plans for a World Centre of Communication in a classicist spirit to bring together "the scientific forces of the world."*

nation state and had to be organized accordingly. The experience of the First World War was the driving force for all related plans. Otlet, who had lost a son in the war, made this very clear in the intensity with which he pursued the Mundaneum project. In this, the Mundaneum also was a direct reaction to the war's challenge to universalism, as it strove to safeguard knowledge from the deliberate destruction seen between 1914 and 1918. For example, the program included a provision to protect libraries threatened during times of war by preserving their contents in the Mundaneum.

Indeed, only a year before the Mundaneum controversy took place, the library of the Belgian University city Leuven-Louvain had been reconstructed, having relied heavily on U.S. funding organized by Herbert Hoover, the later U.S. president (1929–33)—and engineer—through the Belgian–American Educational Foundation, itself conceived in the spirit of universalism during the First World War and established in 1920. In 1914, when German troops burned down Louvain's centuries-old library, the city became a symbol for a new, brutal form of warfare and the deliberate destruction of knowledge. The library's ruins figured prominently in Allied war propaganda. German efforts to justify what had happened, such as a manifesto "To the civilized world," signed by 93 prominent German scientists and intellectuals, actually made their actions look even worse.[11] An alternative manifesto, *Aufruf an die Europäer*, asking for European unity as an answer to the war and signed, among others, by Albert Einstein, found decisively less support. The feeling that German scientists had collectively compromised themselves during the war contributed to their exclusion from international cooperation after 1918. Thus, the Mundaneum plan also reflected an exercise of European self-assurance after the most serious crisis the Continent had experienced in centuries. The problematic role that scientists and experts had played in the war and the deep division among experts from the formerly belligerent countries hovered over the mid-1920s. Because this plan and a number of similar projects were intensely connected to European postwar reconstruction, it is worthwhile taking a closer look at them. The imagined new orders of knowledge and their implications for European societies will be at the center of this chapter.

Orders of Knowledge: Mundaneum

Despite what the grandiosity of Otlet's vision might suggest, its realization was not completely utopian. In many concrete aspects, it was connected to the League of Nations, which took its seat in Geneva at the same time and soon developed into an organization of considerable importance for exchanging knowledge and a focal point for new forms of intellectual organization.[12] Moreover, the League's Committee on Intellectual Cooperation was linked to Otlet and his Mundaneum project. For Le Corbusier, his failure to be entrusted with the task of building the *Palais des Nations*, despite having been ranked first in the competition, was a crucial defeat. Yet, his *Palais* submission caught Otlet's attention and convinced him that Le Corbusier was the obvious man to implement the Mundaneum project. After the experiences of the World War, Le Corbusier's modernism seemed more adequate than more classical architectural approaches.[13] Together with Henri La Fontaine, a lawyer who had won the Nobel Peace Prize in 1913, Otlet had first conceived the idea for the Mundaneum in 1910. Originally envisioned as a "city of knowledge," the project's external structure and internal content were strongly interdependent from the beginning.

Otlet was able to draw on government support. Important players in Belgium, including the king, were keenly interested in placing their nation on the international map next to Switzerland and the Netherlands—or even ahead of those countries—and this project seemed an excellent way to turn Brussels into a *district fédéral mondial*, or at least a *ville internationale*.[14] In this sense, the project offers a telling example of the congruence (for a time, at

Fig. 3.4 **Knowledge Burned:** *In August 1914, shortly after invading neutral Belgium, German troops burned the famous library collection of the University of Leuven/Louvain. The Germans claimed to have been victimized by snipers and had acted only in retribution. The international outcry was nevertheless gigantic. Leuven turned into a symbolic place of a war which turned also against knowledge and institutions of education, and thus the very European values themselves. The Germans attempt to excavate the destroyed books in 1917 (pictured here) came not only too late for the books but could also do nothing to improve the invaders' shattered reputation.*

least) between nationalist, internationalist, and dynastic interests in certain spatial settings, particularly smaller European nations. After their efforts had been interrupted by the war, in 1919 Otlet and La Fontaine secured funding from the Belgian government and space in the center of Brussels, with officials now hoping to improve Belgium's bid to host the future League of Nations. The original Mundaneum was housed in the *Palais de Cinquantenaire* erected for the Brussels International Exposition of 1897.

Apart from the building project, the Mundaneum had a history before and after the "Teige affair." It was not just a plan on paper. Parts of the concept were realized in very concrete ways, both in permanent exhibitions and in various cooperative activities. Otlet and his projects were also elements in a much wider movement of progressivist organizations that flourished immediately before the First World War but did not halt after 1918. This was the environment in which Otlet and Le Corbusier met.[15]

The Mundaneum project was grounded, both directly and indirectly, on a growing desire for international exchange, particularly in the sciences.[16] Its two designers initiated the International Conference of International Associations in 1907, which later led to the Union of

International Associations (UIA). In 1910, they established a Central Office of International Associations, which is still located in Brussels. Its goal was the "international coordination of intellectual work and as such it was one of the first platforms in history for discussing the politics of science on an international scale."[17] The projected Mundaneum, the growth of knowledge, and international organization were intrinsically interwoven. Their advancement, proponents argued, would foster world peace. Although the First World War had proved these assumptions to be rather naïve, they remained key to the development both of the Mundaneum idea after 1918, and for a number of similar and related organizations.

One such was the International Committee of Intellectual Cooperation (ICIC) of the League of Nations. Set up in 1922, it attracted a number of illustrious members, such as Henri Bergson, Albert Einstein, Marie Curie, Béla Bartók, Thomas Mann, Salvador de Madariaga, and Paul Valéry. A so-called Expert Committee of scientific advisers, architects, and library experts aimed to advise the ICIC. The International Institute of Intellectual Cooperation (IICI) established in Paris in 1924 and funded by the French government, became the executive organ of the ICIC in 1926.[18] Two years later, likewise within this framework, the International Studies Conference commenced. This loose association of mainly European actors, albeit having a global scope, focused on international relations and the sciences. Again, this organization predominantly enlisted the "great and the good," prominent figures such as Einstein and Sigmund Freud.[19]

The Committee and the Institute suffered more from the inherent tension between nationalist and internationalist agendas than from financial problems. These tensions were particularly apparent in the question of how to treat German scientists after the First World War. Einstein, for example, left the Committee over this matter for a time and became increasingly doubtful that such an organization could function outside the realm of politics.[20]

Even more perceptive, but even less successful, was a little-known project by the exiled German architect Erich Mendelsohn, who, together with Dutch architect Hendricus Theodorus Wijdeveld and French painter Amédée Ozenfant, conceived a South Mediterranean Academy of Arts (*Académie Européenne Méditerranée*). Mendelsohn had stated in 1929 that it was "unthinkable that we can turn back time, unthinkable that we leave unused the greatly broadened possibilities of technology. That we see the machine as the enemy of humanity, instead of as our powerful tool that we need to master."[21] Albert Einstein headed the advisory committee for the planned academy, which included the architects Frank Lloyd Wright, Heinrich Pertus Berlage, August Perret, and Henry van der Velde, the theatrical director Max Reinhardt, the composer Igor Stravinsky, and the poet Paul Valéry. Along with Mendelsohn, Wijdeveld, and Ozenfant, a number of eminent intellectual figures of the time were to be its future professors before a fire destroyed the building site in southern France. Eventually, growing concerns about the darkening political climate led to the project's abandonment.[22]

Compared to this ill-fated *Académie*, Otlet's goal was more focused and more ambitious. The core of his work was the documentation and dissemination of knowledge. In his attempt to classify existing knowledge, he drew upon historical precursors, Gottfried Wilhelm Leibniz in particular. More important were influential thinkers of his era, particularly Auguste Comte and Émile Durkheim and their attempts to recreate society by enhancing knowledge, bringing together the humanities and the sciences, thereby re-establishing social order and slowing social change.[23] *L'utopie de nos jours deviant scientifique* ("the utopia of our time must be a scientific one"), exclaimed Otlet in 1935.[24] Politically, Otlet never took sides, which enabled him to cooperate with conservatives, socialists, and centrists. So, for example, he established a link to the influential Belgian liberal reformist and internationalist Ernest Solvay, sponsor of a number of high-level conferences on general problems of physics.[25]

In his "utopian universalism," Otlet represented what scholars have termed "cultural internationalism" in its most concentrated form.[26] Otlet lived international cooperation himself and established working relationships with writers like H.G. Wells. Known for his "scientific romances" like *The Time Machine*, by the 1920s Wells was also an eminent promoter of re-educating humanity through the spread of knowledge. Otlet also collaborated with French physiologist Charles Richet and German chemist Wilhelm Ostwald, both of whom were ardent internationalist networkers. They all saw transnational and transcultural understanding and the exchange of knowledge as two sides of the same coin.[27]

What was fresh in Otlet's concept, however, even in his first writings from the early 1890s, was his idea of separating single facts and ideas and documenting them individually on cards.[28]

With La Fontaine, Otlet founded the *Office International de Bibliographie* (OIB) in Brussels in 1895.[29] In 1938, he transformed the OIB into the International Federation for Documentation (FID). This was a reaction not only to the quantitative growth of knowledge, but also to the rise of scientific journals in recent decades.[30] The OIB emerged following the first International Conference of Bibliography, a pattern that institutionalized the ever-growing business of congresses from the late nineteenth century onwards.[31]

Building on Melvil Dewey's Decimal Classification, Otlet developed the Universal Decimal Classification (*Classification Decimale Universelle*) or UDC, as a European, multilingual reaction to the American system—though, of course, the American influence on Otlet's system was obvious. Tellingly, European librarians opposed Otlet's new system due to its mechanical and allegedly arbitrary character that was supposedly unable to cover European complexities.[32] In Europe, moreover, the focus was more on documentation than on librarianship in the narrower sense. The "technology of intellectual work," it was hoped, would also further social, economic, and political goals, such as spreading education, enabling people to engage in current affairs or improving international exchange.[33]

Despite all the criticism, Otlet and La Fontaine developed their own bibliographical database of some 13 million entries by 1927, all indexed according to the UDC. They also welcomed postal inquiries. The collection included letters, reports, newspaper articles, and images, which were kept in separate rooms and indexed. The Mundaneum eventually contained one hundred thousand files and millions of images. Again, the venture's spatial setting and its connection to international communication was decisive, as the following UIA statement made clear:

> The International Centre organizes collections of world-wide importance. These collections are the International Museum, the International Library, the International Bibliographic Catalogue and the Universal Documentary Archives. These collections are conceived as parts of one universal body of documentation, as an encyclopedic survey of human knowledge, as an enormous intellectual warehouse of books, documents, catalogues and scientific objects. Established according to standardized methods, they are formed by assembling cooperatively everything that the participating associations may gather or classify. Closely consolidated

and coordinated in all of their parts and enriched by duplicates of all private works wherever undertaken, these collections will tend progressively to constitute a permanent and complete representation of the entire world.[34]

These projects were new in scope but clearly descended from the tradition of European attempts to gather all written knowledge.[35] Even more remarkable was the role that technology played in Otlet's visionary projects. He was enthusiastic about the combination of textual and visual media and later came up with creative ideas concerning the electronic storage of data.[36] Otlet was all too aware of the hubris inherent in his project. His hope was that new storage and communication techniques would make it possible to overcome the most pressing obstacles for gathering such huge amounts of information. In his 1935 essay, "Monde: essaie d'universalisme – connaissance du monde; sentiment du monde; action organisée et plan du monde" ("World: An essay on universalism – knowledge of the world, perception of the world; organized action and plan of the world"), he imagined "an instrumentation acting across distance which would combine at the same time radio, x-rays, cinema, and micro photography as well as the recently invented television." Making use of existing technology or envisioning new uses for it, Otlet also projected not yet existing devices in order to achieve a "memory" of the world, "its true duplicate." "From afar anyone would be able to read the passage that, expanded or limited to the desired subject, could be projected on his individual screen. Thus, in his armchair anyone would be able to contemplate the whole creation or particular parts of it."[37]

Otlet's close collaboration with an architect such as Le Corbusier was not just the result of a need for structures to house his project. Otlet also described and apparently conceived of books, and the organization of knowledge in general, in terms of architecture, as an "architecture of ideas." To Otlet, this was far more than a way of speaking. He argued that in order to understand this framework of "intellectual data," it was necessary to consider "the enormous revolution architecture itself accomplished in our days."[38] At the same time, Le Corbusier drew heavily on the vocabulary Otlet used and developed, and like him, believed that classification and standardization were essential. Both men shared the assumptions that these were key approaches that not only reached beyond

Fig. 3.5 Organizing Knowledge: *Otlet's plans might seem bizarre, but his attempts to organize all the world's knowledge made quite concrete progress. Here, we see the huge collection of index cards, organized according to a system proposed by Otlet and still sorted the classic way. Otlet believed that new technologies to store and communicate indexed information would be able to solve the problem of the sheer mass of data.*

ordering knowledge or designing architecture, but also referred to human beings, and thus could be instrumental for establishing a new and better society. Not by chance, the Mundaneum was to figure as the core of a new "world city."

Of course, there had long been a tradition of envisioning ideal cities. The model most influential to Otlet, which included strikingly similar features, was by the Norwegian-born U.S. artist Hendrik Christian Andersen and the French architect Ernest Hébrard, published in 1913 and branded the World Center of Communication.[39] While its design followed the Beaux-Arts tradition and stood in the long line of planned ideal cities, this model already featured a number of the institutions envisioned in Otlet and Le Corbusier's later project.[40] More important, it shared the link between the creation of a center of science, a new urban model, and a belief that this would advance world peace.[41] In a similar vein Tony Garnier's *Cité industrielle* also pursued the vision of bringing about a better society by means of an urbanism based on scientific insights and rules.[42] The 1899 peace conference at The Hague was a major step towards implementing such plans, culminating in 1913 with the Peace Palace in the Dutch capital.

Otlet and La Fontaine met Andersen and Hébrard in 1912 and urged them to merge their hitherto largely independent efforts and join forces with the Brussels team. What Andersen and Hébrard had achieved for architecture, Otlet and La Fontaine wanted to match in terms of functionality. Moreover, and significantly, they sought to catapult the project into their envisioned orbit of communication and have the center's plan become "the object of a circulating exhibition in all large capitals." In 1912, backed by a declaration of the UIA, Otlet asked Hébrard to draft a design for land near Brussels, even before the related publication by Andersen came out. When Otlet received the published project, he enthused in a telegram to Andersen: "It is magnificent. Hurrah! On with realization."[43] With the noteworthy exception of the Pope, most of the political leaders Andersen approached granted him audiences. The Belgian King Albert I expressed great interest in the plans, and Wilhelm II of Germany even created a five-person commission to increase the project's chances of realization.

The 1913 publication, which was to be followed by another volume, contained all of Andersen's main ideas and Hébrard's illustrations. While a number of its cultural and sporting structures paid tribute to antique urban models and the newly popular Olympic idea of peaceful competition between nations, the most interesting element was the Scientific Center. In a positivist vein, the designers dedicated it to using scientific communication to improve the world and bring about everlasting peace:

> The scientific forces of the world, if brought together without selfish motive into a harmonious center, would be—beyond all doubt—of immense benefit to the present and would leave to succeeding generations valuable records of their forefathers' love and faith, and of their effort, in spite of all temporarily hindering circumstances,

to raise humanity to a higher level of existence, both physically and mentally.

The Scientific Center was to be accompanied by an International Reference Library and an International Bank or Clearing House, topped by the Tower of Progress, the "central symbolic structure of the World Center: light tower of the Enlightenment, transmission tower of world information, a completed Tower of Babel—the ultimate symbol of human progress."[44] The authors saw the tower as a "marvel of practical engineering" embodying the glory of science and technological progress.

The plan contained state-of-the-art technologies throughout, but carefully avoided appearing utopian, which would have made it seem unrealistically ambitious. While Andersen also planned that his center would take on political functions as soon as the required international bodies were created, its function as a place of international scientific communication was far more important. Andersen believed that new forms of exchanging knowledge would automatically foster peace, thereby providing the center with a direct social purpose and, indirectly, a political one. The city of the past represented a destructive force that was "as despicable as war."[45] Urbanism was thus fundamentally an effort to make peace a concrete reality, and it was seen as the most effective, visible, and important outcome of the scientific approach. This was again strikingly similar to Le Corbusier's arguments. In this view, the World Center of Communication was the apogee in the history of urban design.

As it turned out, the center did not prevent the war, but was itself prevented *by* the war. Interestingly, from the beginning a range of public criticism denounced Andersen as nothing more than "a new project peddler"—an ambiguous term already well established at the time. A socialist newspaper in Germany assailed the capitalist basis of the project, which it said would inevitably lead to the "most scientific exploitation."[46] These attacks anticipated Teige's later critique of Otlet and Le Corbusier, as well as the significant challenges that scientific internationalism would face after 1918.

Still, the essence of the plans for the World Center of Communication lived on in the Mundaneum project. Over the course of a decade, architects presented proposals for establishing it in a number of major cities throughout Europe, not just in Geneva: at

Brussels (Victor Bourgois 1931), or Antwerp (again Le Corbusier with Huib Hoste 1933), on the Chesapeake Bay near Washington, DC (Maurice Heymans 1935), and in Antwerp again (Stanislas Jassinski and Raphaël Delville 1941).

The ubiquity of such attempts to centralize knowledge by creating new cities is telling in itself. There was clearly a political dimension, not in the sense of a particular ideological position, but in prioritizing the act of collecting all knowledge at a central place and thus enabling experts better to conduct politics. One might also see authoritarian tendencies in the inherent desire for information control.[47] Not unlike the visions of technocratic engineers, experts stood at the top of the pyramid in this system.[48]

Although Otlet's political relevance is hard to grasp, he represented a growing movement to promote social change through the diffusion of knowledge. In this, he was close to the Scottish sociologist and town planning pioneer Patrick Geddes. In 1912, both men had planned a so-called *Encyclopedia Synthetica Schematica*. This enormous synthesis, which was to be achieved by an international group of scientists, combined the major characteristics of internationalist science with a social cause. Promoters believed that new forms of graphs and charts would make it possible to interpret the vast body of scientific information being produced, and also to make this information accessible and suitable for exhibition, thereby fostering social improvement.[49]

Geddes innovated ways to make complex information comprehensible through visual representation. He particularly believed in the didactic value of exhibitions.[50] His Outlook Tower in Edinburgh was a "civic museum representing the city as its living counterpart, as a living museum of which the individual visitor was a vital constituent." Otlet took up this concept but dropped the restrictive link to the local. Like Geddes, Otlet believed in preserving the knowledge presented at world exhibitions by creating permanent institutions that echoed their format.[51] Otlet also clearly absorbed Geddes's idea of an Index Museum, that is, a museum perceived as "graphical encyclopedia," and his notion of "thinking machines." At stake was nothing less than the redefinition of the Encyclopedia project and the application of the values of scientific universalism to programs of international education and cultural reform. Education was a key goal here, with the documentation and exhibition of knowledge operating an "educational instrument."[52] Otlet felt that encyclopedias would enable people to gain the skills to acquire general knowledge. This concept expressed itself in a new vision of the role of the museum as a source of information. By insisting on the visual effect and its large impact with minimal effort, Otlet was close not only to Geddes, but also the German chemist Wilhelm Ostwald and later the Austrian economist Otto Neurath.

Working independently of these Western experts, but planning in the same direction was Soviet avant-garde architect Ivan Leonidov. His plan for a Lenin Institute for bibliographic studies was meant to contain all books of the world in a futuristic architecture, including a planetarium and celebrating the Soviet Union's knowledge-based political vision and universalist claim.[53] Otlet and Le Corbusier, Geddes and Leonidov, and in a different way also Otto Neurath, tried to combine the dynamics of knowledge accumulation and urban growth. Science was meant to shape the city and new forms of architecturally arranged knowledge were to heal the wounds of Europe and pave the way for a better future.[54] As the example of Neurath will show, re-thinking the relation between knowledge and society went much further than planning museums and libraries.

Visual Education: Isotype

When he first came into contact with Paul Otlet, Otto Neurath was already an eminent and internationally renowned figure in the developing field of visual education. While the First World War clearly transformed Otlet and La Fontaine's project, the connection between the war experience and ideas for radical reform was even more evident in Neurath's effort to devise a picture-based international language. Born in Vienna in 1882, Neurath founded the *Kriegswirtschaftsmuseum* (Museum of War Economy) in Leipzig during the World War. The museum was a reaction to the introduction of a war-based economy and its centralized attempt at planning, specifically by systematically gathering complex information. During these years, Neurath had developed his *Kriegswirtschaftslehre*, a science of war economy, as a sub-discipline of general economics.

Far from perceiving the war solely as a crisis, Neurath saw opportunities to shape and improve society. He joined the German

Social Democratic Party in 1918–19 and ran an office for central economic planning in Munich's short-lived *Räterepublik* (Munich Soviet Republic). When the Republic was dissolved in 1919, he was imprisoned, but used this period of restricted activity to transform his political experiences and optimism by writing *Anti-Spengler*, a refutation of Oswald Spengler's *Decline of the West* (1918). Back in Vienna, Neurath became one of the leading members of the Vienna Circle, focusing on logical positivism, a "unified science" strictly based on experience and linking different fields of investigation. In fact, the Vienna Circle was given its name by Neurath, who, as philosopher Rudolf Carnap put it, was the "great locomotive," the movement's driving force.[55]

Logical positivism was not only understood as a mode of thinking, but also as a mode of reforming society. In particular, Carnap's *Logical Construction of the World* was important to Neurath as an expression of striving to achieve new forms of communication.[56] In the progressive climate of the 1920s, Neurath moved from political prisoner to the directorship of the *Gesellschafts- und Wirtschaftsmuseum* (Museum of Society and Economy). The wide scope of this institution provided him with the means to develop the Isotype project, a visual dictionary with some two thousand symbols.[57]

The Isotype project drew on the Vienna Circle's attempts to find standardized expressions for ideas and culture and encapsulated most of Neurath's goals and convictions. One main goal was to develop pictograms to approach the "less educated," thereby spreading knowledge about social developments much more widely than had hitherto been possible. Overcoming the gap between the illiterate and the literate became an issue for Neurath when he observed the 60 percent illiteracy rate in Galicia during his travels as a Carnegie Foundation fellow before the war. In addition, Neurath believed that his picture script was destined to be used internationally to overcome cultural borders, a program expressed most clearly in the 1936 text entitled *International Picture Language*.[58] When Neurath referred to "international means"[59] of communication, he meant that the highly abstract graphic information he developed, together with artist Gerd Arntz, would reach back to the "primal beginnings of language."[60] Visualization would transcend linguistic and cultural borders in this "century of the eye."[61]

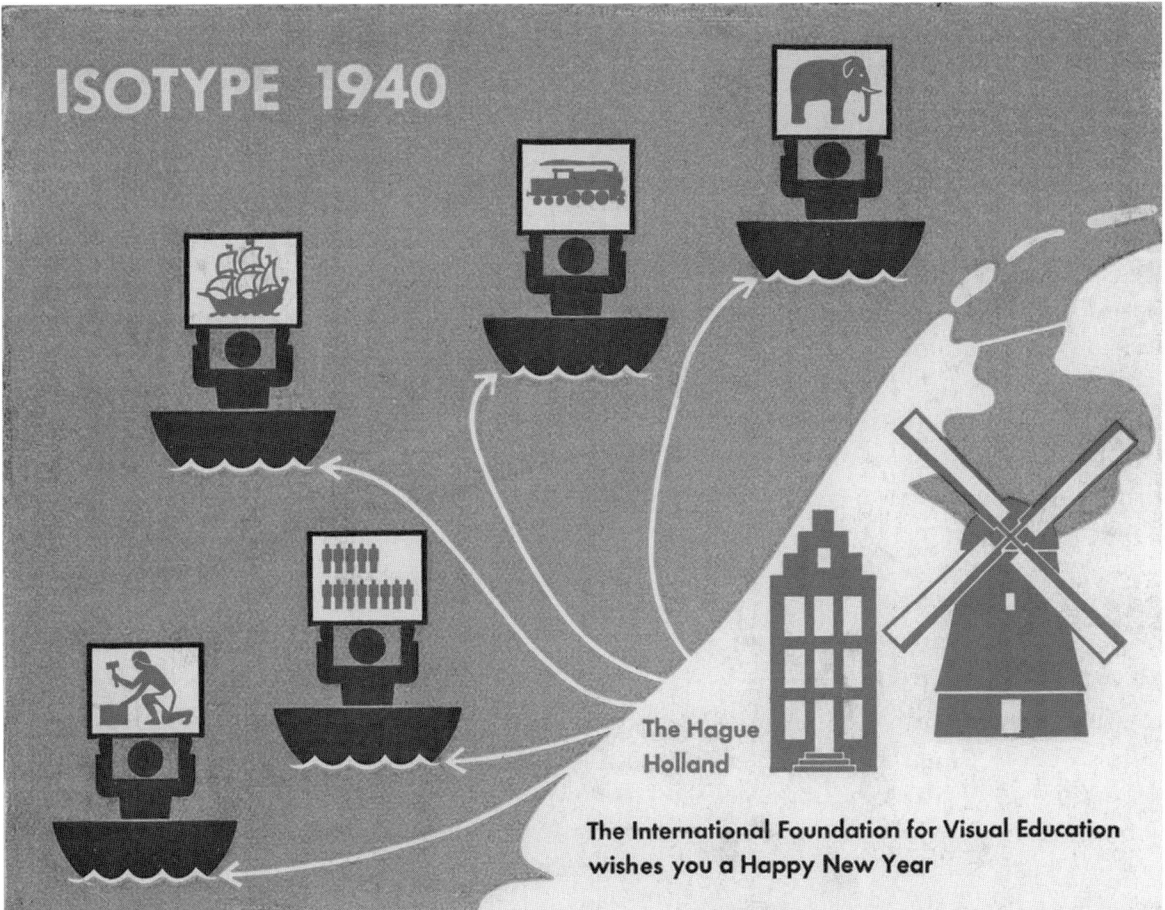

Fig. 3.6 New Year's Greetings in Isotype: *International cooperation was at the heart of the Isotype visual language. This holiday postcard used by the International Foundations for Visual Education displayed the graphic language developed by Otto Neurath and the institute. The year 1940 would see Neurath and his colleagues forced to leave the Netherlands for Britain, where Neurath was able to survive the war and continue his work.*

While this was not, in itself, a new approach, embracing modern storage and communication media was.[62] Under modern conditions, Neurath saw not only an urgency for such an undertaking, but also the chance of implementing it. Neurath directly linked the development of his pictorial language to technological progress. What technology had achieved for industry and hygiene for health pedagogy, visualization would now achieve for social progress.[63]

Neurath and Arntz judged that standardization of graphic information was an essential precondition for improving society. Through strict standardization, it was possible not only to communicate information but also easily compare it, which was particularly important to Neurath. His "Vienna method" (*Wiener Methode*) could provide a simplified illustration of social facts and social relations and developments, and was supposedly so easy that anyone could learn it.⁶⁴ In addition, scientific results should be applicable to everyday life and easy to communicate. This resonated with Neurath's objective; that is, providing workers with the knowledge they needed to understand and change their situations, and thus reform society. Taking up general trends like housing reform with its tendency to teach tenants how to live, Neurath declared that his era was one in which individuals could actively shape their own lives. His graphs delivered "instruments of thinking for daily life." Isotype was a very European undertaking as it clearly referred foremost to the economic and social problems prevailing during the interwar decades. Neurath believed that most people would be interested in the ever-present question of why there was still so much poverty and suffering despite tremendous technical achievements. In this sense, he believed his project was essentially democratic.⁶⁵

While the Vienna Museum of Society and Economy provided the institutional background for developing Isotype, Neurath envisioned a much more important role for the "museum of the future."⁶⁶ Yes, some innovative museums had developed a mission that extended far beyond simply displaying their collections. Geddes's Outlook Tower is an important example of such an "activist" institution. Another is Dresden's *Hygiene-Museum* (Museum of Hygiene), founded in 1912 after the *Internationale Hygieneausstellung* (International Hygiene Exhibition), with its program of social reform, including eugenics, a set of ideas aiming at improving human hereditary traits. With a visual power surpassing Neurath's approach—and with equally democratic aspirations—German-Jewish gynecologist Fritz Kahn published highly successful books depicting the human organism as an "industrial palace." Kahn turned into an international bestselling author; but when the Nazis burned his books in 1933, Kahn was forced to leave Germany, finally emigrating to the U.S. with Albert Einstein's support.⁶⁷

In order to make knowledge more accessible to everyone, Neurath developed the idea of mobile museums, mobile Mundaneums, inspired (as the name suggests) by Otlet's visionary project.⁶⁸ Projected as a kind of mobile drive-in cinema, they would be serially produced according to techniques already developed for traveling exhibitions. Soon they would lead to a new stage in communicating knowledge, including an *Orbis-Eisenbahn Museum,*" a World Railroad Museum. Neurath imagined slogans such as "Every city is a Mundaneum!" "The Mundaneum as a general educational instrument!" "The current order of life in a state of change!" "Where do we come from?" and "Where do we go to?" to promote his idea of linked Mundaneums, or what could be called a Mundaneum chain.

Fantastic as this may sound, such plans were not completely unrealistic. Neurath managed to establish spin-offs of his company in Berlin, Amsterdam, New York, The Hague, London, and, notably, Moscow (ISOSTAT), with the Vienna Museum of Society and Economy as the central institution. He wanted to teach people how to look at the world, to "reach the point where all knowledge can be grasped from one point of view" and to thus shape a "geographical account of knowledge."[69]

Clearly, it was no coincidence that Neurath and Otlet also forged a personal link via the *Maison International* in Geneva, one of many institutions that shared Otlet's approach.[70] They almost immediately started to cooperate in staging joint projects. The *Atlas de la civilisation universelle* ("Universal Atlas of Civilization") was closely linked to the previous work and unique expertise of both men.[71] The atlas was rooted in a bigger idea of creating a "uniform international history textbook," adopted by the League's ICIC and its sub-committee tasked with creating an International Text Book of History.[72] Moreover, this venture was very much in tune with the zeitgeist, becoming the subject of discussion by the World Federation of Education Associations (WEFA), an organization with which both Neurath and Otlet cooperated.[73]

To produce this ambitious multi-volume work, the two men set up the Orbis Institute in Brussels (*Palais Mondial*) and Vienna (Museum of Society and Economy). While Otlet benefited from Neurath's expertise in museum design and experience in graphical presentation of complex information, Neurath's career became much more international as a result of this new contact.[74] However, Neurath also developed his understanding of museums as a "transnational means of communicating, an Esperanto-like language as universal as the laws of nature," after coming into contact with Le Corbusier's and Otlet's design for a world city. Otlet's strategy of "information sharing" among linked centers of cultural production suited Neurath's general ideas, although he favored a decentralized version.[75] Neurath was, in many respects, more realistic than Otlet. Unlike Otlet, Neurath argued against the pyramidal organization of knowledge, denouncing the "system builder as a born liar." That Neurath wrote his 1939 book *Modern Man in the Making* in BASIC English (a simplified version of the language) was an expression of more than just pragmatic choice.

Standardizing Knowledge: Wilhelm Ostwald

Although Otlet and Neurath present two particularly vivid examples of the interwar documentation scene, they were far from alone. The notion of standardizing scientific communication was an urgent topic across Europe even before the First World War. Wilhelm Ostwald, who cooperated with Otlet on different occasions, illustrated the scope of this movement like no one else.

An eminent chemist and 1909 Nobel Laureate, Ostwald had a scientific background that made him particularly suited for taking on projects for organizing and communicating technical knowledge. He derived his standardization schemes from his experience with scientific modes of classification and definition. At least as important, however, was his immense international standing as a scientist and the numerous contacts that this created beyond the sphere of his own professional fields.[76] By the time he received the Nobel Prize, however, Ostwald had already ventured beyond the expertise grounded in his scientific training. He extended then-current theories of energetics (*Energetik*), applying them to social and economic phenomena and bringing in questions of efficiency, conservation of energy, and entropy. Thus, for him, standardization was primarily an attempt to save energy, understood in a broad sense. The constant translation from one language into another in scientific exchange was seen as a waste of energy and a nuisance at international congresses, which were becoming increasingly common. The solution to this, Ostwald believed, as did others, was to create an artificial auxiliary language which would enable international communication. He became an ardent promoter of Ido, together with the French mathematician, editor of the works of Gottfried Wilhelm Leibniz, and pacifist, Louis Couturat, who had developed this language using Esperanto as a base.[77] Esperanto, in turn, derived from work twenty years previously by Lazar Ludwig

Zamenhof, who described his main motivation as transcending the different languages spoken in his native Białystok (in the Russian part of Poland). This multilingual background was not unusual among advocates of auxiliary languages. Martin Schleyer, who invented Volapük, another artificial language, lived in the Swiss-French-German contact zone of Baden, while Ostwald grew up in multilingual Riga, then part of the Russian Empire. Ludwig Wittgenstein's oft-quoted sentence, "The limits of my language mean the limits of my world," was therefore accurate both in terms of how language restricted the intellectual imagination that Wittgenstein had in mind and also in the sense that it imposed certain geographical boundaries.

In contrast with early modern attempts to create a universal language such as Leibniz's *lingua generalis*, which were mostly concerned with scholarly questions, the new auxiliary languages (*Welthilfssprachen*) that started to mushroom around 1900 were directly linked to the rise in economic exchange, new modes of traffic, and the emergence of new communications media. The idea was that such languages would be learned alongside one's mother tongue so that everyone participating in international exchange could, eventually, do so in both their native language and the auxiliary one.

Science was seen as the field most in need of language innovation Pushed by the immense success of standardization and classification in other areas like time, electrical values, and chemical elements, the heralds of new languages believed it was only a question of time until their efforts would bear fruit. "Heralds" like Couturat and Ostwald saw themselves almost as demiurgic system builders, defining the future of communication.[78]

Volapük, Esperanto, and the Esperanto-derived Ido became the subject of countless congresses and frequently-changing international organizations. However, because of the complex interplay of national loyalties, aspirations to being international, and, even more challenging, the problem of setting standards in a realm that was hardly influenced by academia, all efforts ultimately proved futile. Tellingly, a promoter of constructed languages like Ostwald identified the problem as one of insufficient organization. In his view, only a World Office for Language (*Weltsprache-Amt*) would be able to implement his plans, as he explained in a 1910 talk provocatively titled "The Organization of the World" (*Die Organisation der Welt*). Ostwald said there was a "need to organize regular progress."[79] Although he felt officialdom should provide authority, the office itself was seen as an institution of dilettantes operating beyond the existing governmental and administrative structures. *Weltgeld*, that is, an international currency, was another brainchild of Ostwald, as was a World Format for Printed Matters (*Weltformat für Drucksachen*) or, more concretely, the World Organization of Chemists.

More often than not, the prefix "World" actually meant Europe. Couturat made this quite clear when he explained that the European market would reach across the whole world and that other continents would be incorporated into European institutions, given their market power. It is equally telling that the outbreak of the First World War caused Ostwald to radically redirect his efforts and devise a new candidate: *Wede*, an abbreviation for *Weltdeutsch* ("World German"). This was nothing more than the idea that a simplified form of German would ease the introduction of German products—and troops—in Eastern Europe and the Near East. In line with his earlier goals, Ostwald also insisted that the new language would allow speakers of other languages to more easily gain access to the fruits of German science. Instead, the project actually demonstrated the contradictions inherent in internationalization under national auspices. Because the language's relevance was limited to German national interests, its fate was sealed well before German troops retreated from the Eurasian zones, where Ostwald had hoped they would spread the new linguistic standard.

Despite being extremely ambitious and far more locally restricted than their supporters suggested, these projects, like Otlet's, were not mere daydreams. In line with what the nineteenth century seemed to have taught, Otlet assigned science the central role in organizing the world. Science was to serve as the "brain of the whole of humankind." Ostwald profited immensely from his many networks and communication channels, such as the chemistry journal he edited and used to urge standardized paper size, or his collaboration in commissions to standardize the classification of the atom. Ostwald lamented that the results science had produced were not adequately structured and thus only accessible with difficulties.[80] Like Otlet, Ostwald strove for the standardization and centralization of knowledge documentation.[81] Both men aimed to collect all the world's knowledge in one place.

It was this link that Swiss businessman Karl Wilhelm Bührer and journalist Adolf Saager took up in their concept for an Institute for the Organization of Mental Labor (*Institut zur Organisierung der geistigen Arbeit*), which became known by its programmatic name *Die Brücke* ("The Bridge") after 1911. The initiators of *Die Brücke* skillfully managed to use the buzzwords that Ostwald himself had brought into the discussion and, in so doing, convinced the master himself to become the Institute's largest donor; a considerable portion of Ostwald's Nobel Prize money went into the undertaking.

Consequently, given its mission of serving as a bridge, the Munich-based *Brücke* established cooperation with Otlet's project in Brussels, leaving the documentation of books to the players in Belgium and focusing on other kinds of artifacts such as pictures or sculptures. Moreover, by leveraging Ostwald's name, the group convinced prominent contemporaries from different national backgrounds to join forces. Henri Poincaré, Peter Behrens, Ernest Solvay, Lord Rayleigh, Marie Curie, Wilhelm Röntgen, Ernest Rutherford, and even Albert I of Monaco all supported the *Brücke* and attested to the enormous transnational radiance such an institution could broadcast.[82]

A number of sub-projects were started, all of which intended to enable better flows of information, like a *Weltwörterbuch* (World Dictionary), a *Weltwörterbuchinstitut* (World Dictionary Institute), and a *Weltschrift* (World Typeface). However, Ostwald's funds and the division of labor, both internally and between Brussels and Munich, could not overcome the project's inherent infeasibility.[83] Only two years after its foundation, the *Brücke* had to be dissolved due to financial constraints and because its founder, Bührer, diverted the activities of the project away from its original goals, devoting much energy to collecting postage stamps and other rather remotely relevant ephemerae.

In his involvement with the *Brücke*, Ostwald emerged as a typical example of a project-maker. Such schemes were not only intended to solve a particular problem, but also to bring about general progress and to create a better world through more effectively communicating scientific knowledge.[84] Expectations were enormous and extended far beyond better ordering the knowledge of the past in well-documented collections. All of Ostwald's projects, like those of Otlet and Neurath, had a strong imperative for shaping the future. As Ostwald phrased it, the organization of mental labor was the "Demand of our times."[85] The

Fig. 3.7 **The Ambivalences of Technical Progress:** *Building on the immense fascination for technology, the physician Fritz Kahn turned himself into one of the global bestseller authors of the 1920s and 1930s. As in this illustration, Kahn described the human body—the most potent machine in the world as he declared—in technical metaphors. His motivation was to overcome the ills of the past and of human ignorance by employing the new knowledge of his age. Yet, Kahn also embodies the endangerment of progressive expertise in the 1930s. His books were banned in Nazi Germany and Kahn had to leave the country. He continued pursuing his mission in France, the U.S. and Denmark among other places.*

Brücke was to serve as an instrument to advance the "organization of the organizers." As the involvement of prominent scientists shows, these attempts were deeply embedded in a culture of technological progress. Unsurprisingly, these projects were not just concerned with educating the general public, but also with finding new solutions to the problem of educating engineers. Ostwald himself had reflected on how better to present engineering knowledge. In line with these remarks, the German Federation of Technical-Scientific Societies

(*Deutscher Verband technisch-wissenschaftlicher Vereine*) set up a Central Office for Technical and Scientific Teaching Materials (*Technisch-Wissenschaftliche Lehrmittelzentrale*, TWL) in 1921.[86] Embarking on the viral ideas of centralization, rationalization, and organization, the TWL developed new teaching techniques for technical matters, particularly using visual material.

Strongly motivated by Germany's defeat in the First World War, the TWL sought to improve engineering education, thus enhancing general productivity and enabling Germany to regenerate as a powerful nation. Interestingly, the decisively national mission of the project did not prevent the "Archive of Technical Progress," as the TWL later called itself, from enlisting in the Universal Decimal Classification effort. In this context, the TWL was in touch with Otlet's OIB in Brussels, intending to connect with its system. It was also influenced by both the *Brücke* project and Neurath's Isotype visualization. Both were discussed as examples of effective teaching techniques.[87] The TWL indicates how technical education went hand-in-hand with attempts to achieve standardization in other fields and reflected the new place of technological knowledge in society and culture.

The Social Promise of Technology

Projects such as Ostwald's standardization, Neurath's Isotype, and Otlet's Mundaneum were directly linked in various ways. That all of them arose at the same time makes them all the more significant. They all spoke to a seminal change that became full-blown after the First World War. As philosopher Tzvetan Todorov stated, referring to this period, "for the first time in the history of the Western world (Western Europe and North America), people had the impression that innovation was triumphing over tradition and that works of human design counted more than natural phenomena."[88] All these projects reacted to what was perceived as the age of the machine and tried to use the new opportunities this age brought to establish an order of knowledge that fit the new time. This was also the significant period in which industrialization and the first phase of high technology became historicized, that is, when they were understood and described as significant historical phenomena in their own right.

The Mundaneum was thus an interesting case, worth arguing about because it localized and centralized both old and new concepts in the documentation and dissemination of knowledge, both traditional concepts of collecting information and newer ideas of making these collections work for social progress. At the same time, the Mundaneum presented a new vision, of a city solely based on knowledge which it would simultaneously integrate and produce. Otlet's vision was social improvement based on knowledge. In this, though partly drawing different conclusions, Neurath followed Otlet. Ostwald was more interested in the organizational potential of standardizing knowledge itself than in social change.

This conflation of technoscientific and sociocultural factors may be observed in the new prominence that these topics acquired in the public sphere. Documentation (that is, documenting knowledge) was both a prerequisite and a reflection of this blending and served as a "power resource" in its own right. It has been argued that Neurath realized a form of what has been called the "techno-imaginary."[89] In that he anticipated the potential of modern technology, Neurath directly and indirectly promoted visions of social change brought about by technology. This was true also for Otlet. These new, technology-based systems of knowledge documentation and communication were plausible because technology had attained a completely new status. "Technology's storytellers" had successfully redefined the relationship between the public sphere and technoscientific knowledge.[90] New types of technical experts emerged, whose expertise and influence derived at least as much from glossy texts and sketchy pictures as from what they actually built in a physical sense. In reaching out to the wider public beyond the narrower confines of science and technology, Otlet and Neurath seized upon the possibilities offered by scientific innovations, but in doing so hazarded their scientific colleagues' respect. Ostwald lost much of his backing in the scientific community after turning himself into a promoter of energetics and diverse standardization projects which his colleagues regarded as not based in scholarship. Moreover, many established scientists regarded his popular appeal as something suspicious in itself. On the other hand, outsiders like German engineer Franz Feldhaus, who had no standing in the scientific community when he started single-handedly "inventing" the history of technology based on a detailed reference and documentation system, also failed to create a bridge between scientists and the general public.[91] Such attempts also always sought to define who was in and who was out: they attested to the cultural state of advancing knowledge and progress, while identifying its heroes. Sometimes these heroes were the documenters themselves. Ostwald, for example, pursued science historiography, including self-historiography.[92] However, others also tried to denounce European icons such as Einstein by reporting information intended to harm their reputations, in this case with heavily anti-Semitic motivations. Crucially, the prominence of Einstein and the media hype that surrounded him became a major case in point. In order to give more weight to their claims, Einstein's enemies formed their own international associations, for example, an "Academy of Nations," in imitation of the organizations such as the Expert Committee of the League of Nations. [93]

In simultaneously marketing their own cause, touching on current debates, and stressing the social relevance of communicating knowledge through new technical means, the players described here were deeply affected by European traumas and hopes after the First World War. They developed universal systems that were particularly relevant in the European context: that is, suited for a continent connected in many ways through networks of trade, shared knowledge, and history, but divided by the multitude of languages spoken. The emergence of auxiliary languages in Europe was meant to overcome what was perceived as a typically European Babel.

There is no doubt that these projects were intended to have global reach. However, the individuals brought together (such as the members of *Die Brücke*) were still predominantly European—and, in all cases, were closely connected to the positivist tradition.[94] Paradoxically, the scope of the systems developed by figures such as Otlet required reflection upon what was European and how this would relate to the outside world, particularly the U.S.[95] Though meant to be a global undertaking, Otlet's UDC system of classification remained a European answer to a similar U.S. system. On the other hand, pragmatic factors also contributed to the European bias, for distance and substantial travel times made significant exchange outside of Europe largely unfeasible.

These projects reflected a new understanding of technology. Le Corbusier had declared in the 1920s that the "machine is king," a

Fig. 3.8 Gathering the Best Minds: *The Solvay conferences were one of the most remarkable ventures among the many forms of international scholarly exchange developed since the nineteenth century. Following an initial meeting in 1911, they were organized from 1912 on. With most science and technology conferences struggling with their own success—that is, their sheer size—the Solvay conferences assembled a small elite mainly from physics to discuss the discipline's big questions of the field—reaching toward issues far beyond science.*

statement much more elaborately and skeptically formulated by Swiss art historian Siegfried Giedion in his 1948 book, *Mechanization Takes Command*.[96] In the two decades before it was published, Giedion became one of the foremost "translators" between the general public and the technical sphere and promoter of new expert groups, a paradigmatic techno-intellectual. Yet, Giedion was also an organizer of sorts, bringing together, at least for some time, those like Le Corbusier, who believed in evolutionary social improvement through technology and those like Teige, who argued for more radical solutions and prioritized political solutions in a socialist vein.[97]

Le Corbusier clearly embodied the new breed of experts described here, as did Teige. Teige might have been less prominent in the wider public than Le Corbusier, but he had turned himself into the "best-informed expert on contemporary Soviet culture and architecture outside the Soviet Union" as a result of his travels there.[98] When Teige published a book on Soviet constructivism in 1936, arguing once more that architecture had to serve men scientifically and technologically, he was already disillusioned with what had happened in the East.[99] He lamented both Socialist realism in Stalin's Soviet Union as well as anti-modernist politics in Nazi Germany. After surviving the war in Prague, Teige increasingly came under pressure from the Czech postwar government until his early death in 1951.

These ideas and concepts emerged in a European context strongly shaped by the First World War, a World War mainly fought in Europe which left its deepest marks on this continent. These ideas came up in an environment more than ever shaped by a diverse linguistic situation and the competing interests of various nation states. As the argument between Le Corbusier and Teige showed, the discussion about how to apply new technological opportunities transcended all borders and linked European experts. Le Corbusier, Teige and others also illustrate how the debate on the role of technology, with its new social relevance, became increasingly personalized.

As the next chapter will explain, the way technological spill-overs into the realm of the social and political were to be organized was highly contested and increasingly involved taking a position in relation to the new authoritarian and totalitarian regimes entering the scene in Europe.

II
Endangered Experts, New Social Orders

4
Expertise with a Cause

In July and August of 1933, a year not normally associated with progressive international exchange, Le Corbusier, Sigfried Giedion, and many other luminaries of international architecture and urbanism traveled the Mediterranean from Marseille to Athens on board the *Patris II* to discuss the idea of the Functional City. They were attending the fourth gathering of the organization *Congrès internationaux d'architecture moderne* (CIAM), founded in 1928 and guided thereafter by Giedion, its secretary-general. Notably absent in 1933 was Czech Karel Teige, although he had been a key figure in the preceding CIAM discussions. He preferred the *Leva Fronta*, the left-wing political movement of his own country, over what he considered to be a hopelessly bourgeois undertaking.[1] Missing, equally for political reasons, were most of the German members, including CIAM heavyweight Walter Gropius. They believed (correctly, as it happened) that international exchange would not help their position at home. However, the sociologist and political economist Otto Neurath was among the passengers, and this was certainly neither a coincidence nor a consequence of the attractive voyage that had been promised. Neurath had been deeply impressed by his previous exchanges with Otlet and Le Corbusier and understood the city as the main arena of human improvement. The CIAM architects

welcomed Neurath as the first "specialist," which for them meant a non-architect equipped with relevant knowledge for their cause, a key figure who would help them communicate their aims and transpose their agenda into wider economic and social spheres.

Conditions on board the *Patris II* were anything but luxurious. The few participating women shared the only cabin, while the gentlemen spent the nights outside, sometimes soaked by the rough sea. On August 1, 1933, the illustrious passengers eventually disembarked in Athens, and discussions at the local technical university elaborated on what had been sketched out at sea. The university also hosted an exhibition on the Functional City with more than thirty plans of cities, mainly European, grouped into seven categories. The term "Functional City" was based on the idea that cities should be developed according to scientific findings and according to what were seen as their main functions of work, traffic, housing, and leisure. These principles and insights, it was argued, could be applied everywhere but would be particularly beneficial for less-developed countries.[2]

Fig. 4.1 **Experts Cruising the Mediterranean:** *After their unsuccessful attempt to stage a congress in Moscow, the modernist architects organized in the CIAM retreated to the ship* Patris II, *cruising the Mediterranean from Marseille to Athens. In 1933, with the political skies darkening in Europe, this was an impressive demonstration of the optimistic spirit still holding sway among these experts. This spirit is captured in the photos taken on the* Patris II.

Indeed, the Athens CIAM chapter expected a lot from the visitors from the lands of progress. Le Corbusier delivered a special lecture on "Air—Sound—Light" that was widely attended. Tellingly, however, he arrived late for his own lecture and was still in his tourist clothes—he had been distracted while studying the Acropolis and the "timeless" fishing boats that he had spotted at Piraeus.[3] Slightly irritated, Hungarian architect Fred Forbat noted that the modern housing achievements proudly presented by the Greek architects went almost unnoted, while Le Corbusier, Neurath, and the others strove to find European modernism in the remnants of classical Greece.[4] More in line with the new times, Otto Neurath gave a talk on the Vienna Method of pictorial statistics in the courtyard of the Technical University in Athens.[5]

The story of these experts cruising the Mediterranean marks a watershed for a number of the developments that have been described so far. The meeting on the *Patris II* was the apex of a new culture of congresses, led by an optimistic spirit of collaboration.[6] Photographs of sunglass-wearing, half-naked experts bowed over technical plans and graphs convey something of this spirit even today. Hungarian László Moholy-Nagy captured the atmosphere in his film *Architects' Modern Congress*.[7] The enthusiasm found here has to be understood against the background of a shared and far-reaching cause. The *raison d'être* of the Athens Congress, and the CIAM in general, was not so much the exchange of knowledge, but rather collective work towards new social goals, a new society. Sigfried Giedion, CIAM's secretary-general, visionary networker, and agenda-setter, reflected in 1942: "These are congresses, which are based on collaboration, not congresses in which each single participant only talks about his special field like in the nineteenth century."[8] Being modern was part of its very program. The CIAM heads thought about how to bring women to the fore of the organization, still largely dominated by men, in order to stress its progressive profile.[9] Likewise, they took pride in its internationalism, represented by the eleven languages spoken within the organization. From the beginning, CIAM saw itself as global, reaching beyond Europe and discussing problems of worldwide relevance that were to be solved with universal methods. In reality the organization was very European. Its members came almost exclusively from Europe, its congresses took place in Europe, and the topics discussed mostly reacted to the Continent's specific problems.[10]

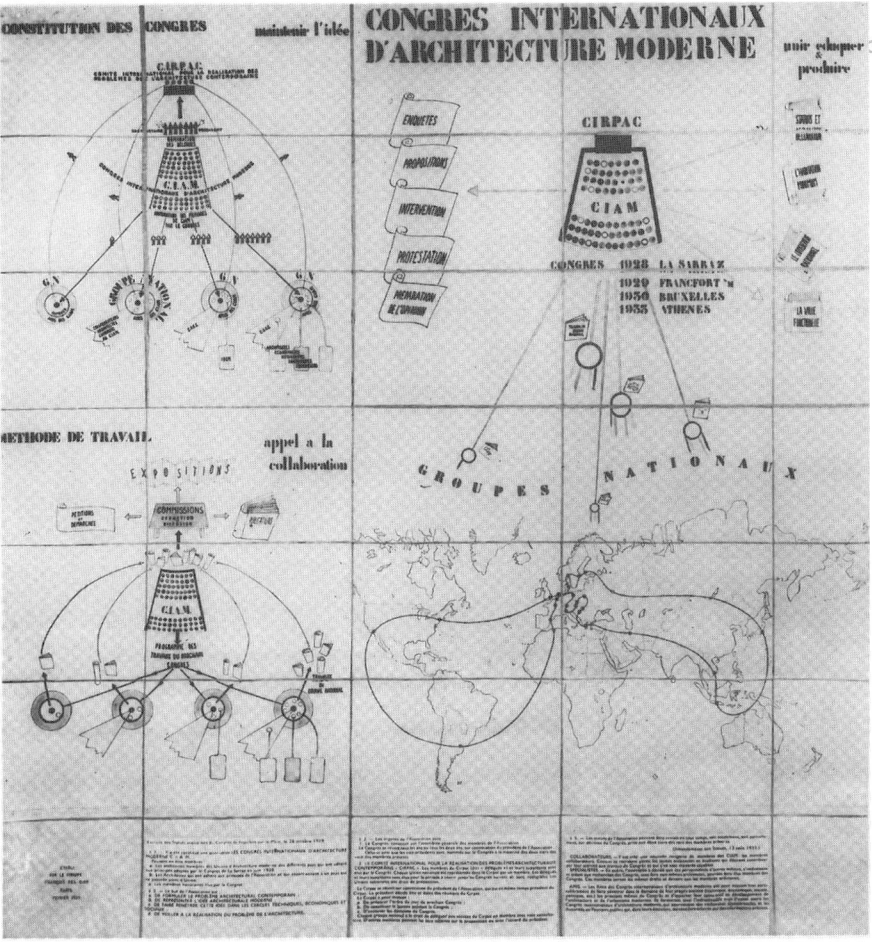

Fig. 4.2 Global Expertise: *The CIAM made much out of its alleged global reach. The universal application of its insights was a key message of this organization. In fact, the members were almost exclusively European and only under the pressure of war did CIAM reach out to the U.S. Thus the map reproduced here represents ambition more than the reality.*

Because of the worsening political situation, there was little room after 1933 for such gatherings as the one in Athens. Moreover, inherent tensions within the CIAM group, established only in 1928, now became all too obvious.[11] What was the place of the technical expert in society? How could the new scientific and technical insights be used to form a new and better society, and should this be done by force? If so, what stance should experts take towards politics, particularly authoritarian politics? Tellingly, although a special committee was formed to come up with a resolution based on the results of the intense shipboard and Athenian discussions, a unanimous declaration was never produced. Only in 1943 did Le Corbusier publish what he, somewhat misleadingly, called the

"Charter of Athens."[12] Previously, a series of documents called "Constatations" had served as a compendium of the group's reflections, although they hardly formed a uniform text. These had been drafted starting in late 1933. In an enlightening way, Giedion had described the most contentious point that September: "Question of principle = Technicians or politicians?" For Giedion, this implied two options: the congress members could either decide to understand themselves primarily as technicians, as they had so far, and thus resolve problems on a technical basis; or they could see themselves as politicians with a clear stance against capitalism, as Karel Teige and the Czechoslovakian group had chosen to do.[13]

Clearly, overall changes in the European political situation, in particular the rise of totalitarian regimes, left its mark. As Giedion saw it, technicians were well-respected and therefore had the ability to exert influence at that moment. However, if the socialist movement won out—as it seemed likely to do—only a politician who clearly positioned himself in favor of the new regime would be able to help bring about the changes he envisioned. In the background loomed what Giedion described in 1933 as the big question: whether in the future all specialists would be required to choose sides politically. Somewhat prophetically, Giedion firmly believed this would be the case. Yet, given the contemporary political chaos, he held that presenting the voice of technical experts outside politics, as CIAM did, was enormously important "to create order."[14] It was this conflict that estranged Neurath from Le Corbusier, who was "wedded to safeguarding the autonomy of the 'master planner,'" while Neurath insisted on keeping in touch with the "common man."[15]

The conflict that formed the background for the rather academic dispute between Teige and Le Corbusier in 1929 pointed quite clearly to a new political relevance of expertise. This soon led to situations in which experts had to decide where their loyalties lay. Another reason why the argument between Le Corbusier and Giedion—and many others—over phrases and formulations was so laborious was because the group's resolution was to be sent to the press and to governments alike and was expected to have a considerable impact. In August 1933, Le Corbusier used dramatic language to underline the urgency of additional action, in this case a resolution clarifying CIAM's goals. Le Corbusier characterized those in executive positions—ministers and other politicians—as waiting for CIAM's statement and its practical action. "It is high time, Giedion," said Le Corbusier, "the world is on fire. There is a need for reinforcement. We are the technicians of modern architecture. In the name of due procedure and of the holy cause I demand that the resolution be published."[16]

Le Corbusier insisted on an "active resolution" that would also tackle the question of economic change. In doing so, he crossed expertise's classic professional boundary. Commenting on Le Corbusier's report for the 1937 CIAM congress, Walter Gropius noted in a memo to Giedion that "to me he [Le Corbusier] seems almost dangerous in his arrogance and partial superficiality...I understand increasingly better why experts who work out special questions based on years of experience turn on their heels in view of this pattern."[17] Indeed, Le Corbusier embodied many of the traits of the "new" expert who was involved in fields beyond material technologies. By 1933, he was undoubtedly a public figure, if not an international celebrity, with his own trademark, the black owl-eye glasses. More important, he turned himself into probably the most influential prophet of the social mission of technical and scientific progress. His daring application of new technology to housing seemed poised to change and improve society or even bring about a completely new social order.

Such far-reaching aspirations called for strong measures. Throughout his career, Le Corbusier had an inclination to rely on powerful men, even authoritarian politicians, in order to mobilize the power to implement his far-reaching concepts.[18] The cruise on the *Patris II* was the indirect result of such an attempt to gain political backing—one that failed spectacularly. Attracted by a land of ostensibly infinite resources taking a progressive political course, CIAM initially planned its fourth congress in Moscow. The extensive preparations had been helped by the fact that a number of modernist architects and urban planners, mainly from Germany, the Netherlands, and Switzerland, had undertaken huge projects in the Soviet Union from 1930 onwards. Yet, soon Soviet leaders decided that it would be dangerous to allow too many foreigners to occupy influential positions. Moreover, Stalin had his own views on architecture. The rise of so-called socialist realism went along with a return to more classic aesthetic forms. This shift became evident in 1932, when the competition for the Palace of the Soviets did not result in a victory for Le Corbusier or another modernist proposal but for a rather conventional neoclassical entry. The CIAM leaders were appalled. Le Corbusier spoke of "treason of the spirit of the modern age," and other prominent CIAM members were harshly critical of the decision of the Soviet jury. They threatened to withdraw their proposition to hold a congress in Moscow and boldly explained their concerns in two telegrams sent directly to Stalin. Giedion managed to soften the messages slightly, erasing Le Corbusier's closing remark that the competition meant the Soviet Union was doomed "to a miserable mediocre, retrograde and decadent end." The chief comrade never replied, of course, but the position of Western experts in the Soviet Union was put at considerable risk.[19] Giedion, Le Corbusier, and the others had also inadvertently demonstrated the experts' overestimation of their capacities, as well as the growing likelihood of clashing with political ideologies. The overreaction of Le Corbusier and others can only be understood against the background of the hopes they had placed in the Soviet Union as a "laboratory of history."[20] Clearly the CIAM experts lacked the clout they thought they had.

Shifting the congress to the Mediterranean and Athens could only briefly displace the tensions inside CIAM about the position of experts vis-à-vis society and state. These tensions were deeply

Fig. 4.3 Battle Plan: *Already in 1928, on occasion of the foundation of CIAM in La Sarraz, Switzerland, the masterminds of the organization planned far beyond just bringing like-minded modernist architects together. They saw themselves as experts commanding the latest technologies and well positioned to bring about a better society. Le Corbusier's "battle plan" illustrates his strategy for winning over the public and the powerful.*

ingrained in the organization's DNA. Its experts dreamed of penetrating all areas of life with new progressive solutions derived from technological progress. They linked technology to a social cause in a way that few other fields attempted. This young discipline used what activists referred to as "scientific methods" to emphasize its credibility, reflecting the importance that science and technology had gained in previous decades. After 1900, statistical comparison

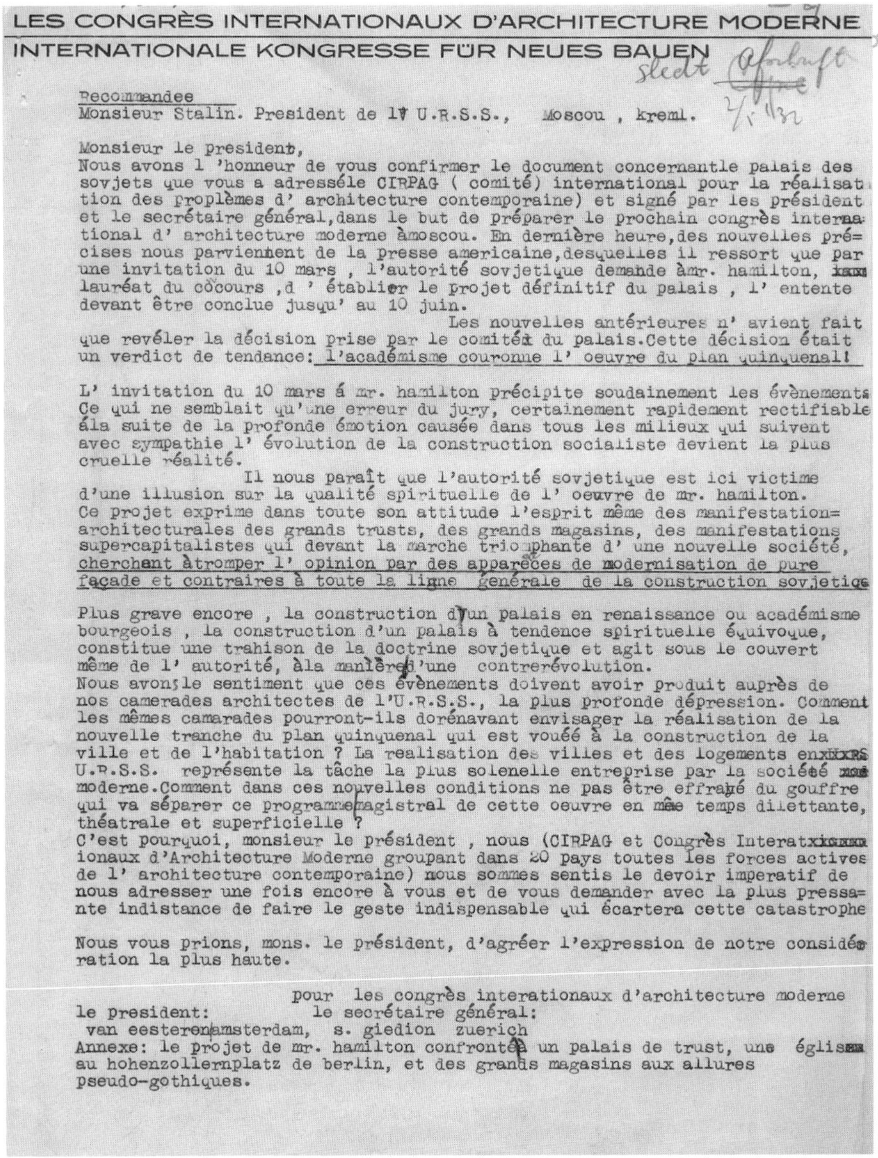
LES CONGRÈS INTERNATIONAUX D'ARCHITECTURE MODERNE
INTERNATIONALE KONGRESSE FÜR NEUES BAUEN

Recommandee
Monsieur Stalin. President de l' U.R.S.S., Moscou , kreml.

Monsieur le president,
Nous avons l 'honneur de vous confirmer le document concernantle palais des sovjets que vous a adresséle CIRPAG (comité) international pour la réalisation des proplèmes d' architecture contemporaine) et signé par les président et le secrétaire général,dans le but de préparer le prochain congrès international d' architecture moderne àmoscou. En dernière heure,des nouvelles précises nous parviennent de la presse americaine,desquelles il ressort que par une invitation du 10 mars , l'autorité sovjetique demande àmr. hamilton, lauréat du côcours ,d ' établir le projet définitif du palais , l' entente devant être conclue jusqu' au 10 juin.
Les nouvelles antérieures n' avient fait que revéler la décision prise par le comité du palais.Cette décision était un verdict de tendance: l'académisme couronne l' oeuvre du plan quinquenal!

L' invitation du 10 mars á mr. hamilton précipite soudainement les évènements Ce qui ne semblait qu'une erreur du jury, certainement rapidement rectifiable àla suite de la profonde émotion causée dans tous les milieux qui suivent avec sympathie l' évolution de la construction socialiste devient la plus cruelle réalité.
Il nous paraît que l'autorité sovjetique est ici victime d'une illusion sur la qualité spirituelle de l' oeuvre de mr. hamilton. Ce projet exprime dans toute son attitude l'esprit même des manifestation= architecturales des grands trusts, des grands magasins, des manifestations supercapitalistes qui devant la marche triomphante d' une nouvelle société, cherchant àtromper l' opinion par des apparèces de modernisation de pure façade et contraires à toute la ligne générale de la construction sovjetique

Plus grave encore , la construction d'un palais en renaissance ou académisme bourgeois , la construction d'un palais à tendence spirituelle équivoque, constitue une trahison de la doctrine sovjetique et agit sous le couvert même de l' autorité, àla manière d'une contrerévolution.
Nous avons le sentiment que ces évènements doivent avoir produit auprès de nos camerades architectes de l'U.R.S.S., la plus profonde dépression. Comment les mêmes camarades pourront-ils dorénavant envisager la réalisation de la nouvelle tranche du plan quinquenal qui est vouéé à la construction de la ville et de l'habitation ? La realisation des villes et des logements en U.R.S.S. représente la tâche la plus solenelle entreprise par la société moderne.Comment dans ces nouvelles conditions ne pas être effrayé du gouffre qui va séparer ce programme magistral de cette oeuvre en même temps dilettante, théatrale et superficielle ?
C'est pourquoi, monsieur le président , nous (CIRPAG et Congrès Interationaux d'Architecture Moderne groupant dans 20 pays toutes les forces actives de l' architecture contemporaine) nous sommes sentis le devoir imperatif de nous adresser une fois encore à vous et de vous demander avec la plus pressante indistance de faire le geste indispensable qui écartera cette catastrophe

Nous vous prions, mons. le président, d'agréer l'expression de notre considération la plus haute.

pour les congrès interationaux d'architecture moderne
le president: le secrétaire général:
van eesteren amsterdam, s. giedion zuerich
Annexe: le projet de mr. hamilton confronté à un palais de trust, une église au hohenzollernplatz de berlin, et des grands magasins aux allures pseudo-gothiques.

Fig. 4.4 a–b Pitfalls of Political Engagement: *The heads of the CIAM organization believed it most natural that they would turn to the powerful directly with their concerns. This letter to Stalin voiced the architects' criticism over the decision to build the Palace of the Soviets in a rather traditional way (reminiscent of a Berlin department store as an attached collage stated). The CIAM's threat not to organize a congress in Moscow hardly touched Stalin, but the letter endangered the position of Western architects in the Soviet Union considerably.*

and the use of newly available visual material like aerial photography, surveys, and sociological analyses contributed to the idea that planned development of all aspects of the city—if not of society as a whole—was not only desirable, but also achievable. Urban planning, however, particularly on the Continent, often carried more expectations than simply improving the organization of a city.[21] In tackling the ills of the modern city, it strove to address the ills of modernity itself. Radical urban planners envisioned a new society and the rise of the "new man." This excess of expectations was also a reaction to the possibilities offered by new technology, whether real or imagined. Almost inevitably, urban planners became technoscientific experts with a close relation to the state. They were strongly tied to the political, social, and cultural developments and debates of their time. These experts embodied a particularly European expertise and European answer to pressing problems—or at any rate, problems perceived as pressing, even if the urgency was only in the eyes of those who believed they had a solution to offer.

While the notion of the Functional City was only one of many concepts, it was the one that best exemplified the dynamics of the

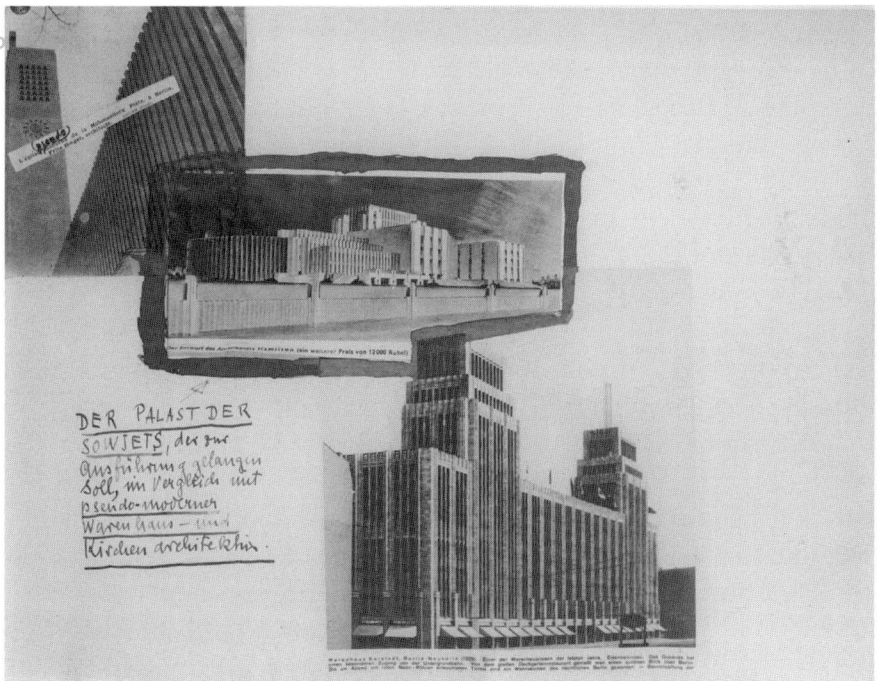

Fig. 4.4b

movement. Talking about the Functional City provided a framework within which to compare and standardize what was, by its very nature, something extremely diverse. This was not only true for far-reaching plans to bring such a city to life; it was also the case for the more prosaic present, at the center of which was the analysis of the city. A common language that allowed for effective comparison was developed, based to a large extent on graphic symbols developed by Neurath.[22] The great effort put into these developments seemed justified, as this common language would eventually enable the experts to transcend local shortcomings in favor of a uniform conception of a better future. Theirs was a common language that not only technically facilitated comparison, but was also attributed to the universal and cosmopolitan aspirations of experts in general. Cornelis van Eesteren, the Dutch president of CIAM in the 1930s, coined the term "one-hour urbanism," claiming to provide a universal system that was applicable in every place with minimal planning time—one hour or even just five minutes.[23] Urbanism promised to compare any place in the world with another, regardless of how different they were. In fact, however, its usefulness was mainly limited to Europe.[24] CIAM members systematized the politics of comparison, which had a much older tradition in Europe, and raised them to the rank of a stand-alone analytical method. They also radicalized this phenomenon in calling for action. Le Corbusier saw "the force of our congress" in the thirty-four city plans presented during the congress.[25] From today's perspective, this reveals a surprising combination of a belief in statistics and rational analysis of what was to be found on the ground and an almost unlimited faith in the capacity to promote change for the better, using exactly these "scientific" methods.

The plans of Polish CIAM members Szymon Syrkus and Jan Olaf Chmielewski for a *Warszawa funkcjonalna* (functional Warsaw), one of the few attempts of CIAM to apply the idea of the Functional City in practice, are symptomatic of the urbanist approach. Firstly, Warsaw was conceived of as almost a *tabula rasa* at the intersection of major European traffic arteries, an ideal place to apply the universal insights of CIAM urbanism. Secondly, and characteristically, Chmielewski and Syrkus declared that their conclusions were in no way derived from local conditions, but rather from their collaboration with CIAM and its earlier congresses in 1928, 1929, and 1933. In this sense, they took CIAM's universalism, based on the applicability of technology and "scientific" principles, to its logical extreme. At the same time, they accepted the reality check posed by the current global economic crisis, which was still strongly felt in the early 1930s. Their proposal, they claimed, should not lead to the neglect of social conditions:

> We do not want, like the technocrats, to get carried away by technical enthusiasm in order to forget the crisis, unemployment and the homelessness of the masses. We know all too well that in this very moment, when production and consumption are in such disorder, and when the ground-breaking social forces unload such a dynamic, we may only theoretically prepare the Warsaw of the future—the functional city.[26]

Rationalizing the Factory & Beyond

In fact, the decades that generated so many visions based on technology were not particularly welcoming to their realization. Only a few years after the congress in Athens, CIAM's universalist stance was challenged, as was its belief in ongoing progress. External conditions had worsened considerably with the rise of totalitarian and authoritarian regimes. Organizational tensions grew as well. Virtually pushed to one of the few corners of the Continent still not dominated by authoritarian regimes, a subgroup called CIAM-Ost gathered in Czech Brno in late April 1937. The group met there because the delegates from a number of Eastern European countries believed the region was in need of particularly radical measures in housing and regional planning. Brno had been chosen not least because of its proximity to one of the rare examples of an attempt to implement functionalist concepts in reality. The nearby town of Zlín was the headquarters of a flourishing shoe-making company founded by Tomáš Baťa (1876–1932). Baťa had turned a small family business into one of the leading shoe producers in Europe, and at the same time transformed Zlín from a marginal small town in Moravia into the pulsating urban headquarters of a worldwide network of technical and industrial expertise. The company embodied not only many of the opportunities inherent in planning expertise for all areas of life, but also its inherent dangers, which became apparent in the years that followed.

In 1921, Baťa had welcomed the first emissaries of "knowledge tourism," lured to Zlín by the news of something new happening and further spreading this news themselves. Many more would follow. Recapitulating his visit to this showcase city in 1935, Le Corbusier recorded his impressions in a letter to the company's owner and director, Jan A. Baťa (1898–1965), Tomáš' brother. Le Corbusier was enthusiastic about this "resplendent" place, saying: "Zlín is, when everything is told, one of the hot places of the new world: there is life in it! In Paris I have not met anything like Zlín."[27] The correspondence between Le Corbusier and Baťa is also notable for the fact that it happened at all. Apparently, Baťa saw Le Corbusier as kind of meta-expert who combined international fame, drive, and critical organizational knowledge. Le Corbusier hoped to find in Baťa the great "enabler," someone whose resources would allow him to put his ideas into action. Previously he had hoped to win over the French and Italian industrialists Citroën and Olivetti to support his plans, but with limited success. Although Le Corbusier had flirted with the left and the Soviets, he was perfectly willing to turn to someone whom the leftist avant-garde—and Teige in particular—detested as a capitalist exploiter.

The Baťa factory was, in many respects, a direct outcome of the First World War. The company had been one of the major manufacturers of shoes in the Austro-Hungarian Empire prior to 1914. However, it only started to overtake its competitors after Tomáš Baťa managed to obtain a huge commission for army boots at the start of the Great War. This commission allowed him not only to reach new scale in production, but also to save his workers from conscription. Moreover, Baťa drew on prisoners of war in the production process. By the end of the war, the firm employed almost 5,000 workers who produced 10,0000 pairs of army boots a day. A few years later, shoes were still the company's main product, but far from the only one. Baťa had diversified into flooring, toys, car tires, socks and stockings, mining, maritime transport, air transport and, later, even light airplanes. After the death of Tomáš Baťa, who crashed his company airplane into a factory chimney on July 12, 1932, Jan A. Baťa took over. Under his leadership the company increased to almost 6,000 retail shops, and eventually it employed more than 100,000 workers worldwide.

By 1923, Tomáš Baťa had also become the mayor of Zlín, by then a medium-sized town. Yet, Zlín was soon only one of many production sites, although it remained the central one. In an attempt to survive the economic crisis and new tariff barriers, Baťa established factories abroad, often in conjunction with the creation of new towns according to the Zlín model, most of which had their own company publication.[28] These included Chełmek in Poland, Batadorp Best in the Netherlands, Ottmuth in Germany, Borovo in Yugoslavia, Möhlin in Switzerland, Bataville, and Hellocourt in France (with an urban plan by Le Corbusier), Martfü in Hungary, and East Tilbury in the United Kingdom. The company later erected branches in Batanagar (India), Beirut, and Baghdad, among many other places. "I want to build copies of Zlín all over the world!" Jan A. Baťa exclaimed, as he exported his model towns to North and South America, Africa, and Indonesia. When visiting Zlín in 1935, Le Corbusier is reported to have said "Ah, a self-duplicating town."[29] In this drive for modernization, the old town center of Zlín was held in low regard and its partial destruction met little resistance. Flying over the region of Moravia, Tomáš Baťa expressed his wish for an earthquake that would make ample space for a better and hygienic new world.[30] He complemented the existing structures with an innovative health department focusing on preventive medicine, a welfare department with a wide range of activities established in 1924, followed by a Baťa hospital (1927). These measures reflected demands of the workers' movement and even included May Day celebrations organized by Baťa, but in many respects they also resembled fascist leisure organizations such as *Dopolavoro* in Italy or *Kraft durch Freude* in Nazi Germany.

Right from their first day in the company, Baťa workers—referred to as "young men" and "young women"—became part of a system that reached far beyond the factory. Special schools, similar to those used by Ford in the U.S., initiated the new generation of employees. The most gifted and the most committed entered an elite corps called *Tomášovici* in honor of the company's founder. They underwent special training as leading employees, including assignments abroad. They learned foreign languages, acquired perfect manners, exercised self-control, and trained in horsemanship. In order to highlight their special status, these men wore top hats and lived in a separate building of their own.[31] The *Tomášovici* formed an elitist modern, knowledge-based order, reminiscent of older clerical or knightly orders in Europe. In response to the company's growing dependence on specialized and scientific knowledge, Baťa founded a Center of Science with research institutions that dealt with

Fig. 4.5 **An Order of Knowledge:** *Bataism was based on the idea of developing the full potential of modern knowledge, and also the bearers of such knowledge. The best young employees of the company entered an elite corps called* **Tomášovici** *in honor of the company's founder. They underwent a comprehensive form of training and wore top hats to signify their special status.*

airplane construction and mechanical and chemical engineering. High-profile scientists who had suffered from the closing of the universities of Prague and Brno after the German invasion entered these institutions.[32] Although industrial Zlín was first and foremost a shoe-town, postcards and other visual material presented it as "a new world that provided seemingly endless possibilities identified with modern technology."[33] Modernist architecture served as a brand that contributed to corporate identity.[34] That Le Corbusier designed the Baťa pavilion at the 1937 World Exhibition in Paris should come as no surprise.

Everything about the company was standardized in the name of efficiency—from the production process to the factory architecture and the company housing in Zlín. In order to supervise his employees more easily, Jan A. Baťa even had his office constructed in a 6.15 by 6.15 meter elevator, a "mobile throne room," as reporter Egon Erwin Kisch called it, which allowed him to move up and down his

17-story high-rise, one of the first in Czechoslovakia. The elevator, which came to be known as the "Baťa skeleton" was an adaptation of a 20-foot model used in U.S. projects and inspired by the famous Ford architect Albert Kahn. Baťa had apparently brought back the idea from a 1919 trip with his father to the U.S.[35]

In an extremely consistent way, Baťa had transposed the logic of industrial production into the logic of architecture. During the 1930s, Baťa's housing division constructed between 300 and 600 apartments each year for the ever-growing army of workers and their families. The central importance and symbolic meaning that he gave to norms, also illustrates his conviction that it was essential to apply new technological solutions to every area of life. Baťa's chief planner František L. Gahura later explained that from the very beginning he and his colleagues strove to have the city grow organically out of industrial architecture and, in so doing, to develop a new concept of work and life.[36] In this, Zlin had much in common with a number of other interwar cities founded or rebuilt in the spirit of "planning for national regeneration."[37]

The identification of the Baťa company's modernist mission with modernist architecture was probably best expressed by the glass-concrete structure that Jan A. Baťa erected to memorialize his older brother. In addition to celebrating the man who had made the company great, it celebrated an "aesthetic of rationalization"[38] at the city's highest spot, in between the boarding schools for the new Baťa men and women.

Just as Baťa's city model and aesthetic mission were expanded far beyond the boundaries of Zlín, transport planning, too, involved altering an entire region. Gahura initiated the regulation of the Moravia River, making it navigable as a first step to link the Black Sea to the Baltic via the Danube and Oder rivers.[39] Jan A. Baťa envisioned nothing less than transforming Czechoslovakia into a country of 40 million inhabitants, which made it very clear that his ambitions were no longer just business-related. In accordance with Baťa's dictum "We are not afraid of the future,"[40] master plans for the region around Zlín anticipated enormous future growth and tried to seize the opportunities of modernist planning applied to a backward district.[41]

What came to be known as "Bataism" had been highly controversial almost since its beginnings. Pamphlets like *The Unknown Dictator Thomas Baťa*,[42] and terms such as "Dictatorship of the boot" or "Shoe Mussolini,"[43] set the tone. Egon Erwin Kisch was Baťa's most prominent critic, focusing particularly on the excessive control used in the factory system. As Czech writer Ludvík Vaculík remarked sarcastically: "I wanted to say that BŠP [Baťa School of Work] is an assembly line and that the Young Men are the shoes, that everything is mechanized here (even culture is manufactured here as if on the assembly line)."[44] In 1930, an investigation by the International Labour Organisation (ILO) looked more deeply into the concerns the critics had raised.[45]

Still, there were at least as many positive voices. French technocrat Hyacinthe Dubreuil published a brochure in 1930 entitled *L'Exemple de Baťa*. The Belgian CIAM group enthused about the good lighting and ventilation and the safe traffic in Baťa's factories.[46] Joseph Vago, a prominent Hungarian mainstream architect, described Zlín as an exceptional phenomenon due to its exemplary social infrastructure. He contrasted the European example with the noisy and monotonous American Detroit, which he claimed lacked any welfare structures. Vago described in great detail the superiority of the social model in the European knowledge-based economy over its American competitor. An Italian publication placed "the city without courtyards" in the same group as other successful European industrial cities and the latest example of a fascist city, Sabaudia in Italy.[47] Not surprisingly, Le Corbusier also joined in. Having criticized the U.S. as the "land of the timid," a land not prone to enact his visions, Le Corbusier explained that Baťa had achieved much more than just effective rationalizing—a man guided by his heart. Baťa, he believed, had created a "homogenous social group" and thus had overcome class conflict.[48]

In short, Baťa seemed to provide an example of how the admired American traits of productivity and innovation could be tamed and "humanized" in a European spirit. The "American anthill" that Le Corbusier had discovered in remote Moravia was overseen by a European Ford who, unlike his American counterpart, was presented as a loving patriarch. This contrast between America and Europe was also emphasized when discussing the city's master plan, in which Le Corbusier defended his own vision of a condensed city center, accusing Baťa and Gahura of Americanism, leading, he argued, to urban sprawl—"the awfully bad side of America, the true cancer."[49]

Fig. 4.6 Memorizing Modernism: *Merging the cult of the late Tomáš Baťa and the cult of technology, Jan A. Baťa hung a replica of the Junkers airplane in which his brother had crashed in this strikingly modern memorial. Baťa's belief in the value of new technology was as apparent as his unwillingness to stick to traditional ways of doing things—in this case established forms of memory.*

While some like Le Corbusier tried to present Baťa as a project in which everything positive was associated with Europe, the reality was much more complex. In 1937, a film was shot, later entitled *Baťa Flies around the World*, which captured the decisive features of Bataism. Filmed and edited by some of Czechoslovakia's most gifted cinema professionals, the film differed somewhat from the blockbusters of the age. Business meetings, production processes, factories, and transportation systems intermingled with coverage of daily life in exotic settings and tourist attractions. Jan A. Baťa functioned almost as a second director, intervening even in seemingly minor scenes.

Baťa was not just interested in the advertising potential of the film, however; he was concerned that it effectively communicate knowledge about, and the corporate identity of, his company. The film presented his vision of standardization and training in a highly condensed form and covered subjects such as the workings of new machinery distributed in the satellite-town system. Within the Baťa universe, print, telegraphy, telephone, photography, slide projectors, cinema, gramophone, radio, teletypewriter, phototelegraphy, and pneumatic tubes were in use, and television had already been discussed. An employee notice from 1938 demonstrates the degree to which these media were part of the company's work processes:

> Fifteen hundred telephones have been installed in our factory and they are used for 200,000 calls every day! If only one minute were wasted in each individual call, the total loss would reach 30,000 per day, which means 138 days, i.e., more than a quarter of the year. How big the loss in money per year would be—that's impossible to calculate.[50]

Transatlantic Fordism

The other side of Bataism—the obsessive control that was essential to the whole system—clearly comes to the fore here. Examples include a system of sound and light signals that directed the workers, and Baťa's habit of eavesdropping on his employees' telephone conversations by using monitoring devices attached to his elevator office. Baťa's particular version of Fordism took American ideas to the extreme. Rudolph Philipp, Baťa most obstinate critic, exclaimed in 1936: "Baťa beats Ford. Zlín beats Detroit."[51]

Of course, Fordism, Henry Ford's all-embracing reorganization of his car factories and of benefits for his workers, was initially an American phenomenon. It could even be argued that it was the quintessential American phenomenon for most of the twentieth century. Yet, as Antonio Gramsci has pointed out, Fordism may say more about a receptive Europe than about the U.S. itself.[52] Fordism, seen as the epitome of what was more generally referred to as Americanism, presented an opportunity as well as a challenge for a Europe undergoing a severe postwar crisis. The factory provided a structure full of opportunities, offering solutions to the causes of the crisis and touching much more than just the economy. Europeans had exhibited a strong fascination with U.S. technological progress since the mid-nineteenth century, and the sweeping success of the U.S. in the First World War enhanced its status as a great power, increasing European interest in what was happening on the other side of the Atlantic.

Fig. 4.7 Shoe Mussolini: *The revolutionary new production methods used by Czech shoe manufacturer Bat'a found many admirers and almost as many fierce opponents. What some saw as bringing to the fore Europe's best traditions of an economy based on social principles, others saw as the ruthless rule of commercial goals over human needs. This book cover tellingly links Bat'a's attempt to control everything to the political threats of the period.*

Fordism became a highly charged and omnipresent concept across the whole of interwar Europe; the term was often used almost as a synonym for Americanism. Whenever or wherever one positioned oneself in the ongoing debates on the economic and social crisis, ignoring the relentless talk about Ford was almost impossible.[53] Fordism clearly resulted from an intensified techno-intellectual exchange between Europe and the U.S. that had started by 1900. Shortly after Henry Ford established his innovative production system based on standardization and moving assembly lines, experts flocked to see this miracle of productivity. In 1905, well before Ford Motor Company built its famous Highland Park factory and experimented with moving assembly of automobiles, legendary Fiat founder Giovanni Agnelli traveled to Detroit to carefully extract and adopt what he believed useful for his thriving business in Torino.[54] German visitor Franz Westermann reverently reported that the spirit of the Ford company "simply draws people into its orbit."[55]

Critics and admirers alike recognized that Fordism was a phenomenon that could be ignored only at one's peril. First-hand evidence was easily available, even for those who were unable or unwilling to cross the Atlantic. Ford's book *My Life and Work* became a huge success on the European market, with translations into most European languages. The gospel of Ford appeared to fall on particularly fruitful ground in those countries that had lost the Great War and gone through a painful transitional phase. By the end of the 1920s, more than 200,000 copies had been sold in Germany alone. Though many commentators criticized the inconsistencies of Ford's self-made world view, enthusiastic reactions came from every part of the political spectrum, including from an admiring Adolf Hitler.[56] In victorious France, too, Fordism had an appeal far beyond the workplace, with car manufacturer André Citroën fashioning himself as a sort of French Ford.[57] Both Le Corbusier and Karel Teige could agree about the immense potential of Fordism for almost all areas of life.

Though the "ism" suggests otherwise, Fordism was anything but a clear-cut system. In part, this openness and flexibility was a precondition for its success. Taylorism, a system devised by the American engineer Frederick W. Taylor to organize labor for greater efficiency, had been widely dismissed immediately before and briefly after the Great War as too mechanical. Fordism, on the other hand, carried overtones that made it an attractive concept beyond the production site. Inter-European cultural prejudices and political rivalries also made it more appealing for many European nations to emulate American models rather than those of their neighbors. For example, while industrialization in Britain was also highly advanced and successful, Germans preferred to look to the U.S. for inspiration. The cultural distance of the U.S. also seemed

Fig. 4.8 Hope of Salvation: *Henry Ford, besides becoming the world's leading car maker, turned into a remarkable techno-celebrity. In 1920s Europe he was believed to be a relevant thinker far beyond his actual business. His insights on organizing the economy were hailed as offering a solution to the Continent's immense economic problems. Here, Henry Ford is depicted in another context dealing with the hope of salvation, the Oberammergau Passion Play of 1930.*

to promise more space for critical analysis and an opportunity for Europeans to consider what was worth transferring. Although—or perhaps because—it was often rejected as "un-European," Fordism appealed deeply to a continent in a severe crisis of identity, with Henry Ford's notorious anti-Semitism being hardly irrelevant for his appeal to right-wing circles.[58]

In concrete terms, this was also true for Bataism as a European adaption and development of Fordism. Europeans placed great expectations on Fordism as a European modernization program that was particularly attractive to the countries like Poland or Czechoslovakia, that had newly emerged on the European stage after 1918.[59] The belief that modernization offered a solution to these countries' social and economic ills contributed greatly to Baťa's communicative success. The local conditions were also advantageous to Baťa's endeavor: a rather underdeveloped region offered cheap labor and the traditions of Czech technical knowledge and education could be used to the best effect. Away from the Continent's main political centers, Baťa seized the opportunity to exert influence beyond the factory gate. From an early stage, Jan A. Baťa strove to introduce the American models, namely Fordism and Taylorism, that he learned about when traveling the USA in 1904–05 and 1919.[60] Baťa was deeply impressed by the assembly-line production and conveyor belts, but also by the self-management of workshops at use in Detroit. Fordism infused every aspect of production at Zlín and many aspects of the organization of social life. However, none of this would have been successful if it had not been compatible with ideas already circulating around Europe. The idea of the Functional City is symptomatic of the trends of the time. As sources of inspiration, both Fordism and so-called scientific management were closely related to the idea of the Functional City.[61] These concepts and belief-systems became powerful because they promised to use superior knowledge in the context of daily social relations.

In this, Bataism was certainly a European variety of Fordism, with very distinct features. Its extreme internationalization strategy was a reaction to the obvious limits of a small country, as was Baťa's obsession with transport. The Europeanness of Bataism was particularly evident in what could be called its social dimension. Its urbanism, health care, and advice on healthy living (communicated to the employees in radio broadcasts during breaks), social welfare, and cooperation between employer and employee were all part of traditions and discourses that also existed in the U.S. but had a much greater urgency in crisis-ridden interwar-Europe.[62] Moreover, Bataism was strongly inspired by Fayolism, a management theory developed by the French engineer Henri Fayol.[63] This was in many respects another European adaption of Fordism and Taylorism, yet placed more stress on the responsibilities of the individual worker.

As the example of Bataism shows, Fordism was part of the extremely intense transatlantic flow of ideas and concepts, providing both a shared language for specific ideas and a concrete method of production. The examples that crossed the Atlantic in both directions were deeply influenced by their context and their place of origin. In the years before the First World War, Werner Hegemann, one of the earliest urbanists and a sought-after transatlantic traveler, was marketed in the U.S. as a harbinger of the latest knowledge from Europe. Back on the old Continent, he rose to prominence as an eyewitness to American progress in all things urban.[64] The transatlantic exchange intensified after, if not during, the Great War, with U.S. relief work setting in even before the conflict's end.[65] In a reverse flow, during the depression years and

Franklin Roosevelt's New Deal, European ideas for overcoming the crisis by "social expertise" gained immense influence. For example, when German architect and Bauhaus founder Walter Gropius, a proponent of social housing based on Fordist production processes, accepted an influential position at Harvard University in 1937, this mechanism was certainly at play. In many respects "modernism was the aesthetic wing of the...growing technocratic movement in interwar Europe."[66] As the decision makers at Harvard made clear, they chose Gropius mainly because of his "social edge."[67] They hailed him as the creator of ideas that had been accepted worldwide and, at the same time, that had triggered high expectations.[68] Tellingly, *Time* magazine welcomed the newcomer to the U.S. with the headline "Bauhaus Man."[69] This testified both to the rise of transnational expert celebrities (in this case Gropius), but also of a trademark European modernism (in this case the Bauhaus), thought of as poised to solve deep social problems using technology and effective production. What the Americans perceived as European had been, for Gropius, who traveled the U.S. early in his life, essentially American—the hope of providing mass housing using Fordist methods. Housing production was one of the least evident—but for contemporaries most remarkable—signs of the machine age, with some three hundred thousand homes built in Germany by the late 1920s and an almost equal number in Great Britain.[70] Not by chance did Walter Gropius state in 1923 that architecture was now walking "hand in hand with technology."[71]

An important interlocutor between both sides of the Atlantic was Lewis Mumford. Mumford, a self-professed "disciple" of Patrick Geddes, had studied extensively in the library of the *Deutsches Museum*, Germany's foremost museum of technology, and became an advocate for community building using new technologies in housing.[72] Mumford is an example of a fruitful misunderstanding that found Americans admiring the social standards of housing in central Europe, while architects in the Netherlands or Germany prided themselves on their Fordist approach towards housing, their "dwelling machines," or as Dutch modernist architect Jacobus Oud described their products, their "dwelling Fords."[73]

Mumford became a key figure in bringing this Americanized housing back to the U.S. via suggestively illustrated articles for *Fortune* magazine, but he ran into trouble with his editors when he stressed the importance of the state as an actor in social housing achievements. This flow of ideas was largely centered on modernist visions, which were sought and found on both sides of the Atlantic.[74] Yet, it also went hand in hand with "the politics of lag"[75]; that is, social progress in Europe invoked by groups pressuring for similar measures in the U.S.

As Bataism also shows, the transfer of Fordism to Europe largely altered its character. In Europe, Fordism turned into a wide-ranging concept that was seen as potentially capable of overcoming a deep social, economic, and political crisis.[76] At its core an organizational concept, Fordism spurred in Europe what could be called "expertise with a cause," that is, systematic use of technology with the goal of improving society. Advocates saw in it a possibility for overcoming class conflict. Fordism, which made it possible to offer high wages and low prices for consumer goods, seemed a balm directed at the point where Europe's crisis was most severe. In this, Fordism also seemed to have the potential to offer an alternative path of development beyond the old liberal systems and the threat of right- or left-wing authoritarianism.[77]

Fordism offered an attractive justification for all kinds of technocratic visions. American technocratic models, which were derived initially from Taylorist scientific management and heralded the "era of knowledge" overcoming "the era of force," thrived in the dysfunctional European postwar systems more than in the rather successful U.S. Thus, ironically, technocratic models that were developed within a parliamentarian system were often used to legitimize authoritarian solutions.[78] Technocratic solutions became popular with those who believed in the central importance of technoscientific experts and saw technological progress as a neutral medium for bringing about improvement without necessarily transforming politics or society. In spite of this supposed neutrality, however, they became key starting points for various technocratic political models. Technocracy became the most important framework in which the arbiters of knowledge and the state could meet and join forces—be it the *Groupe-X-Crise* in France, consisting of graduates of the *École Polytechnique*, the think tank Political and Economic Planning in Britain, or the plan for Tectology, a "systematization of organized experience," developed by Alexander Bogdanov in Russia.[79] Perhaps the most illustrious example of the adoption of management methods developed in the U.S. was that of Frenchman Charles Bedaux, a disciple of Frederick W. Taylor and major contributor to the development of scientific management.

Bedaux saw no problem in working together with the Nazis and the Vichy regime in France. In fact, he even regarded authoritarian political regimes as a precondition for effective business.[80]

Self-Empowerment & Social Engineering

Europe's experience of extremely fast technological and scientific progress in the decades leading up to the 1930s had given rise to high expectations of change—change that was not even necessarily tied to any concrete form of scientific progress, but was of a general nature. At the very least, after the destructiveness of the First World War, technological change was also seen as a symptom of a deep crisis by very different groups. Criticism of technology and science that had existed since the beginnings of industrialization was now expressed more radically. Simultaneously, in what has been labeled "reactionary modernism," many right-wing intellectuals embraced technology, combining it with romantic and undemocratic political concepts.[81] Even those who wholeheartedly supported technological development and the social change which came with it were concerned, for they believed that Europe was falling behind the U.S. As has been shown for the European adaptation of Fordism, technology and scientific progress had become part and parcel of European self-definition. The long-standing equation of knowledge with Europe was being severely challenged.

However, this crisis of science and technology did not prevent technoscientific experts from claiming and gaining new ground. After the First World War, as Tzvetan Todorov rightly put it, science "was no longer content to restrict itself to acquiring knowledge of the world but sought instead to transform the world, in keeping with Marx's tenet, in order to achieve an ideal that, it claimed, was rigorously deduced from scientific observation."[82] While this development was certainly not restricted to Europe, it was far more pronounced there. Europe's troubled transformation from traditional societies into modern, more egalitarian orders was the most important background development. The traumatic Great War, with the internal upheavals and fears of Europe losing its dominant global position, added to the attractiveness of visions of transforming the world through knowledge.

Although they were not always fully appreciated by the experts involved, technical problems turned into very human ones. If, for example, experts started to define what exactly the minimal dwelling needed to comprise, they had to define what Teige called the "existential minimum," the subsistence level of human beings.[83] Experts had to make judgements that increasingly touched the realm of the social. This form of intervening in individual lives has often been labeled "social engineering." Technology played both a symbolic and a more practical role in this. The enormous progress of technology triggered a general belief in the feasibility of organizing and structuring even the most complex affairs. Yet, applying concrete technological innovations beyond the laboratory or the factory also seemed more and more attractive in areas such as housing and hygiene.

Social engineering became a particular mode of "problematizing modernity."[84] It was used to frame problems that would otherwise have remained invisible or diffuse, and it carried the promise of solving these problems once and for all. Experts, who occupied center stage in such visions, focused on the family as the key unit from which to transform not just individuals but whole societies. The example of Baťa shows how every worker was seen as a potential expert, as profiting from the fruits of systematically applied knowledge through the superior organization of factories and through direct training.

Thus, problems such as housing acquired great relevance, and other fields, particularly health, entered the horizon of technoscientific experts. Concepts of efficiency were now increasingly applied to humans and the human body, and governments began to dedicate time and resources to public health projects.[85] There was no structural logic for the increasingly intense attempts to improve the human condition via science that would lead it to eventually transgress ethical borders. Yet more than just a few morally questionable scientists entered into biologistic schemes like eugenics or racism. In fact eugenics became the obvious merger of science and politics. Self-empowerment by experts played as much of a role here as the dynamics of expertise with a cause. Their activism yielded "technologies of race," the use of scientific knowledge and pseudo-scientific knowledge to establish techniques and measurements that reinforced eugenics and racism.[86] Such technologies were communicated throughout Europe. They fell on particularly fertile ground in those countries that felt a need to catch up after the Great War, the newly emerged states, or those that, like Germany, had lost the war.[87] Moreover, the economic crisis, which dramatically worsened after 1929, played an important role in encouraging experts to embark on radical measures. However, the many schemes for disease control and improvement of public health were not, in and of themselves, necessarily the seeds of criminal and eventually genocidal politics. While we see that the self-mobilization of experts and their political radicalization was of key importance, particularly in Germany, a change of political regimes was needed to turn ideas into crimes.[88]

Transatlantic exchange and the "international of racists" helped to legitimize the new strand of radical social experts.[89] Nevertheless, it was in Europe that eugenics and racial politics led to disastrous results. Paradoxically, traditional society and the obstacles to modernization that it presented seemed to have made proponents of European modernization particularly willing to take radical measures during the interwar period.[90] This was obvious among CIAM architects and urbanists, as well as for the increasing number of experts who joined forces with totalitarian regimes advocating completely different goals and very different ideas of change. In Europe, and particularly in Germany, experts demonstrated their totalitarian potential very strongly in the course of the next several years, and not just in hygiene. Their belief in authority was often justified by the need to implement long-term planning and overcome technological shortcomings.[91] What made many experts perfectly suited to becoming indispensable proponents of National Socialism or Stalinist communism was, of course, in most cases opportunism. But many also longed for stable and durable political structures, which they saw personified in authoritarian regimes. What is more, many experts identified with the new regimes for reasons that were inherent to their professional fields.[92] Some fields like eugenics obviously were sympathetic towards the Nazi ideology. Yet, with ever more resources necessary to put through expert schemes and to make people conform to what experts regarded as the necessary steps to achieve what they saw as a better society, an intrinsic logic developed for merging a substantial group of experts with the radical agendas of Fascism, Stalinism, and National Socialism. In hindsight, the year 1933, when this chapter opened, was a turning point for this amalgamation of expertise and politics. As the discussions within CIAM on board the *Patris II* and after showed, the role of experts in the emerging political regimes was still being negotiated. How and why experts entered into ultimately catastrophic bargains with European authoritarian regimes will be our next topic.

5
Faustian Bargains in Totalitarian Europe

By 1940, the fortunes of the experts described in the previous chapter had completely reversed. The Warsaw University of Technology (WUT)—the pride of the new Polish nation state—had been closed immediately after German troops conquered and partially destroyed the Polish capital in October 1939. The German occupants now restricted higher education more tightly than ever, even under Russian rule before 1918. Tellingly, this situation did not change until 1942, when increasing pressure on resources forced the Germans to introduce an extremely reduced version of the WUT, so as to produce a small number of technical experts with basic qualifications for those Eastern areas that the Wehrmacht aimed to conquer but where German engineers were unwilling to go.[1]

In the meantime, at least in Eastern Europe, the German occupiers had made plain their new policies regarding knowledge production, or rather, the eradication of knowledge. In 1939, Krakow's ancient and famous Jagiellonian University became the site of an act that was both brutal and sinister, the so-called *Sonderaktion Krakau*: having invited all university professors and faculty from other academic institutions to a lecture in the Collegium Novum, the Germans then used this as a trap to imprison 183 professors in concentration camps. Only due to severe international pressure were most of them released, although some eventually died for committing the "crime" of holding an academic position. Two years later, after Polish Lviv (which had been part of the Soviet Union since 1939) fell into the Wehrmacht's hand, the Germans no longer even tried to maintain a semblance of decency. Only a few days after the army arrived, more than forty Polish professors and their families were executed. Technoscientific expertise had become a prominent target in warfare, as the so-called *Intelligenzaktion*, a genocidal sweep targeting Polish elites, had shown.

It was not by chance that some of the Lviv professors had also been on Soviet prosecution lists.[2] Like Nazi Germany, Stalinist Russia intended to partially integrate and partially annihilate technoscientific expertise in Eastern Europe. The history of the Ukrainian Physical Technical Institute (UFTI) in Kharkiv illustrates how a single research center could suffer from both Sovietization and Nazification. Founded in 1928, the institute exemplified the Bolshevik policy to colonize the Soviet periphery by establishing kernels of industry and, at the same time, subjugating the local "kulak" peasantry. Despite obstruction by Soviet party officials, the institute developed an excellent international reputation under the leadership of the renowned Lew Landau, who in 1962 received the Nobel Prize for physics. During the purges of 1937–38, Kharkiv "fell directly into the Stalinist maelstrom" and lost most of its research personnel, who were arrested, if not executed.[3] During the Second World War, the institute experienced another major blow. When the Wehrmacht invaded, the remaining Kharkiv experts evacuated the institute and tried to relocate parts of their laboratory equipment to Almaty in Kazakhstan, where they helped take the first steps in the Soviet atomic bomb project. Returning to Kharkiv in 1943 after the city was freed from Nazi rule, they found the institute ransacked and in ruins. Reconstruction suffered from an ideological struggle in Soviet science and the institute again experienced Sovietization and marginal access to resources. Although the Kharkiv institute was able to develop novel ideas in solid state and nuclear physics, most of its talented young specialists were transferred to Moscow after graduation. In Stalinist Russia, knowledge production in the Ukraine was seen as peripheral, even if it had the potential to further Soviet power.

While Krakow, Lviv, and Kharkiv represented extreme cases of the totalitarian politics of obstruction and destruction, many other higher technical education and engineering research institutions in the West also suffered under Nazi occupation. A noteworthy example is the *Technische Universiteit Delft* (Delft Technical University) in the Netherlands, which the Nazis closed in 1940 after 3,000 students went on strike to protest against the forced removal of Jewish civil servants from the Dutch administration. Typical of the Nazis' mixture of suppression and pragmatism in the occupied West was their policy towards aeronautical research centers like the French *Institut Aérotechnique de Saint-Cyr* in Paris and the Dutch *Nationaal Luchtvaartlaboratorium* in Amsterdam. These organizations were able to preserve a high degree of institutional autonomy by claiming that their technical equipment and personal resources would feed the German war machine. Thus they became "an integral part of a European-wide network of outposts of the Nazi regime" to mobilize expertise of potential military use.[4]

A more specific case was the *Reichsuniversität Strassburg* (*Reich* University of Strasbourg) which Nazi Germany established in 1941 after annexing Alsace. The *Reich* University was conceptualized as a continuation of the Kaiser Wilhelm University, which had existed between 1871 and 1919 when Alsace-Lorraine was under German control. In the view of Werner Best, the infamous mastermind of annihilation in the *Reichssicherheitshauptamt* (*Reich* Security Main Office) and by then a kind of Super-Minister of the Interior in occupied France, the *Reich* University would help "integrate the neighbors in the West into the new European order and to secure them for the emerging community of peoples under German leadership."[5] Nazi plans for a huge new university quarter in Berlin or the gigantic "House of Knowledge" within the *Ordensburg Vogelsang*, a center to educate future Nazi leaders, attest to the high symbolic value that institutions of knowledge and education enjoyed. Yet, the Nazis stripped away the universalism inherent in science, subjugated knowledge to Nazi ideology, and made it accessible only to a racially defined group.

Meanwhile Paul Otlet's small version of the Mundaneum in Brussels' *Palais de Cinquantenaire* closed after the German conquest in May 1940. By that time, the installation's home, the *Cinquantenaire* Museum, housed a display of Nazi art diametrically opposed to Otlet's internationalist vision of knowledge. Simultaneously, Otto Neurath, the other great visionary of universal knowledge organization, escaped the Wehrmacht under dramatic circumstances, abandoning his exile residence in The Hague in an open boat headed to Britain. Only after nine months of internment and an intervention by Albert Einstein was Neurath able to resume his work.

In East Central Europe, the Baťa works had already been confronted with the harsh new realities, although the company met a very different fate than many universities. Before German troops invaded the remainder of what had been Czechoslovakia in March 1939, Jan Baťa had met Hermann Göring in Berlin. What Baťa and the *Reich* Minister for Aviation actually discussed remains open to debate. In any case, the factories in Zlín were closely integrated into the German war economy. Hitler even sent specialists to study their scientific management methods in order to enhance the productivity of comparable industries back in the *Reich*. Baťa's emblematic hyper-modernist showroom—the Baťa Palace at Prague's Wenzel Square—turned into a particularly sinister scene. On Nazi orders, Baťa filled the huge glass storefront with a bicycle and other remnants from the site where Reinhard Heydrich (known as "Hitler's hangman") had been assassinated by Czech paratroopers. A poster offered an immense reward for capture of the assassins.[6]

With the beginning of the Second World War, the *Congrès internationaux d'architecture moderne* (CIAM) had been dissolved in everything but name, as it had become practically impossible to take part in free exchange and travel within Europe. This was, of course, also true for all other scientific and expert congresses and associations. The luckier CIAM members managed to escape to the U.S., while many from Central and Eastern Europe suffered from persecution, some ending up in concentration camps. Le Corbusier, however, who was always eager to find a great sponsor who would allow him to realize his grandiose schemes, had fewer scruples. After the French defeat one could find him in Vichy, literally next to the strongmen of the French collaboration regime.[7]

These cases make it dramatically clear how, all across continental Europe, experts were forced to make choices by the late 1930s, regardless of how neutral and apolitical they considered themselves. While Le Corbusier, for example, had other options than supporting the Vichy regime, those in the way of the Nazi war machinery faced potentially fatal consequences even if they did not actively oppose the regime. Continuing to work without taking sides was not an option. Moreover, both the Soviet Union and the Nazis pursued totalitarian politics, demanding loyalty from everyone and striving to control both the professional and private spheres for expertise. Either for racist reasons (in the case of the Nazis), or ideological reasons (in the case of the Soviet Union), both regimes were willing to sacrifice political, economic, and technological advantages to consolidate control.[8] For the experts, this meant that they had to demonstrate involvement with the regime not only personally but also professionally. In many cases, this could save their lives. Those who were more privileged, from the viewpoint of the regime, still had to make decisions that often amounted to Faustian bargains. We have deployed three biographical cases to explore and document what was at stake.

Hitler's Mastermind of Annihilation: Konrad Meyer

In Germany, National Socialism ushered in the "hour of experts."[9] Firstly, the mass expulsion of Jewish and democratic academics in 1933 and the *Gleichschaltung* (forced alignment) of the universities, research institutions, and academic associations left no room for the traditional figure of the autonomous academic intellectual—the German "mandarins," in Fritz Ringer's words. The National Socialist regime asked for scholars who were loyal to, if not in full solidarity with, the new political elites and their goals; they should support the racist *Volksgemeinschaft* (community) and expansion of German *Lebensraum* (living space). Secondly, the regime favored the rise of academically trained experts and provided attractive career opportunities in the rapidly growing bureaucratic apparatus of administration and planning. National Socialism was not, as historians used to argue, an anti-science movement that hindered the rise of a modern knowledge society in Germany. On the contrary, the political and ideological context of the Nazi era accelerated the secular rise of expert cultures that characterized the inner dynamics of industrial societies in the first half of the twentieth century. In that sense, the Nazi regime was a very modern political system: particularly during the Second World War, it provided plentiful resources for newly emerging expert communities and enabled these communities to flourish as long as they adhered to the regime's political agenda.

Framing the relationship of experts and the state as a mutually beneficial exchange of resources, as one historian of science has convincingly proposed, makes it easier to understand why scientists and experts were neither coerced nor exploited, but actually collaborated willingly with the Nazi regime.[10] Anticipating scientific findings and technical innovations of political, ideological, and military utility, the regime offered far more funds and personnel than had been the case during the Weimar Republic. Researchers rushed to meet the regime's expectations. Similar relational exchanges of resources can be found in other totalitarian countries and, again particularly after the outbreak of the war, also in Western capitalist democracies. What was unique about National Socialism, and what remains difficult fully to understand, was the extent to which experts actively self-mobilized their personal energies for the regime's felonious goals, right until the very end. In Nazi Germany, the Faustian bargain between experts and the state paved the way for annihilation, genocide, and ultimately the Holocaust.

If there is one thing academic experts are particularly good at, it is creating plans and concepts. In Nazi Germany, many experts worked to design refined plans to support the expansionist and racist goals of the regime, and, in so doing, developed into "architects of annihilation."[11] Others went even further, leaving their drawing boards and actively engaging in executing their theoretical designs for the barbarian practice of mass extinction. The agricultural scientist and spatial planner Konrad Meyer was one such expert and was particularly skillful in designing grand schemes. The plan for which he became renowned in the early 1940s, when Nazi Germany aimed to reshape Europe under racist conditions, was the infamous *Generalplan Ost* (Master Plan East). The scheme, which he introduced to SS Leader Heinrich Himmler in June 1942 aimed largely to cleanse Eastern Europe of its Slavic people in order to create space for "Aryan" settlers.

The case of Meyer and the *Generalplan Ost* highlights three interrelated processes of crucial relevance to this book. Along with underlining the rise of academic experts in general, the case points to the importance of expertise in European spatial planning, and it spotlights the continuities and ruptures in academic concepts about designing Europe's spatial order under changing political contexts.

Who was Konrad Meyer?[12] Born in 1901 in rural Lower Saxony, Meyer was part of the "generation of the absolute."[13] Like many other members of his cohort, he considered the defeat of the First World War and the Treaty of Versailles as a national humiliation, to which he responded by becoming an early proponent of National Socialism, unconditionally supporting the regime's politics of expansion and racial extermination. Meyer studied agricultural sciences in Göttingen and quickly developed an astounding academic career. In 1934, he took over the newly founded chair of agriculture and agrarian politics at the Friedrich Wilhelm University (today, Humboldt University) in Berlin, while also heading simultaneously the department of biology and agricultural and veterinarian sciences at the *Reich* Ministry of Education. In 1936, he also became vice president of the German Research Foundation and led the *Reich* Consortium of Spatial Planning. On the eve of the Second World War, Meyer commanded a true reseach empire that coordinated all relevant academic and political positions in agriculture and related fields.

Meyer had long envisioned a new spatial order for Eastern Europe, but only after the end of the campaign against Poland in autumn 1939 did he obtain the resources needed to realize his visions on a grand scale. Heinrich Himmler, now also *Reichskommissar für die Festigung deutschen Volkstums* (*Reich* Commissioner for the Strengthening of German Nationhood), authorized Meyer to develop a general scheme to Germanize the annexed territories and to select the staff of his newly founded *Hauptabteilung Planung und Boden* (Main Department of Planning and Land) in the *Reich* commissariat. Meyer and his team of experts enlarged the scope of their plans as the war's course unfolded. In earlier versions of the *Generalplan Ost*, from January 1940 and July 1941, Meyer limited it to the Germanization of Poland and the adjoining territories in the East. In early 1942, however, when Nazi Germany still had high hopes of victory on the Eastern front, Meyer thought big. He envisioned that over the course of the subsequent 25–30 years, 45 million non-Germanizable people would be either physically eliminated or deported, while another 14 million people would remain and be treated as slaves. To populate the emptied *Lebensraum*, so-called settlement marches and a chain of settlement bases would be erected to house some 10 million Germanic colonists.

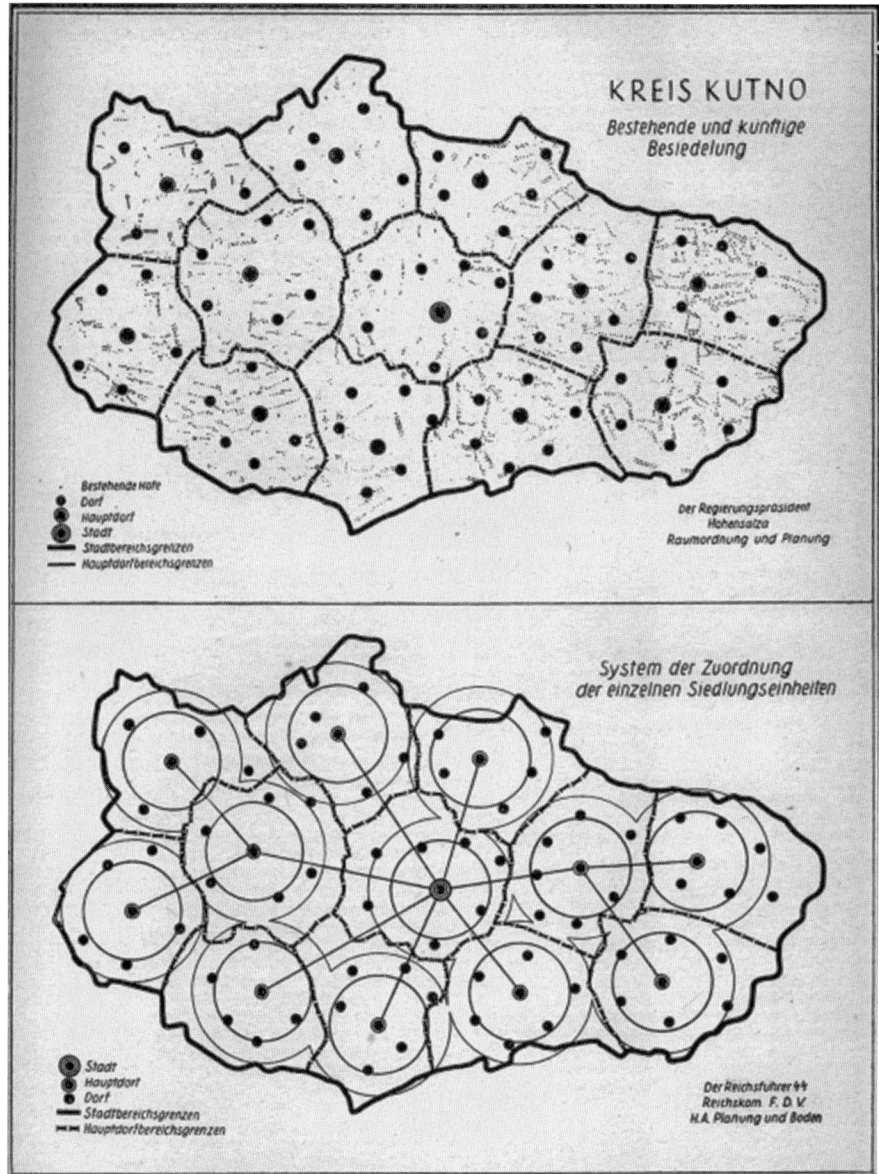

Fig. 5.1 **Experts of Annihilation:** *With the Master Plan East (Generalplan Ost) Konrad Meyer and his interdisciplinary team of experts designed a new spatial and ethnic order for East Europe that would have eliminated tens of millions of people. The map from 1941 depicts the master plan for the Kutno district in occupied West Poland ("Warthegau"). The plan was based on the "Central Place Theory" of the German geographer Walter Christaller, who argued that settlements as central places provide services to surrounding areas. During the Cold War, Christaller's theory became a key concept for spatial planning both in East and West.*

Himmler appreciated the new version of the *Generalplan Ost*, although he felt it was not expansive enough. He advised Meyer to extend the concept to a Europe-wide settlement plan, including the Western occupied territories, and he demanded that the plan be realized within twenty years after the war. The large-scale

Generalsiedlungsplan (General Settlement Plan) was never finalized, although parts of it were already realized during the final years of the war. Unlike many other experts in spatial planning who served the regime as armchair strategists, Meyer occasionally left his office in Berlin to oversee the implementation of his plans on-site.

The *Generalplan Ost* originated from the office of an expert who was not only a member of the Nazi party for opportunistic reasons, but also an unconditional supporter of the regime's racist ideology. In that sense, Meyer represented a minority of deeply politicized academics and stood at the fringe of the scientific community, which still generally adhered to the ideal of research as an apolitical undertaking. However, it would be misleading to understand the *Generalplan Ost* as having emanated from academic backwaters. On the contrary, the plan emerged from the very center of the scientific community. Meyer received substantial financial support for the research project—more than half a million *Reichsmarks*—from the country's most traditional and most prestigious funding agency, the German Research Foundation.[14] Furthermore, the plan involved hundreds of experts from multiple disciplines, which points to the emergence of large-scale, truly interdisciplinary programs. Geographers, historians, ethnologists, anthropologists, spatial planners, literary scholars, linguists, agricultural sciences, botanists, plant-breeding scientists, and lawyers all collaborated in the so-called *Volksdeutsche Forschungsgemeinschaften* (Ethnic German Research Associations) to legitimate Germany's claim for supremacy in Europe through scientific research. Technocrats and social engineers joined forces and mobilized their scholarly resources for the racist goals of the Nazi regime. Their irreplaceable expertise helped develop and refine the "technologies of racism."[15] They gilded the arbitrariness and chaos of annihilation practices in the annexed territories with the authority of scientific rationality and order—and they framed their ideas deliberately in a European context.

European views of the rest of the world were often deeply rooted in colonial visions. German expert visions for spatial planning to Germanize Europe, particularly Eastern Europe, can be traced back to the late nineteenth century. Colonial expansionism on a global scale in Imperial Germany aimed at acquiring "a place in the sun" in Africa and Asia. For *Reich* Chancellor Otto von Bismarck and the Prussian government, however, the primary goal was to secure German influence and culture in Poland. The agrarian crisis caused by falling grain prices and the flight of Germans from the eastern lands nurtured deeply rooted political and cultural anxieties of a "Polonization" of Germany's eastern territories. In 1882, the Prussian government sent the young agrarian economist Max Sering on a fact-finding mission to North America to study overseas experiences with inner colonization. Starting in New York in February 1883, Sering traveled through the Midwest, up and down the West Coast, northeast to Winnipeg, and westward towards the frontier. Sering noticed a "government-sponsored, seemingly rational allocation of land to settlers—settlers who then ventured forth, ordering the wild, 'empty' landscape, colonizing the land and fundamentally strengthening the Canadian government and the Canadian race."[16] For Sering, North America's Wild West became the blueprint for Germany's "Wild East." Germany would be rejuvenated and strengthened by colonizing the emptying lands in the East. He put his ideas into his influential book entitled *Die innere Kolonisation im östlichen Deutschland* (The Inner Colonization in Eastern Germany, 1893), which paved the way for his successful academic career.[17] Generations of agrarian scientists and spatial planners followed in his intellectual footsteps, and so did politicians. German imperialist policy in the East during the First World War reads like a massive effort to implement Sering's ideas.

Sering promoted spatial planning on a grand scale in interwar Germany. After the Nazi seizure of power, however, a new generation of experts entered the scene, including Konrad Meyer, who now dominated the field of agricultural and spatial research. Meyer forced Sering to retire and dissolved his institute for settlement science in Berlin. While Meyer did not proceed exactly according to his predecessor's ideas, there is a definite line of intellectual continuity linking Sering's concept of inner colonization with Meyer's *Generalplan Ost*, which laid the groundwork for the genocide in the East. These continuities support the argument made by some historians that the Holocaust took place within a colonial context.[18] On the other hand, there was no such linear path from Bismarck to Hitler, from colonial imperialism to racist genocide, from Manitoba to Auschwitz. While Sering tried to transfer the North American case for providing incentives for settlers colonizing the East, it took the radicalism of a new generation of technocratic experts such as Meyer to develop grand schemes of Germanizing the East by eliminating its settled people to clear uninhabited spaces for "Aryan" in-migrants.[19]

Meyer and his fellow spatial planning experts thought in a European dimension and conceptualized a new European order under German hegemony. With the collapse of the Nazi regime, Meyer was smart enough to adjust his spatial ideas to the integration-oriented newspeak of postwar Europe, on both a conceptual and semantic level. He stripped his ideas out of the totalitarian context and framed them as purely scientific. "The truth is," he claimed in 1960, supported by the West German community of spatial planners, "that neither spatial planning nor spatial research have had anything to do with National Socialism."[20] After rehabilitating himself as a well-respected academic and intellectual leader of his discipline, Meyer quickly adopted the new framework of European integration. In the early 1940s, he advocated a Europe-wide alliance of the Germanic nations under German leadership, which he argued would bring about a "healthy sense of European solidarity" and the "constitution of a more fortunate Europe."[21] In the late 1950s, Meyer remained convinced that the "Germanic nations" (Germany, Belgium, and the Netherlands) were ahead of their neighboring states in the South, but he welcomed European integration in the European Economic Council (EEC), as it would facilitate a Europe-wide order of spatial planning.[22] By semantically replacing *Lebensraum* with "sustainability" or "nature protection," Meyer could claim that he had always been a pioneer of regional planning in Europe and an advocate for a true European collaboration based on scientific and technical expertise.

Again, there is no linear path from Nazi spatial planning to regional planning in the age of European integration. However, there are many continuities, both personal and conceptual. Expert cultures are obviously quite willing to adapt themselves to changing political circumstances—in this case, being probably more adaptable than was desirable for postwar Germany and Europe.

Fig. 5.2 **Whitewashing Nazi Experts:** *The picture was taken on the very last day (March 10, 1948) of the trial against the SS Race and Settlement Main Office in Nuremberg. Konrad Meyer (third from left in the dock) and his defense counsel convinced the American judges that the* Generalplan Ost *was solely based on theoretical expertise in spatial planning. He was absolved of the charges of war crimes and crimes against humanity and only received punishment for his SS membership. Later, Meyer even claimed that the Nuremberg Trials had proven the* Generalplan Ost *was a purely scientific program.*

Best Man: Guglielmo Marconi & Benito Mussolini

It was a wedding ceremony and more. Italy's pre-eminent inventor-entrepreneur, Guglielmo Marconi, a pop star both at home and abroad, married Maria Cristina Bezzi-Scali, the only daughter of Vatican official Francesco Count Bezzi-Scali. The couple had Benito Mussolini, *Il Duce*, as the best man at their wedding. The marriage of the young and demure Maria Christina to Gugliemo, famous for his long record of love affairs, developed into a happy relationship, if we are to believe Maria Carla's account in her book entitled *Mio marito Guglielmo* (Marconi My Beloved).[23] It was also a happy marriage for Fascist Italy. The wedding ceremony, held on June 12, 1927, was designed to show the world that Fascism was fully backed by Italy's scientific and technological elite, with Marconi at the top. The Nobel Laureate and ingenious inventor was a true godsend for Mussolini, as he allowed the Italian leader to "prove" that technoscientific experts with undisputed international credibility had legitimated Fascism.

| Faustian Bargains in Totalitarian Europe | 155

When Benito Mussolini and his Fascist movement seized power after the infamous March on Rome in October 1922, Guglielmo Marconi was already internationally famous. Born in Bologna in 1874 as the second son of the Italian landlord Giuseppe Marconi and his Irish wife Annie Jameson, Guglielmo developed an interest in physics in general and in electricity in particular. He was

Fig. 5.3 Best Man & Close Friend: *Mussolini (left) served as Marconi's (right) best man, when the latter married his second wife Maria Cristina Bezzi-Scali in 1927. When this picture was taken (around 1936), the close friendship was of mutual benefit to both men. Unlike what the image suggests, Il Duce did not have to tell Marconi what to do. Marconi willingly supported Mussolini's fascist cause as much as he could. He even did not refrain from proposing infamous weapons of mass killing for the Italian Army.*

introduced to the subject of radio waves by Augusto Righi, a well-known physicist at the University of Bologna and the family's neighbor. The young Marconi began to experiment with electromagnetic radiation in his home at the Villa Griffone in Pontecchio with the aim of realizing a practical system to transmit telegraph messages without using connecting wires. After tinkering with his apparatus outside the villa, he was able to send wireless signals over distances of more than a mile.

How this creative inventor-entrepreneur managed to deliver what numerous investigators had been exploring for half a century—a technically reliable and commercially successful system of wireless telegraphy—has often been told as a straightforward success story.[24] However, it took more than a decade of trial and error, improvements and setbacks, before Marconi gained credibility both as an inventor and an entrepreneur. Furthermore, Marconi depended on a wide network of supporters, including his mother Annie and Italy's London Ambassador, Annibale Ferrero, who paved the way for Marconi's to demonstrate his technical innovation

Fig. 5.4 Young Father of Wireless: *Based on James Clerk Maxwell's electromagnetic theory of light, the German physicist Heinrich Hertz developed a radio wave transmitter and an antenna receiver. Publications on Hertz's experiments after his death in 1894 sparked young Guglielmo Marconi to realize a practical system for the wireless transmission of telegraph messages. The picture shows Marconi and his technical system that included a spark-producing radio transmitter, a coherer receiver, a telegraph key, and a telegraph register to record messages received in Morse code.*

to British government officials. In 1896, Marconi took his apparatus to Great Britain, where he was granted the world's first patent in wireless telegraphy. He founded the Wireless Telegraph & Signal Company the following year, renamed Marconi's Wireless Telegraph Company in 1900.

In January 1903, Marconi launched a promotional campaign that helped him to create a market for wireless transatlantic telegraphy. He transmitted a message of greetings from U.S. President Theodore Roosevelt to King Edward VII from a station which the company had started to build in 1901, located near South Wellfleet on Cape Cod, Massachusetts. International visibility also resulted from the ultimate technical disaster of the early twentieth century, the sinking of the *Titanic* in May 1912. The radio operators aboard the *Titanic* were employees of the firm's subsidiary, the Marconi International Marine Communication Company. In the legal inquiry following the disaster, the British postmaster-general officially stated that the survivors were only saved through Marconi's invention.

The biggest impetus to his international reputation, however, resulted from the Nobel Prize in Physics that Marconi received jointly with Karl Ferdinand Braun in 1909 for his scientific contributions to wireless telegraphy. Marconi's home country was keen to profit from the prestige of the first Italian Nobel Laureate. On the eve of the First World War, Marconi became a member of the Italian Senate and served during the war in the Italian Army and Navy. In 1919, he was appointed Italian plenipotentiary delegate to the Paris Peace Conference.

When King Victor Emmanuel III made him a Marchese in 1924, Marconi's support for Mussolini was not a pretense. He had joined the Fascist party in 1923 and could later even jokingly claim to have been the "first fascist in Italy," for his beam of defined radio waves had been named *fascio* in Italian.[25] Thus, Marconi's wedding ceremony in June 1927 with *Il Duce* as his best man offered another public demonstration of the close alliance between politics and science and technology in Fascist Italy. Just three months later, the international scientific community witnessed a similar demonstration of this alliance when the members of the international Volta Congress of Physics in Como were invited to Rome. They were received by *Il Duce*, who, after expressing his appreciation for the physicists' scientific achievements, passed the ball to his supporter Marconi, who lectured on Volta's life and work in the Capitol.[26]

Fig. 5.5 **Staging Radio Technology:** *Guglielmo Marconi at work in the wireless room of his yacht* Elettra *in 1920. The steam yacht was built in 1904 for the Archduchess Maria Theresa of Austria and was originally named* Rovenska. *During the First World War, the British Admiralty used it as a minesweeper. Marconi purchased the yacht in 1919, renamed it* Elettra *and converted it into a floating laboratory. The yacht had cabins for Marconi and his wife Maria Cristina, three guest cabins, four bathrooms, and an oak-paneled study.*

The mutually beneficial alliance of Italian leaders in politics and science became even closer when Mussolini appointed Marconi to the position of president of the *Consiglio Nazionale delle Ricerche* (National Council of Research) in 1927 and then in 1930 as president of the *Reale Accademia d'Italia* (Royal Academy of Italy). Until that time, Marconi had no experience in Italian academic circles. He was only chosen by Mussolini because of the "splendor" he could add to the two institutions, thereby helping to consolidate the "Fascist revolution."[27] A historian has recently demonstrated that Marconi clandestinely prepared the dictator's anti-Semitic campaign years before the persecution of Jews came into the open in the form of Mussolini's race laws of 1938. Marconi systematically blocked all Jewish candidates, on Mussolini's instruction, at a time when the regime still denied implementing policies based on religious prejudice. Marconi himself added the letter "E" (for Ebreo, the Italian word for Jew) beside the names of Jewish scientists his colleagues had shortlisted for academy membership. During Marconi's tenure, no Jewish scientist entered the ranks of the Academy. Even such brilliant scholars as the internationally famed physicist and mathematician Giancarlo Vallauri and Italy's most celebrated archaeologist Alessandro Della Seta, an ardent Fascist despite his Jewish origins, were excluded.[28]

Marconi went further than most of his academic colleagues in allying technology and politics. In October 1935, Mussolini's troops invaded Ethiopia and started the brutal Abyssinian war. In massive aerial bombardments of both military and civilian targets, the Italian army used poison gas and burned villages to ashes, followed by mass slaughters of local communities. Targeted attacks against military hospitals were based on maps that the Red Cross had provided with the aim of avoiding unintentional assaults. Europe responded to these violations of international law with a trade embargo. Marconi was among the first to heavily criticize the League's decision to impose sanctions. He broadcast from Rome to the U.S. to argue Italy's claims, justifying the Italian Army's war crimes by calling the invasion a civilizing mission to backward African people. At Mussolini's request, Marconi undertook a long propaganda tour to South America, France, and Great Britain and developed into something of a quasi-ambassador. However, when he met with British officials, including the future King Edward VIII, he was surprised that the British Broadcasting Corporation (BBC)—1927 successor to the British Broadcasting Company he had helped to create in the early 1920s—refused to let him speak in defense of the Fascist case. Despite serious health problems, Marconi did not tire of justifying Mussolini's regime and even provoked the international public with his aggressive view that Italian workers and colonists would always be ready to defend their country at any cost. Moreover, Marconi hurriedly sent a National Council of Research commission of experts to Ethiopia to study the best way of exploiting local resources for the Italian nation.

Totalitarian regimes in Europe were keen to exert control over the media. In Italy, coercion aiming at establishing political control over the radio came only after Marconi's death in July 1937. Until then, Mussolini had at his side the inventor of the radio, who enabled him to utilize broadcasting for propagating Fascism. Moreover, radio was a powerful tool that helped Mussolini become omnipresent in public media, penetrating the private lives of Italians "to lay the foundations of Fascist modes of behavior in the spheres of daily life."[29] Thanks to Marconi, the voice of *Il Duce* was audible even in rural Italy.

Marconi was certainly not alone in his efforts to support the Fascist regime. Many, if not most, of his peers in science and engineering did the same in one way or another and, in return, profited from the state's policy of providing tangible benefits and resources. Members of the Academy, for example, received a high salary, free first-class travel, and other privileges, including the right to dress in uniforms with a plumed hat and gilt sword. For Nicola Parravano, the renowned president of the International Union of Chemistry, Marconi personified "the genius of the [Italian] race." Parravano stated that science in Fascism should no longer be "an unreal abstract creation" in a search for truth, but "mainly an instrument for directing and utilizing all the resources of the country to secure life to its citizens and to give the maximum strength to its commanding will."[30] Most technoscientific experts took Parravano seriously. They were keen to foster the regime's goals of making Italy self-sufficient and responded by focusing their agendas on developing substitutes for imported products such as hydrogenized fuels, synthetic rubber, and chemical fertilizers. Marconi was probably right when, in 1932, he concluded a long article on *Scienza e Fascismo* (Science and Fascism) in the newspaper *Il Populo d'Italia* by saying that "in Italy there is harmony of purpose between the Fascist Government and the scientists."[31]

As was the case with the self-mobilization of experts for the Nazi regime in Germany, the close alliance of large parts of the scientific and engineering community with the Fascist government in Italy was not based on coercion or exploitation, but rather on intentional collaboration. As the case of Guglielmo Marconi exemplifies, technoscientific experts and the Fascist regime exchanged resources and mutually profited from each other. In contrast to Nazi Germany, the collaboration between experts and politics in Italy did not directly result in war crimes and genocide. However, the Faustian bargain did enable Italy's totalitarian regime to augment its power by improving its political credibility at home and abroad and, in this way, also to establish a brutal colonial regime in Ethiopia.

Stalin's Rainmaker: Trofim Lysenko

When the Lenin All-Union Academy of Agricultural Sciences (VASKhNIL—the leading scientific agency in this field in the Soviet Union) met in summer 1948, it certainly sponsored one of the most momentous gatherings in the history of twentieth-century scholarship.[32] This assembly was the culmination of Trofim Lysenko's two-decade campaign against Mendelian genetics, as a result of which Gregor Mendel's ideas were practically banned from the Soviet Union by VASKhNIL under Lysenko's leadership. The consequences for scientific development in the Soviet Union (and in many of its satellite states in the emerging Soviet bloc) were severe and far-reaching.

Reacting to renewed scientific debates in the Soviet Union, the decision that VASKhNIL made was—as Lysenko made clear from the beginning—politically backed. Lysenko's paper opened with the following forthright declaration: "The question is asked in one of the notes handed to me: What is the attitude of the Central Committee of the Party to my report? I answer: the Central Committee of the Party examined my report and approved it."[33] Stephen Jay Gould called this rather clumsy introduction "the most chilling passage in all the literature of twentieth-century science."[34] That Lysenko felt it necessary in the same paper to add "Glory to the great friend and protagonist of science, our leader and teacher, Comrade Stalin!" clarifies just what he meant.

In fact, Lysenko was not so much pointing to the future, but rather sealing what he had achieved in earlier years. Remarkably, the final editing of his 49-page manuscript had been done by none other than Stalin himself. The Soviet leader made manifold and detailed changes, including the side note "HA-HA-HA!!! AND WHAT ABOUT MATHEMATICS?" where Lysenko had declared that any science was class-biased. In fact, Stalin went to great lengths to have Lysenko's text appear less political and more universal and scientific. In general, however, Stalin fully agreed with Lysenko's condemnation of Mendel's insights, adaptation of the teachings of Jean-Baptiste de Lamarck, and his ambition to keep Mendelian genetics out of Soviet academia by all means.

The striking gap between the potentials of "scientific" agriculture and the much more modest reality had already stirred the imagination of experts during the Imperial period. Technology and science, experts claimed with ever-increasing urgency, could transform Russia's huge natural resources into a source of abundant wealth and propel the backward country into modernity. The many new agrarian experts educated in late Imperial Russia believed that their hour had come in 1917 when the Bolsheviks took over and ousted what they perceived as a hopelessly deadlocked political system.[35] Russia's dramatic situation after the lost First World War, the ongoing civil war, and a political transformation of unprecedented dimensions meant that agricultural policy-makers had to fight for a share of scarce government resources. Furthermore, it meant that experts had to bind their fate to a political party in a way that had not existed anywhere before.

The situation presented enormous opportunities and equally enormous risks for the experts. Lysenko was clearly willing to seize the opportunities. While Russian scientists faced numerous challenges, they still remained among the best in the world. Lysenko, however, held almost none of his colleagues' credentials. Nonetheless, he successfully climbed the ladders of the Soviet Union's scientific hierarchy and became one of the most influential scientific figures in the twentieth century. Who was this man?

Fig. 5.6 Man on the Ground: *Trofim Lysenko (right) successfully stylized himself as the man in tune with simple farmers and able to come up with practical results, which egg-headed traditional scientists ignored or could not deliver. In fact, Lysenko instrumentalized the science system of the Soviet Union, after having removed a number of his opponents with Stalin's backing. For years he was able to control scientific institutions in the Soviet Union far beyond his field.*

In many respects, Lysenko was a typical Soviet "new man," born in 1898 from a modest peasant background in Ukraine. Despite the impression given by what has often been written about him, however, he was not a complete autodidact. Lysenko graduated from the Kiev Agricultural Institute, which placed him among the quickly growing number of experts in *the* critical battlefield of the Soviet Union, agriculture. Moreover, he completed his studies at a particularly significant moment, in an intermediary phase of a highly contested change in Soviet elites. This shift resulted in the Bolshevization of the Soviet Academy of Sciences between 1928 and 1932.[36]

Lysenko's first commission took him to the outskirts of the Soviet Empire, to Azerbaijan, where he undertook research into the "vernalization" of plants, a process on which he published in 1928. This method exposed wheat seeds to cold and moisture, attempting to make winter seedlings follow the pattern of spring seedlings and thereby increasing the number of potential harvests. What might seem at first glance like a rather obscure topic was of central importance to the Soviet Union. Lysenko's approach—which was generally in line with similar research taking place elsewhere in Europe, and hardly original—promised to solve a number of problems haunting the Soviet Union. First, the productivity of agriculture desperately needed to be improved in a country with huge amounts of territory but facing difficult climatic conditions. Secondly, Lysenko's methods did not require fertilizer, which was important given scarce resources. Thirdly, Lysenko claimed to be able to grow specific plants in regions where this had not been thought possible before, such as peas in Azerbaijan.

The crisis of Soviet science in the 1920s affected many fields, and agriculture in particular. Furthermore, even if scientists had developed more productive methods than they did, it would not necessarily have been possible for farmers to apply them because of traditional convictions or a lack of knowledge and financial means. What was needed instead was literally a down-to-earth approach, although it was difficult substantially to improve upon what farmers had already achieved on their own by trial-and-error and with their mongrel seeds. The failures of agrarian collectivization efforts in the late 1920s and a dramatic drop in crop yields increased the output problem's urgency. Lysenko's method did little, if anything, to increase production directly. However, some indirect effects could be observed. It brought farmers hoping for better harvests back to their land, which they often had left, protesting measures taken by the regime. Focusing on seedlings' problems allowed the regime to shift discussions away from collectivization, which remained a strict political taboo. Not least for this reason, the regime publicized Lysenko's experiments prominently in *Pravda*, the Communist party's newspaper, as well as in many other publications. Press and radio campaigns spread Lysenko's claims, making him a model celebrity of socialist science. With his modest peasant background, he was well suited to symbolize much more than just a specific scientific success. He stood for the accomplishments of socialist science itself and could be presented as a true hero from the lower classes, embodying native genius of the new era, for now anyone with talent could make it. Expectations soared, fueled by Lysenko's ever more ambitious claims. In the late 1920s, agricultural output was expected to double within ten years. Moreover, Lysenko was provided access to other big Soviet projects, such as planting gigantic shelter belts in the steppes.[37]

Therefore, the regime's gamble on Lysenko rested, on the one hand, on its belief in the superiority of science under communism, which itself claimed to be based on scientific principles. On the other hand, the emerging Lysenko cult helped to convince innumerable lay technicians, who were instinctively suspicious about science, to dedicate their energies to the Soviet cause. Lysenko's self-chosen mentor, Ivan Michurin, a plant cultivator who had gained his expertise in the field, had called their academically trained competitors the "caste priests of jabberology." Lysenko constantly expressed his contempt for the "fruitfly lovers," that is, the university-based geneticists occupying themselves with *drosophila*. "It is better to know less but to know just what is necessary for practice," Lysenko explained, not forgetting to mention that his method was based on using the insights of the many, the truly socialist rural classes. Lysenko's claims appealed to Stalin, who fantasized about *spetseedstvo*, literally "eating of specialists."[38]

With his rather non-conventional approach, Lysenko was able to deliver the quick "solutions" that classical science could not, by its very nature, provide, but which the Communist party demanded. Lysenko's true talents lay in using to his own best advantage the improvised, but ideologically grounded, pattern of rule and administration that was the essence of Stalinism. Lysenko had political

Fig. 5.7 Soviet Celebrity: *Lysenko, the man and Lysenkoism, the theory, fitted perfectly in the Stalinist ideology. The man who rose from a simple background seemed to show the genius of the people. Lysenkoism formed an alternative to genetics, suspected for ideological reasons. Moreover, Lysenkoism promised to revolutionize Soviet agriculture and deflect the problems of collectivization. While scientists despaired, Soviet propaganda extensively spread the image of Lysenko.*

momentum on his side. In a speech during the 1929 All-Union Congress of Genetics and Breeding, Stalin had declared that praxis had to be favored over theory, which played right into Lysenko's hands. From this point on, Lysenko could authoritatively fend off any criticism that his methods were not useful for the practical progress of the Soviet people.

Until 1938, there was still considerable tolerance for scientists critical of Lysenko. Thereafter, however, the situation changed considerably. Initially used to counter competitors in academic battles, Lysenko's arguments assumed an ever sharper edge. From 1934 to 1939, Lysenko headed the All-Union Institute for Genetics in Odessa. In 1935, he was placed in charge of the Academy of Agricultural Sciences (VASKhNIL). In 1940, he became head of the Institute of Genetics of the Academy of Sciences, succeeding Nikolai

Vavilov, a scientist of international standing. A number of lesser-known scientists climbed the hierarchy in Lysenko's slipstream, filling many posts and dominating more and more influential scientific institutions. It became relatively easy for younger biologists to denounce their older colleagues by referencing Lysenko and then climbing into the newly vacated positions.[39]

This would have been problematic enough had Lysenko had not started to develop his rather modest insights into a scientific field called "agrobiology," an all-encompassing creation that combined aspects of genetics, plant physiology, and Darwinism. With his journal, the *Bulletin of Vernalization* (Яровизация), founded in 1932, Lysenko acquired a forum to popularize his Lamarckian-based theories. Until the mid-1930s, it had still been possible for geneticists to conduct research, but Lysenko increasingly controlled the field.

For a long time, Soviet propaganda and Lysenko's brand of science were so closely connected that it became virtually impossible for the propaganda to set Lysenko aside. The Lysenkoist way of doing "science" and the systemic flaws of socialism interacted in disastrous ways. Few of Lysenko's findings could be reproduced by other researchers, but this could not be expressed in the Soviet narrative of ever-increasing success and productivity. Rather, each astonishing victory in the battles of agricultural production had to be outdone by ever more impressive examples. Political and scientific actors both had a strong interest in overstating their achievements, and on principle neither would allow any questioning or scientific discussion in the narrower sense. Lysenkoism became part of the official agenda. Any skepticism regarding the merits of the new teaching was considered defeatist and thus treasonous.

Indeed, much has been made of ideological overlaps between Lysenko's pseudoscience and Stalinist ideology. Lysenko had clearly left the terrain of established science when he claimed that plants could inherit characteristics by a forced exposure to new conditions in an extension of his work on making seeds more winter tolerant. This belief accorded well with the communist ideology's emphasis on environmental factors and the notion that re-education was both possible and desirable—instead of stressing rather fixed conditions in the genes. Genetics, again, was presented as bourgeois science or even Fascist science because many geneticists also held eugenicist convictions. One of the highest administrators in Soviet agriculture referred to genetics as the "handmaiden of Goebbels' department."[40]

The obvious shortcomings in harvests stemming from Lysenko's flawed theories were explained as caused by sabotage. Using the many public and hidden channels Stalinism offered, Lysenko himself denounced dozens of scientists who had actually or allegedly opposed him, as reactionaries, as bourgeois, or simply as being unhelpful to the cause of the Soviet people. In this way, he became both directly and indirectly responsible for hundreds of scientists' detainment and death.[41] Among the most prominent of these was his predecessor as director of the Institute of Geneticists, Nikolai Vavilov. A geneticist of international renown who had initially been friendly towards Lysenko before becoming one of his fiercest critics, Vavilov was arrested in 1940 and died three years later in prison of malnutrition. He had been accused and found guilty of reactionary tendencies, conspiracy, and many other fabricated charges. By 1936, biologists Israel Agol, Max Levin, and Solomon Levit, all communists working in biological theory, had all been accused of being "enemies of the people." All died in prison or were shot, like many of Lysenko's other victims. The luckier ones only lost their positions and, therefore, any chance of continuing their scientific work. Apart from an estimate of perhaps a hundred scientists who suffered directly from Lysenko's ambitions, immeasurable harm was done to scientific and agricultural development in the Soviet Union. Lysenko caused institutes and journals related to genetics to shut down and libraries to be "cleansed" of the offending scholarly publications. In particular, international contacts suffered, including stopping by political order the Seventh International Congress of Genetics, which was scheduled to take place in Moscow in 1937. No Soviet scientists were permitted to attend the 1939 International Congress of Geneticists in Edinburgh, which reflected how sweeping Lysenko's success had been and how devastating his influence on scientific development and exchange proved to be.[42]

Lysenkoism attracted considerable attention and even some followers in Western Europe. In Belgium, Lysenkoism challenged university professors' loyalties vis-à-vis the *Parti Communiste Belge* (PCB) which made it a litmus test. In fact, a number of academics and intellectuals left the PCB because, although willing to follow a communist course, they were unwilling to accept a theory that was at odds with scientific insights. In France, too, Lysenkoism served as an important yardstick for determining the relation of left-leaning scientists to Stalinism. Between 1945 and 1948, Soviet opponents of Lysenko used international networks to build a "second front" against it in the West. In Soviet bloc states the introduction of Lysenkoism also met opposition from scientists but was generally successful due to dependencies on the Soviet Union. In some cases the political motives of individual scientists played a role. An example of this was Poland's Stanisław Skowron, whose imprisonment in a concentration camp apparently motivated his shift from genetics, which he associated with Nazi eugenics, to Lysenkoism.[43]

Though the rise of Lysenkoism was bizarre in many respects, we should not over-emphasize its strangeness. Lysenko may present the most spectacular case of Stalinist purges in the academy and science, but certainly not the only one. Moreover, Lysenko's research was, at least in its early phase, not that far removed from the academic mainstream.[44] Also, the notion that a particular science culture, to use a modern phrase, would need to fit a national or ideological culture was not confined to the Soviet Union and to agriculture; one only has to think of the rise of a *Deutsche Physik* (German physics) or a *Deutsche Biologie* (German biology) in Nazi Germany and the hype promoting a Soviet cybernetics in the 1950s and 1960s.[45]

However, it would be equally problematic to conclude from the Lysenko case that the Soviet science system was completely defunct. We should not forget that Lysenko's position in the Soviet Union remained fragile. His advance in 1948 had a defensive motive. Lysenko had come under pressure, not least because his brother had defected to the Germans during the war.[46] After the Second World War and Stalin's death in 1953, Lysenko's fate remained closely linked to politics. Lysenko remained in reasonably high esteem even under the reformer Khrushchev, who had denounced Stalin's excesses. Finally, in 1964, the physicist Andrei Sakharov criticized Lysenko in the General Assembly of the Academy of Sciences, which triggered his downfall.[47] International advances in genetics as well as the insistence on scientific principles in the Soviet Union—essential for the long-term survival of the political system—ended Lysenkoism. By the time of Sakharov's speech, Western nations were already spreading the Green Revolution through wheat, rice, and other grain breeding. In Europe, experts in science-based agricultural promoted the idea of a unified (Western)

Europe based on modern concepts of plant genetics. Viewing Lysenko as a mere charlatan, however, misses the point.[48] His case shows not only what was typical about Soviet science, but also the enormous opportunities that individual scientists could seize in a period dominated by economic crisis, war, and political transformation, when a totalitarian regime put science policy at the top of its agenda and entertained far-reaching visions of a new man and society shaped by science.[49]

Experts in the Age of Extremes

As the old European tale of Doctor Faustus and his pact with Mephistopheles suggests, dreadful outcomes from scientific ambition are nothing new. Stepping over moral boundaries or committing fraud to advance one's own position has always been a risk inherent in expert cultures. Yet, as the biographies of Meyer, Marconi, and Lysenko demonstrate, the end of the golden age of European science brought with it a new form of involvement between scientists and politicians: the Faustian bargain. Even before the "Great War," criticism of a mechanized and dehumanized world dominated by technology was gaining ground. Disillusion and insecurity among experts themselves grew dramatically after the experience of technology unleashed to its full extent in warfare.[50] Interwar Europe saw the rise of nations busily preparing for another war they believed to be inevitable, one that would blur the boundaries between military and civilian spheres. Like never before, European powers demanded technoscientific expertise.[51]

This shift was especially true for experts in the nations that lost the Great War. Defeat in a war characterized by completely new roles for technology undermined the losing nations' established technoscientific structures. The shock of loss was amplified by the exclusion of German and, to a lesser extent, Austrian and Russian scientists from postwar international collaboration. Conversely, that ban increased the willingness of the "outcasts" to cooperate with one other. Often enough, scientists who saw themselves as unfairly treated and victimized used this rejection as a justification for engaging in research of questionable morality.[52]

With the consolidation (as in the Soviet Union) and rise (as in Italy and then Germany) of Communist, Fascist, and National Socialist authoritarian regimes, new roles expanded markedly for experts willing to step across the boundaries of traditional science and uninhibited by established moral standards. All these regimes adhered in varying degrees to radical agendas backed by (pseudo-)scientific support.[53] Eugenics and race theories enjoyed wide popularity. Only in Nazi Germany, however, did racism turn into an official state doctrine, which established completely new fields of expertise that, as the example of Konrad Meyer demonstrates, offered remarkable career opportunities.

Apart from their ideological preferences and the need to staff newly created fields, it made sense for totalitarian regimes in Germany, Italy, and the Soviet Union to put their faith in newcomers like Lysenko. In Germany, the older generation of experts was unlikely to be republican, but also not necessarily pro-Nazi. Those who gained new positions through political support, often at strikingly young ages, were most likely to be more loyal, radical, and willing to follow directive from the new regime to the bitter end. Wernher von Braun, who designed the V 2 rocket, and Albert Speer, Hitler's favored architect and Nazi Minister of Armament, are the best-known examples of technical experts who committed themselves and their talents to the alluring promises of Nazism, embracing its ideological consequences with few inhibitions. The self-mobilization of experts for extreme political ends is still best captured by Speer's memoir *Inside the Third Reich*. "For the commission to do a great building, I would have sold my soul like Faust. Now I found my Mephistopheles," Speer stated, albeit in a way which rather flattered himself.[54] Leaving his self-pitying excuses aside, Speer spoke frankly about the lethal combination of extreme opportunism and blind ambition interacting with the power of technical expertise and ideological commitment. This lethal combination, we believe, explains much about the three great wars in Europe's "Age of Extremes" (the First World War, the Second World War, and the Cold War), each with its technoscientific obsessions.

The rise of men like Meyer—and much more dramatically the likes of Lysenko in the Soviet Union—also built on a widespread European discontent with the dominance of traditional elites in academia. In Germany and in many other European nations, the new generational cohorts of those born during the new century's first decade often responded to the First World War and the crises of the 1920s and 1930s by unconditionally supporting totalitarian regimes that seemed to offer new opportunities for the young and gifted. In this way, Faustian bargains of science, technology, and politics reflected fundamental changes, spurred by deep generational and cultural conflicts, taking place among the intellectual elites in interwar Europe.

Lysenkoism lacked the direct, brutal consequences inherent in Nazi race theories and eugenics. Yet, the fact that this dilettantism gained professional status through political backing and the use of force shows the disastrous consequences that ideologized science acquired—disastrous both for those scientists competing with Lysenko or not fitting in his worldview, and for the many others who were dependent on effective agricultural practice. It should not be forgotten that the field of agriculture was also particularly prone to political influence in Nazi Germany, albeit with far less criticism of genetics.[55] Politicized agriculture also formed an important precondition for Meyer's rise.

Marconi was a newcomer too, but in a somewhat different sense. When he entered the grand stage of Italian science policy in the late 1920s, he was already an internationally famous inventor and entrepreneur, but had not been involved in the Italian scientific community's scholarly world. As an outsider, he was best suited radically to transform the institutional framework and culture of research according to Mussolini's Fascist ideology. In the case of Marconi and Mussolini, the Faustian bargain was expressed as a mutual exchange of resources that helped both individuals increase their political power and social prestige.

Lysenko, Marconi, and Meyer represent very different manifestations of Faustian bargains, reflecting their respective biographical backgrounds, the different scientific cultures from which they had come, and, in particular, the regimes for which they worked. However, they, each and all, actualized and embodied a decisive change in interwar Europe. Of course, science and technology have never been neutral and self-sufficient. Also, as we have shown earlier, technoscientific experts in Europe always depended on states that expected loyalty, and they often voluntarily entered into such dependencies.

Yet the regimes in Berlin and Moscow that emerged after the First World War asked for much more than loyalty. Experts had to choose allegiances and take an overt position for or against the regime. This new pressure was by no means confined to experts in Germany, Italy, and the Soviet Union. Political pressure on scientists and engineers was also present in the young and therefore rather vulnerable nation states of Central and Eastern Europe, the Baltics in particular, not to mention the Fascist regimes in Spain and Portugal. Germany, however, went even further and was especially efficient in using hundreds of technical experts to spread the "Nazi gospel" in Central-Eastern Europe.[56] Moreover, technoscientific experts in occupied Europe—by 1941 the greater part of the Continent—also had to make decisions that cannot always be clearly defined as collaboration or resistance. Even those who were not directly affected were drawn into the ideological battles of the twentieth century. Beginning with those who devoted their expertise deliberately to the Communist cause in the Soviet Union or, in far fewer cases, to the Fascist states, the intense love affair of technoscientific experts with political ideologies had repercussions. It certainly did not end with the post-1945 debates about Lysenkoism in Western Europe.

Fascist states were not the only places where social movements flirted with authoritarian, if not Fascist, ideas to control society by means of technocratic expertise. European democracies experienced this too. A striking example is France in the 1930s, where engineers advocated a kind of French "New Deal" to bring order to the politically and socially fragmented nation by modernizing its economic organizations and technical infrastructure.[57] The engineer Jean Coutrot, for example, epitomized a widespread belief that collaboration between engineers and experts in various branches of science was needed to secure the future of the French nation. A strong believer in Taylorist concepts of industrial rationalization and mechanization of work, he founded various institutions of scientific management such as the *Centre national d'organisation scientifique du travail* (National Centre for Scientific Management) and became vice president of the *Centre d'études des problemes humains* (Centre for the Study of Human Problems). He also joined the *Groupe X-Crise*, founded in 1931 by former students of the *École Politechnique* to advocate technocracy and "planification," a government-based system of economic planning. Coutrot regarded political autocracy as a precondition for effective business; in this way he resembled his compatriot Charles Eugène

Bedaux, who collaborated with the Vichy regime in order to further his schemes for Taylorist scientific management. Like many other French experts, Coutrot shared the "engineer's preoccupation with systems and efficiency" and thus paved the way for authoritarian visions of an orderly society that the Vichy regime favored.[58]

Collaboration also meant aligning oneself with Nazi Germany. Coutrot committed suicide in 1941 under mysterious circumstances. However, many other engineers were willing to take his place and join in supporting the autocratic Vichy regime. Just as Vichy entered into a Faustian bargain with Hitler, so did many French technoscientific experts. Elite French public engineers, trained to serve the nation's interest, continued to seek the transformation of state and society through grand technical projects. The technocratic orientation of French engineers remained a constant during the twentieth century, from the Third Republic to the Vichy regime and finally the Fourth Republic of the postwar era.

Obviously, the year 1945 marked a break, at least for experts who had placed their bets on Fascism and National Socialism. The era of totalitarian regimes in which experts had seemingly unending resources and vast power to implement their plans was over, at least in Europe. Even Lysenko had a much harder time in the Soviet Union after Stalin's death in 1953 and eventually fell from favor in 1964. However, the rupture was far from complete. Many Faustian bargainers were able to continue their careers because of their expertise, which was deemed critical for the reconstruction of Europe during the Cold War. Numerous studies have emphasized the striking continuities between wartime and postwar Europe in practically every field of knowledge, on the level both of persons and cognitive concepts, methods, ideas, and even institutions. Experts like Meyer, whose knowledge was largely bound to the specific field of Nazi racial politics, managed to continue their careers and to reframe their infamous ideas as having paved the way for a true integration of Europe. No less telling is the case of Meyer's close colleague Hans-Jürgen Seraphim, who with his deep knowledge of *Ostforschung* (Research of the East) in general and of Poland in particular was a sought-after man in the Cold War era, both in the U.S. and in West Germany.[59]

The general, and remarkable, willingness to overlook the role of experts in the Nazi era such as Meyer and Seraphim was an important precondition for their continuing careers. Moreover,

Fig. 5.8 **Another House of Knowledge:** *In stark contrast to the universalist concepts of Le Corbusier and Otlet, the Nazis conceived their own vision of organized knowledge. The* Ordensburg Vogelsang *was built to educate the future Nazi elite and was to be crowned by a* Haus des Wissens, *a house of knowledge also architectonically expressing an authoritarian approach. Unlike large parts of Vogelsang, that building remained at the design stage. Only the foundations and a small part were begun.*

the Western Allies opportunistically utilized experts among their former enemies both for European reconstruction and to strengthen their military-technical complex in response to the threat of war with the Soviet Union. As we will see, the Soviets were hardly less active in harvesting experts from Germany and Central Europe for their postwar technoscientific projects. Experts in Western Europe also profited from the societal autonomy that had grown during the war, in spite of the autocratic regimes, and from the ensuing need for specialized knowledge. Moreover, the heavy human losses, particularly among those between 20 and 40 years of age, indirectly improved the positions of those who had survived and made for a greater degree of continuity than one would expect. As Meyer quickly realized, European integration opened up a new fields for experts and offered new meta-narratives to cling to.

Like a mock version of the grand progressivist schemes of Le Corbusier and many others who built self-consciously on the technoscientific progress achieved in the nineteenth century, the Meyer and Lysenko cases also tell us something about experts' belief in their own importance: a subject dealt with in the two preceding chapters. These men turned into self-proclaimed shapers of a society based on scientific and technical expertise. Also in this respect, the three individuals presented here have specifically European biographies, but not just because they happened to live in Europe in the age of extremes. Rather, their agendas increasingly reflected widespread European dreams of pushing the old, crisis-ridden continent into a bright future by making unlimited use of modern expert knowledge.

The multiple variations of ethics-crushing, morally-bankrupt Faustian bargains in totalitarian Europe, repeated thousands of times (although usually less dramatically) beyond the cases presented here, should not obscure the equally high number of experts who lost their jobs, had to leave their countries, or were persecuted and killed, thereby vacating many positions used to lure others into Faustian bargains. Starting just after 1930, Europe experienced a fundamental reconfiguration of its knowledge societies and expert networks, derived largely from the political interventions and racist motivations of Nazi Germany (and to a lesser extent the Soviet Union), although the crude mixture of ideology and charlatanism characteristic of Lysenko also contributed to this restructuring. Next we will probe more deeply into the often-forced circulation of expert knowledge in the age of extremes.

6
Experts in Exile

Stalin's Operation Osoaviakhim

Stalin was not amused—not at all. His army had finally defeated Nazi Germany and his country had suffered no less than 27 million casualties from the Great Patriotic War. The emerging Cold War and the political division of Germany limited Soviet influence in Germany to the Eastern occupation zone. The Soviets' high expectations of being compensated for the war efforts were restricted to "their" zone, which actually experienced enormous extractions. Stalin's military administration in East Germany filled train after train with production facilities, machinery, infrastructure equipment, and goods, shipping them to the Soviet Union.[1] However, Stalin's former Western allies, his new enemies in the Cold War, were more successful at capturing the most valuable legacy of Nazi Germany: the technoscientific experts who had worked hard to design a multitude of innovative weapon systems until the very last days of the war. In the immediate postwar period, the Western allies sought to keep these experts for themselves as "intellectual reparations," often by deporting them from territories that the inter-allied treaty of Yalta of July 1, 1945 had designated as part of the Soviet occupation zone.[2]

The Soviets also recruited technoscientific experts from Nazi Germany immediately after the war, particularly nuclear scientists, to foster the development of atomic bombs. Yet, the primary Soviet response to the Western policy of shipping German experts abroad was to set up special design and production bureaus in the Eastern occupation zone to exploit "their" experts' knowledge. In September 1946, the four special design bureaus (*Osoboe konstruktorskoe bjuro*, OKB) of the Soviet Ministry of Aviation in Dessau, Stassfurt, Halle, and Berlin employed more than 8,000 "specialists," as the Soviets called such experts.[3]

The Soviets then decided to change strategy. On the night of October 22–23, 1946, operation Osoaviakhim was carried out. In a coordinated action, the secret police and the Red Army captured some 2,200 experts, along with their families and laboratory equipment, and transported them to the Soviet Union. The operation forces awoke the Germans between 3:00 and 5:00 AM and helped them pack their luggage. They had orders to treat the experts well. They even allowed the spouse of the leading rocket engineer Helmut Gröttrup to bring two cows with her. The majority of the deported specialists consisted of the rocket groups around Berlin and at Nordhausen in the Harz Mountains, where the Nazis had located their infamous V-2 rocket fabrication facilities, reliant on forced labor that annihilated of tens of thousands. Many of the remaining specialists came from design bureaus of the former aircraft companies Siebel and Junkers.

The operation was planned well in advance. Over a period of months, the Soviet industrial ministries drew up lists of experts whom they wished to claim for the plants under their control. In March 1946, for example, the Ministry of Armament asked Stalin for permission to request 400–500 engineers and 1,500–2,000 skilled workers from the optical industry. The Ministry later reduced its claim to 300 optical experts.

Operation Osoaviakhim was top secret until the very last minute, for the Soviets were well aware that the German experts would not leave their country voluntarily or, if they did, would prefer to go to the West. The operation was originally planned to take place one week earlier but was postponed for political reasons. Soviet leaders were afraid that the operation would seriously affect the prospects of the *Sozialdemokratische Einheitspartei* (Social Unity Party, SED), which they favored in the state elections being held in the Eastern occupation zone on October 20. In his memoirs, the famous rocket designer

Boris E. Chertok reported that two days after the elections, he and his comrades at the rocket institute Nordhausen invited their German colleagues to an opulent dinner at the local restaurant, "Japan." On that evening, the two hundred Germans at the institute, who still had no clue what would happen to them only a few hours later, enjoyed themselves at the open bar until one o'clock. However, the Russians, banned from drinking, "were in a gloomy mood."[4]

Intellectual Reparations

The Soviets' response to the Western intellectual reparations programs was based on the idea that advanced scientific knowledge and technical skills could be easily transferred and exploited in another context. However, the next few years revealed all kinds of problems in fusing the diverse knowledge cultures of German and Soviet specialists. Most Germans were quite willing to demonstrate their ability to develop innovative technologies of military importance. The Soviets offered the Germans a second chance to prove their sophistication and ingenuity after their country's defeat. The German cultural tradition of understanding science and technology as apolitical allowed the experts to adapt to their new environment. However, the Soviet military establishment was not ready to allow a mutual exchange of knowledge between the two expert groups. The structural problems of the Soviet knowledge economy, which favored secrecy over open circulation, constrained benefitting from German expertise. The Stalinist regime kept the Germans in isolation and wanted them to transfer expertise to their Soviet colleagues without receiving knowledge about how their innovations would be used within the armament sector. Furthermore, language difficulties, professional rivalries, and resentment from Soviet experts led to various social problems.

Still, all in all, German specialists did contribute to Soviet efforts to catch up with the U.S. in nuclear, missile, and aircraft technologies. An example is the rapid introduction of the swept wing concept that enabled supersonic aircraft design, as is the development of the Junkers Jumo 222 turboprop motor into a family of powerful engines for intercontinental bombers and transport airplanes. The overall technology transfer, however, remained limited. The Soviet Union was never willing to integrate the Germans into its

Fig. 6.1 **Experts for Hot and Cold Wars:** *This image shows the Pirna 014 jet engine at the Spring Leipzig Trade Fair in March 1958 with delegates from various East European airlines. The engine was developed to power an aircraft that derived from the Soviet Samoljot 150 jet bomber designed by the former Junkers engineer Brunolf Bade. Bade and his team of German aeronautical experts were forced to move to the Soviet Union in October 1946 during operation Osaviakhim. After returning to East Germany in 1954, the team built and tested the jet passenger airliner Bade 152 (pictured in the background of the image), which, however, failed to enter service.*

military-industrial complex. As early as 1948, they were practically excluded from Soviet innovation efforts. After their knowledge was essentially put into cold storage for a few more years, they were finally sent back to East Germany.[5]

The pattern of absorbing German technical talent was somewhat different in the U.S. Immediately after the war, the U.S. also sought to make use of the resources of a conquered Germany. Unlike the Soviets, who were initially mostly interested in material resources, the U.S. focused on intellectual resources. During the war, first the British and then the U.S. armies had installed special units to identify scientific and technical assets to be exploited. These units followed in the wake of military troops, dismantled research laboratories and industrial facilities, captured patent libraries, and interrogated thousands of scientists and engineers, often in great haste to avoid the Soviets getting a grip on them. Ton after ton of laboratory equipment and technoscientific papers were shipped out of Germany in order to enrich the knowledge base of the Allied countries. More efficient than standard interrogations was the strategy of commissioning review reports from German experts. After 1945, the joint U.S.–British Field Information Agency, Technical (FIAT) generated more than 1,300 reports. The FIAT reports became an invaluable source of knowledge for the Allies, both in ending the war against Japan in the Pacific, and for the emerging Cold War, as well as later for historians aiming to gain a better understanding of experts' technoscientific achievements during the war.

Hand-in-hand with extracting knowledge by commissioning reports were efforts to integrate German experts into the Allied knowledge societies and innovation cultures. The decades following the war witnessed a true exodus from Germany: by the 1960s, half a million Germans had moved to the U.S., and tens of thousands to Canada, to escape their devastated country. Among these emigrants were numerous technoscientific experts. Thanks to a number of recent studies on the Allied programs that sought to capture intellectual reparations, we have been able to approximately map the transnational flows of expertise within and beyond Europe.

The two pre-eminent programs of the Western Allies to engross the elites of Nazi Germany's technoscientific-expert system were the Alsos Mission and Operation Overcast, later renamed Operation Paperclip. The Anglo-American Alsos program aimed to capture and interrogate the community of nuclear experts, including the Nobel laureates Werner Heisenberg and Otto Hahn. The U.S. Paperclip program was directed towards rocketry and aeronautics, most notably to capture the Peenemünde team of Wernher von Braun plus a number of aeronautical experts, including the aircraft designers Alexander Lippisch and Hans Multhopp and the jet-engine pioneer Hans Pabst von Ohain.

Table 6.1 The global circulation of knowledge: German technoscientific experts abroad after the Second World War.[6]

Country	Number of experts (source)	Major fields of expertise
U.S.	900 (Neufeld)	Rockets and missiles, aeronautics, medicine, electronics, nuclear physics
Soviet Union	2,900–3,000 in 1948 (Mick)	Rockets and missiles, aeronautics, nuclear physics, optics, chemistry, electronics
Britain	800–1,050 (Glatt)	Rockets and missiles, aeronautics, nuclear physics
France	800 in 1946 (Ludmann-Obier)	Rockets and missiles, aeronautics, chemistry
Australia	150 (Jones)	Chemistry, physics, metallurgy, geology, aeronautics
Argentina	108 (Stanley)	Rockets and missiles, aeronautics
Spain	100 (Presas I Puig)	Aeronautics, nuclear physics, chemistry
Egypt	100 (Neufeld)	Aeronautics, nuclear physics
India	50 (Neufeld)	Rockets and missiles, aeronautics
Canada	41 (Margolin)	
Turkey	2,000 (Erichsen)	Aeronautics, physics, chemistry, architecture, social sciences
Brasil	27 (Stanley)	Helicopters
Total	**8,000–8,350**	

Wernher von Braun, the "dreamer of space" and "engineer of war," is perhaps the most famous case of the transatlantic migration of expertise.[7] In the mid-1930s, the charismatic young engineer cut a deal with National Socialism that gave him the opportunity to work on designing rockets that he hoped would eventually fly to the moon. In return, he and his team at Peenemünde on the remote Usedom peninsula in the Baltic Sea devised the deadly V-2 combat rocket. After the war, von Braun and his core Peenemünde team traveled to the U.S. as Paperclip scientists. There von Braun became a public hero, probably the most prominent technical expert in postwar America. First at White Sands Proving Ground in New Mexico and then, after 1950, at the Army's Redstone Arsenal rocket development center in Huntsville, Alabama, von Braun designed intercontinental ballistic missiles (ICBMs). These rockets allowed the U.S. to gain technical superiority over the Soviet Union in an early phase of the nuclear race. "The missileman," as von Braun was named in a 1957 *Time Magazine* cover story, brought not only techoscientific expertise with him from Germany, but also skills in managing complex, large-scale technical projects. Redstone Arsenal was run as an institutional copy of Peenemünde. However, the limits of transferring managerial principles from one place to another, from a large project to a super-large project, became evident when von Braun and his team moved to NASA to operate Huntsville's Marshall Space Flight Center, tasked to develop the Saturn V rocket for the Apollo program. The more the Apollo program expanded, the less von Braun succeeded in managing the Center by applying his Peenemünde principle of designing everything under one roof. Von Braun encountered increasing technical and political difficulties. As the Saturn V program became increasingly complex, it needed the Air Force's experience in hiring industrial contractors to keep it on its feet.[8]

Cold War America showed an insatiable hunger for technical talent. The expanding scientific-military-industrial-complex not only absorbed the first wave of Paperclip specialists, but also continued to recruit experts from former Nazi Germany until the 1960s. The aerodynamicist Bernhardt Goethert presents another interesting case of how a European mode of producing knowledge was copied and transferred to the U.S. At the *Luftfahrtforschungsanstalt Hermann Göring* (Hermann Göring Aviation Research Center)

Fig. 6.2 **Americanizing Nazi Technology:** *With former SS officer Wernher von Braun and his Peenemünde team, expertise in rocketry traveled rather smoothly from Nazi Germany to the U.S. This image shows von Braun (seventh from right) and his core team of German experts at Fort Bliss, a large U.S. Army installation north of El Paso, Texas, where von Braun headed the guided missile development unit. In 1950, the team moved to Huntsville, Alabama, where von Braun and his collaborators developed ballistic missiles for the U.S. Army and later designed the gigantic Saturn V rocket for the Apollo project.*

in Brunswick, Lower Saxony, Goethert had successfully brought together researchers from different disciplinary fields and designed innovative high-velocity wind tunnels to undertake experiments in supersonic flight. After the war, the FIAT interrogation teams were stunned by the quality of the test-works at Völkenrode near Brunswick and two smaller aerodynamics research centers at Kochel and Ötztal, both south of Munich. U.S. wind tunnel expert Frank L. Wattendorf sent a "transatlantic memo" back home proposing a "new engineering development center to consolidate the test facilities and the talents of the nation's best civilian and military scientists—a center to properly test and evaluate weapon systems needed to guarantee superiority of American airpower and thereby the national security."[9] In Cold War America, Wattendorf's ideas fell on fertile ground. In 1951, after years of preparation, President Harry S. Truman formally opened the Arnold Engineering Development Center (AEDC), which quickly developed into the largest complex of flight-simulation test facilities in the Western world. The Center was located near Tullahoma in Tennessee, the region with

Fig. 6.3 a+b Global Circulation of Nazi Expertise: *The Nazi party had presented the aeronautical expert and test pilot Hanna Reitsch as a National Socialist hero and public celebrity. More than most other technical experts, Reitsch was an ardent Nazi. She incarnated the self-mobilization of experts in the final phase of the war, which in her case went so far as to become a founding member of the SS suicide unit Leonidas. The image depicts Reitsch greeting her home town Hirschberg with the Hitler salute in April 1941. Unlike other experts, she did not break away from Nazism after the war. In the 1950s and 1960s, Reitsch set numerous world records in gliding and was internationally highly recognized and honored. The picture opposite (6.3b) shows Reitsch in 1959 in India with Prime Minister Jawaharlal Nehru, who invited her to set up a gliding center, preparing for a sightseeing flight over New Delhi.*

the most waterpower energy in the country. American engineers had learned from Nazi Germany that large technological installations like Tullahoma depended on the availability of large amounts of energy. Walter Dornberger, the former head of Peenemünde-Ost, helped to plan the center on the basis of his experience in Nazi

Fig. 6.3b

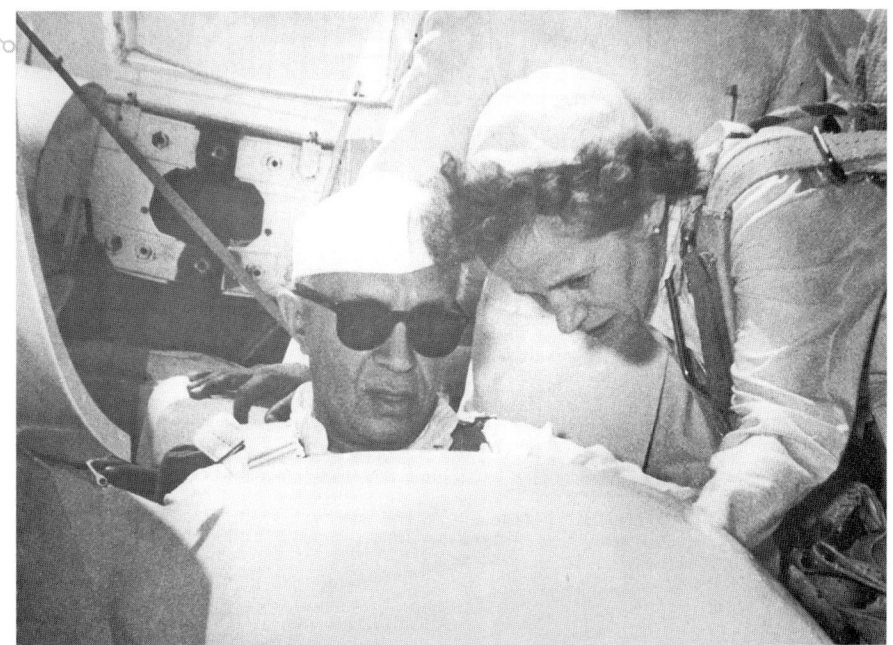

Germany, and the Center's founding director was Goethert, who had been brought to the U.S. as a Paperclip scientist. One historian of technology correctly concluded that AECD combined both the material infrastructure and the expertise of "Peenemünde, Völkenrode and Kochel and Ötztal."[10]

The aeronautical engineer and test pilot Kurt Tank represents the global dimension of the postwar circulation of experts. In Nazi Germany, he headed the design bureau of Focke-Wulf, where he developed multiple new airplanes including the famous single-seat fighter-bomber Focke-Wulf Fw 190 Würger. After the war, this highly confident expert evaluated a number of different opportunities to continue his design work. He negotiated with the U.S., Britain, the Nationalist government of China, and the Soviet Union. Finally, he accepted an offer from Argentina's president Juan Perón to head the Instituto Aerotécnico at Córdoba, where Tank (under the pseudonym Pedro Matthies) and his team of former Focke-Wulff engineers produced the jet fighter Pulqui II. After Perón fell from power in 1955, Tank moved to India, where he became chief designer of Hindustan Air Industries in Bangalore. He created the fighter-bomber HF 24 Marut, India's first military aircraft and part of Prime Minister Jawaharlal Nehru's program to modernize the country by adapting

Western technologies.[11] In the 1970s, Tank returned to Germany and worked as a consultant for the aircraft company MBB.

In the Cold War setting, allegedly "apolitical" experts like Tank were easily able to switch between regimes with diametrically opposed ideologies. The Communist Soviet Union was as eager to engage the former Nazi engineer Tank as was Perón's proto-fascist Argentina, socialist India, or democratic Britain. The U.S. was ready to grant citizenship to SS officer von Braun and even to the infamous physician Hubertus Strughold, whose human experiments had tortured to death inmates of the Dachau concentration camp. The "father of space medicine," Strughold was honored with the Americanism Medal by the Daughters of the American Revolution. Only after the end of the Cold War did Western democracies begin to reassess their policy of unconditionally integrating the intellectual booty from Nazi Germany into their scientific-military-industrial complexes and their civic societies.

Historians have also recently reassessed the postwar intellectual reparations. The resulting picture points to ambivalence and is, in itself, ambivalent. On one hand, the Allies' seizure of German technical knowledge and talent was probably "the largest, best organized, and most thoroughly coordinated international transfer of knowledge ever."[12] On the other hand, its impact was much more limited than was commonly claimed, both during the Cold War and by later historians. FIAT estimated that it had saved U.S. government and industry the huge sum of five billion dollars, which seems "a fair global figure" for one historian, adding that another five billion dollars were saved in the British case.[13] The FIAT figures are methodologically questionable and all the more suspect as they served to justify the organization in the face of growing public criticism in the U.S. No less questionable is the assumption that the loss of know-how massively hampered the economic recovery of postwar Germany.[14] When the *Wirtschaftswunder* (economic miracle) did take place later in the 1950s and 1960s, it was not based on high-tech knowledge resulting from the wartime legacy, but on proven low-tech engineering expertise and a traditional culture of production spurred on by a steady stream of incremental innovations.[15]

The customary accounting of gains and losses is all the more questionable as it assumes a linear transfer of technical knowledge. In particular, Wernher von Braun's Peenemünde team of rocket experts has been hailed as the technical forerunners of practically every Cold War missile program, and direct lines have been drawn, for example, from the V-2 in Peenemünde to the French postwar rocket Véronique and further to the European rocket Ariane.[16] However, as we will see, the road to Ariane was a crooked one, paved with numerous technical and political problems. Expert knowledge rarely travels easily. Likewise, the rather limited and short-term impact of the German experts on Soviet postwar military technologies and the hermetically sealed working culture of von Braun's Peenemünde team in the U.S. both point to the numerous obstacles to transnational transfer and appropriation of technical knowledge. Postwar intellectual reparations boosted technoscientific globalism, but the effects of this boost were limited and differed both from technology to technology and from place to place. Contextualizing the intellectual reparations from Nazi Germany within the larger geopolitical circulation of expert cultures in the postwar period produces a different picture than merely calculating losses and gains. On the one hand, it stresses the ruptures and frictions of experts' circulation between different knowledge societies, while on the other, it points to processes of technoscientific, economic, and political integration. The Allied information-gathering activities helped to enmesh scientists, technicians, and industrialists in transnational networks, where they served as a "conveyor-belt for future business connections."[17] In rapidly expanding global technology markets, this migration gave strong bargaining power to those experts in Germany who possessed knowledge that could not be codified in reports. Likewise, from a European, transatlantic, and global perspective, the postwar circulation of experts fostered both the multinationalization of knowledge and social integration.

Émigré Experts as Agents of Modernity

The global circulation of experts who left Germany after the Second World War was not the first exodus of intellectual capital from a European country in the twentieth century, nor was it the most dramatic. Germany had experienced an enormous loss of expertise after the Nazi seizure of power. Based on the infamous *Gesetz zur Wiederherstellung des Berufsbeamtentums* (Law for the Reconstitution of the Professional Civil Service) of April 7, 1933, the Nazi government started to dismiss scholars from universities and research institutes for political and racial reasons. What has been termed a "cultural decapitation" was only the beginning of mainly-Jewish researchers' forced exodus that led to "the most sweeping and enduring cultural transfer in modern history."[18] German academia, which had previously led the world in many fields, lost 14 percent of its scientific personnel by 1936 (1,145 out of 7,979 professors); this figure rises to approximately 2,000 when researchers from non-university institutions like the Kaiser Wilhelm Society are added. The flourishing field of theoretical physics suffered the loss of such gifted minds as Albert Einstein, Erwin Schroedinger, and Leo Szilard. The Nazi regime expelled twenty-six of Germany's sixty theoretical physicists from their positions and forced them to go abroad. The Göttingen University physics department, internationally recognized as the premiere institution in the field, experienced the eviction of half its academic talent, including the Nobel laureates James Franck and Max Born.[19]

Most of the expelled scholars were Jewish. For racial reasons, however, many European nations were not eager to welcome Jewish emigrants. The comparatively receptive U.S. profited most. Anti-Semitism and prejudice against Jewish immigrants was certainly widespread in the U.S., but dense transatlantic personal networks and interventions by influential figures who advocated rescuing émigré experts allowed many German scholars and scientists to find suitable positions on the American academic market. The Rockefeller Foundation helped fund programs such as the Emergency Rescue Committee and the University in Exile. The latter brought over 300 emigrants to the U.S. who its founder Alvin Johnson described as "Hitler's gift to American culture." Johnson's no less prominent colleague Walter S. Cook, founding director of the Institute of Fine Art at New York University, put it another way: "Hitler shakes the tree and I gather the apples."[20] In science and technology, as well as in the social sciences and humanities, the newcomers stimulated the cross-fertilization of knowledge and enabled the transformation of intellectual ideas, concepts, and practices. A recent study suggests that "U.S. invention increased by 30 percent after 1933 in the fields of U.S. émigrés," such as chemistry.[21]

Countries other than the U.S. exploited the sudden availability of experts on the global intellectual market as well, some even more deliberately. Particularly interesting was the Republic of Turkey, the young nation at Europe's Eastern periphery. Mustafa Kemal,

Fig. 6.4 **Dismantling the European "Mecca" of Physics:** *After the Nazi seizure of power thousands of experts fled Germany. World centers of excellence such as Göttingen suffered heavily, when bright minds like the Nobel laureates James Franck (second from right) and Max Born (second from left) were forced to go abroad. The personal network displayed at this picture from a ceremony in 1923 quickly vanished. Max Reich (left) seized the opportunity to further his career by joining the Nazi party and Robert W. Pohl (right) refrained from protesting against the dismissal of his former friends.*

later known as Atatürk, was the republic's founding father in 1923 and its first president until his death in 1938. Atatürk envisioned a nation state based on secular laws and customs rather than on ethnicity or religion. He and his elite "Young Turks" aimed to modernize state and society along Western European models.[22] European expertise was very much in demand in Atatürk's post-Ottoman Turkey.

Atatürk had high expectations that the *Dar-ül Fünun* (House of Knowledge), the old Istanbul university, would spearhead his modernization policy, but these expectations were not fulfilled. In the early 1930s, the university, based on the traditional Islamic *madrassa* educational system, was at the top of the agenda for reform. The Turkish government hired Albert Malche, a professor of pedagogy from Geneva, as an external advisor and assigned the newly installed minister of education Resit Galip, Atatürk's personal friend, to transform the *Dar-ül Fünun* into a modern university. The official reconfiguration took place on the last day of July 1933, when it was closed and reopened the next day under its new name: Istanbul University. Two days earlier, Resit Galip fired ninety-two professors who had rejected the reforms, thereby providing space for new people hired from abroad.

The coincidence between closing the *Dar-ül Fünun* and the expulsion of mainly Jewish researchers from German academia is striking. Malche mediated a meeting between Resit Galip and Phillip Schwartz, the head of the Emergency Organization for German Scientists, who was looking for places to employ German scientists abroad. On July 6, 1933 Schwartz reached an agreement with the Turkish government to hire a substantial number of expelled Germans. On the same day, the Turkish minister dissolved the self-administrative bodies of the *Dar-ül Fünun*. The agreement was soon known as "the German-Turkish miracle."

In the first wave, the Turkish government appointed thirty émigrés to Istanbul University, which was soon considered "the best German university," although Nazi officials described it as having "turned Jewish."[23] Many more appointments followed until the end of the 1940s, when Turkish higher education employed some 300 immigrant professors and fifty technical experts. Albert Einstein even considered leaving for Turkey while he waited for a response from Princeton, rumored not to hire Jews.

Turkey had longstanding good relations with Germany that continued under the Nazi government. The latter even allowed the Turkish government to rescue some Germans from concentration camps. When a son of the chemist Fritz Arndt from Istanbul University was caught fighting the Germans during their invasion of Poland, the Turkish government used its bargaining power and brought him safely to Istanbul. German émigrés not only transformed Istanbul University into a Western institution following the German model, but also helped to establish the Istanbul Technical University in 1944 and Ankara University two years later.

The modernizing impact of the refugees was not limited to revitalizing the Turkish system of higher education. Before fall 1939, the Turkish government hired approximately 2,000 experts from Germany as well as from Austria after the *Anschluss* (annexation) of 1938.[24] Most helped to develop Turkey's industrial and agrarian sectors. They supported Atatürk and his modernizing elite in fashioning a society based on the secular religion of scientific and technical progress.

One key group was the large cohort of architectural experts who escaped from Nazi Germany to Turkey, which depended on Western architectural expertise in at least two respects. The new state, and Atatürk in particular, strove for aesthetically modern representations.

Social modernization was to be mirrored in architecture. These reforms incorporated housing designs reflecting new family concepts and new roles for women. At the same time, everything reminiscent of the Ottoman period was held in low esteem—merely tolerated but not encouraged. For Turkey, forced modernization after the First World War also underscored the need for technical knowledge of modern building methods, then available in Turkey only to a limited degree. The new elites regarded the predominant Ottoman architects as not qualified for the tasks of modernization. European knowledge was equated with modern knowledge. Due to a lack of expertise and practical skills in traditional building methods, there were great expectations related to rationalized building methods. This was also true for the new discipline of urbanism (as discussed in chapter 4), introduced at Turkish universities in the 1930s. The scientific character of modern architecture was the most compelling argument—avoiding branding the new designs as "Western," in what was still a largely Islamic context.

In "nationalizing the modern," the new Turkish state also strove to sell its "national renaissance" abroad and to gain international standing and acceptance.[25] Against this background, Turkish architects saw being part of an international professional community as a matter of national pride. In the new Turkish narrative, contemporary civilization—modernity—was no longer a European monopoly, although it had originated there. "Being Western in spite of the West," was the new watchword.[26]

For this reason, Western architects who had made their names with modernist solutions seemed predestined to tackle the new state's manifold challenges. Forced modernization with architects in the driver's seat seemed an attractive arrangement for Western architects. Some were drawn by an "Oriental" culture that offered an alternative to a crisis-ridden Western Europe. Yet Turkey was also able to exert a pull on so many architects because it refrained from adhering to just one ideological style which had become *de rigueur* in Nazi Germany, the Soviet Union, and, to a lesser extent, Italy. Generally, it seems that German experts adapted themselves quite well to Kemalism because of their anti-capitalist, yet statist orientation.

It is unsurprising, then, that the circulation of architectural expertise had set in long before the first German architects arrived in Turkey. Many Turkish architects had studied at Western universities. Moreover, Turkish journals devoted significant space to modernist achievements, particularly Le Corbusier's projects. This movement was mostly one-directional. For example, there was no significant Turkish participation in CIAM, and few examples of Turkish modernism were published in Western European sources. Also, other Western architects arrived in Turkey before 1933. In 1927, Clemens Holzhäuser and Ernst Egli of Austria entered, attracted by the potential of building in Turkey rather than by a desire to escape the tense political situation at home. In the 1930s, with building activity and educational reform on a new scale and the huge wave of mostly political emigrants from Germany, the transfer of expertise gained an enhanced quality. Paul Bonatz, Bruno Taut, Margarete Schütte-Lihotzky, Wilhelm Schütte, Martin Wagner, Hermann Jansen, Martin Elsässer, and Gustav Oelsner were among those who came to Turkey at this time. Most of them belonged to the modernist strand of architecture. Luminary Hans Polezig would have joined them, but his health was poor and he died in 1936.[27] For Taut, who had already worked in Russia and Japan, Turkey was his third exile phase, testifying to the astonishing mobility of experts in the age of extremes.

These experts were engaged directly by the state, which was almost the only way to enter the country. Istanbul and the new capital Ankara were beginning to struggle with problems typical of urban centers, but this also led to the formation of important arenas for urban planners and architects alike. Both cities also served as showcases, proving the capability of the new Turkish state and its political system. Finally, the economic development and urbanization of the countryside presented important challenges for the Western experts. Urbanization was crucial for the overall modernization of the country. Moreover, universities served as important transmission belts, particularly in the case of the School for Political Science (*Siyasal Bilgiler Yüksek Okulu*, SBO), but later also for the technical universities in Ankara and Istanbul. Much as had happened in European nations during the nineteenth and early twentieth centuries, technical education in Turkey and the building of a new nation went hand-in-hand.

Ernst Reuter, who later became famous as the mayor of Berlin after the Second World War, is a good example of the opportunities that Western architects and urban planners could seize during this period of rapid change and forced modernization. Reuter, a prominent municipal administrator in Berlin of the 1920s and briefly mayor of Magdeburg, left Germany after having been interned in a Nazi concentration camp. Following a difficult time in Britain and helped by the intercession of agrarian economy expert Fritz Baade, Reuter arrived in Turkey as a special advisor for transport tariffs in 1935. He had developed and proven his organizational talent in Soviet Russia after the First World War and thus became an influential teacher at the SBO in Ankara and an advisor to the state in all urban affairs. He understood better than others that housing, urban planning, and building were always also a matter of creating a civil society and sustainable governance structures, even at the local level.[28] The SBO was a practically oriented school for administration, intended to train those who would implement ambitious state reforms. Many of Reuter's disciples rose to leading positions in the Turkish political hierarchy, an influence that is honored at the Ernst Reuter Center for Urban Studies today.

In addition to his contacts in Turkish politics and administration, Reuter could draw not only on a vast network of other emigrants, architects, and urban planners, but also on individuals such as the economics scholar Fritz Neumark and the conductor Ernst Praetorius. There were two reasons why Reuter was one of the most successful emigrants in Turkey. The first was his experience in local politics, which he regarded as a laboratory for creating a modern state and society, and the second was his willingness to adapt to a new culture and language. Like many other emigrants, Reuter entered Turkey with a strong interest in what was to him a foreign culture. However, also like many emigrants, he was not without a certain sense of superiority having come from a more "advanced" European background, which led to him having an almost ethnographic approach. After some reflection, the experience of the German urban planners and architects in Weimar Germany turned out not to have been so different from the Turkish situation many now faced. In Weimar, they also had to fight severe opposition and win support for a new political regime trying to prove its legitimacy through achievements in housing and urbanism.

However, cases like Ernst Reuter should not detract from the fact that many other emigrants had great economic difficulties and not all managed to adapt to their new environment. When Hungarian architect Fred Forbat traveled to Turkey, he met both Reuter and Martin Wagner, who had been Berlin's chief planning executive before 1933. Forbat concluded that Wagner was clearly not able to find his place in Turkey and, not coincidentally, soon left for Harvard

University.[29] Reuter later wrote to Walter Gropius, then a professor at Harvard, that while not wishing to complain, he was longing to be "useful" again.[30] He described himself as a "modern Job," caught in a "golden cage" and doubtful about the social value of his work.[31]

Beyond the self-doubts and problems of acculturation that haunted many of Reuter's colleagues, by the late 1930s the general climate worsened for the foreign architects. Though certainly a unique story of the immense impact that a group of emigrant experts exerted in a comparatively short time, little was actually built by most of the decisively modernist architects. The large, representative commissions mostly went to conservatives like Bonatz and Holzmeister. Sometimes the categories used in Western Europe changed completely in this country. Someone who counted as an abstract artist among his Western fellows, such as sculptor Rudolf Belling, could succeed in Turkey as a builder of representative state architecture. Modernist Martin Wagner, on the other hand, failed with a plan he made for Atatürk's residence in the seaside resort Florya, by deliberately choosing a structure inspired by Ottoman palaces. Instead, Atatürk chose a modernist concept by Turkish architect Seyfi Arkan, who had studied in Germany.

Moreover, Western experts became increasingly caught in the tensions within authoritarian modernization. Starting in the late 1930s, growing nationalist sentiments led to criticism of immigrants, a reflection of the influence and success of these architects. For the Germans and Austrians the situation came to a head with Turkey's (mostly symbolic) entry into the Second World War on the Allied side in February 1945.

Although it was not directly involved, Turkey suffered economically from the war. Serious inflation set in, and food was being rationed by the end. Even highly paid professors felt the hardship, and after 1945 most refugee academics tried to secure positions in the U.S. Others left for Palestine (later Israel), while some returned to Germany. By 1950, only a small number of these academics still lived in Turkey, the departure of the others having been hastened not only by economic conditions but also by jealousies on the part of some Turkish professors and the opposition of Turkish nationalists to renewing their contracts. With the Nazis' defeat, the tasks awaiting architects returning to Germany and Austria were welcome and potentially attractive alternatives to the restrictions of working in Turkey. Ernst Reuter, for example, was all too eager to use experience gained in Turkey back in devastated Berlin.

Émigré Experts as an Endangered Species

Turkey is a particularly unusual case in that for the first twenty years after it became a state, its national architecture was largely dominated by emigrant rather than local expertise. The emigrants came for very different reasons; some were pulled by professional chances, while others were pushed by persecution.

The scope for exploring new ideas that Turkey offered to foreign experts was significantly different from another destination on the periphery of Europe in the early 1930s: the Soviet Union, which at this time experienced a flood of emigrants very different from the (largely involuntary) one that took place several decades later, after the end of the Second World War. The numbers of emigrants turning to the East were even larger than those voyaging to Turkey, as were the expectations of these often leftist intellectuals. After the Soviets established a bureau for "foreign consultation" as part of the Committee for Construction (Госстрой, Gosstroy), a huge stream of experts, along with less qualified personnel, tried to enter a country that seemed to them to be the promised land. They aimed to fill the immense gaps caused by experts' emigration during and after the civil war. The newcomers were mostly motivated by ideological convictions, but as least as important were both the economic crisis at home and rumors of huge projects and apparently unlimited resources in the Soviet Union. The Soviets' smart PR initiatives attracted more than one million applications from foreign workers and experts between 1930 and 1932. In 1930 alone, approximately 4,000 foreign specialists arrived in the Soviet Union. In 1933, by the end of the First Five-Year Plan, they numbered closer to 35,000. The press reported that German planner and architect Ernst May, who had been granted a major contract with immense responsibilities, received more than 1,400 applications from almost all over Europe, asking to collaborate in what May believed to be "possibly the greatest task an architect ever faced."[32]

The attraction of a realm of unlimited opportunities was not confined to the old continent. Soon up to one thousand experts from the U.S. trained Soviet technicians in the secrets of Fordist production and helped erect factories modeled after those that had made Detroit famous. In Stalingrad alone, some 380 American experts received contracts, with an annual pay of ten thousand dollars being the rule rather than the exception. Albert Kahn Associates, a leading factory designer, for complexes such as Henry Ford's famous Highland Park and River Rouge plants, established a planning office in Moscow that was intended to eventually employ 4,500 architects and engineers. Moritz Kahn noted with astonishment that their Detroit office, which normally housed 400–500 employees, was considered the largest in the world. The Moscow branch quickly exceeded 600 staffers but fell short of the workforce initially planned.[33]

Impressive as the huge influx of foreign experts into the Soviet Union was, its limitations quickly became evident. In contrast to Turkey, political pressure was immense and soon was felt in all aspects of professional and private life. Competition between European and American experts in the Soviet Union grew, but tensions between locals and foreigners were sharper.[34] By the mid-1930s, those experts with strong political convictions were clearly in the lead, while others had either left the country or run into severe problems.

A German journal had warned the first emigrants that the political pitfalls in the East should not be disregarded and noted sarcastically that, of course, a "global celebrity" like Ernst May was not subject to such threats.[35] But even May, who believed that he would be received by Stalin himself, if he wished, had to learn the hard way how limited the relevance of both international standing and professional excellence was when the Stalinist regime increasingly followed its own internal logic shaped by terrorism, paranoia, and annihilation. Hardly voluntarily, May left Moscow at Christmas 1933.

One of the aims of the Great Purge orchestrated by Stalin from 1936 to 1938 was to repress and persecute the so-called intelligentsia, which included many immigrant experts. They often paid with their lives for envisioning their roles in shaping a new society. The Soviet case drastically signaled what was at stake for experts in the age of extremes. Even more dramatic, it soon transpired, was the situation in Nazi Germany, where experts suffered persecution on political and, primarily, racial grounds. Some architects, who could call themselves lucky, managed to escape to what was then Palestine. In 1939, more than one hundred architectural experts who had attended a German university or school of higher education lived and worked in Tel Aviv alone, including nineteen Bauhaus alumni.[36]

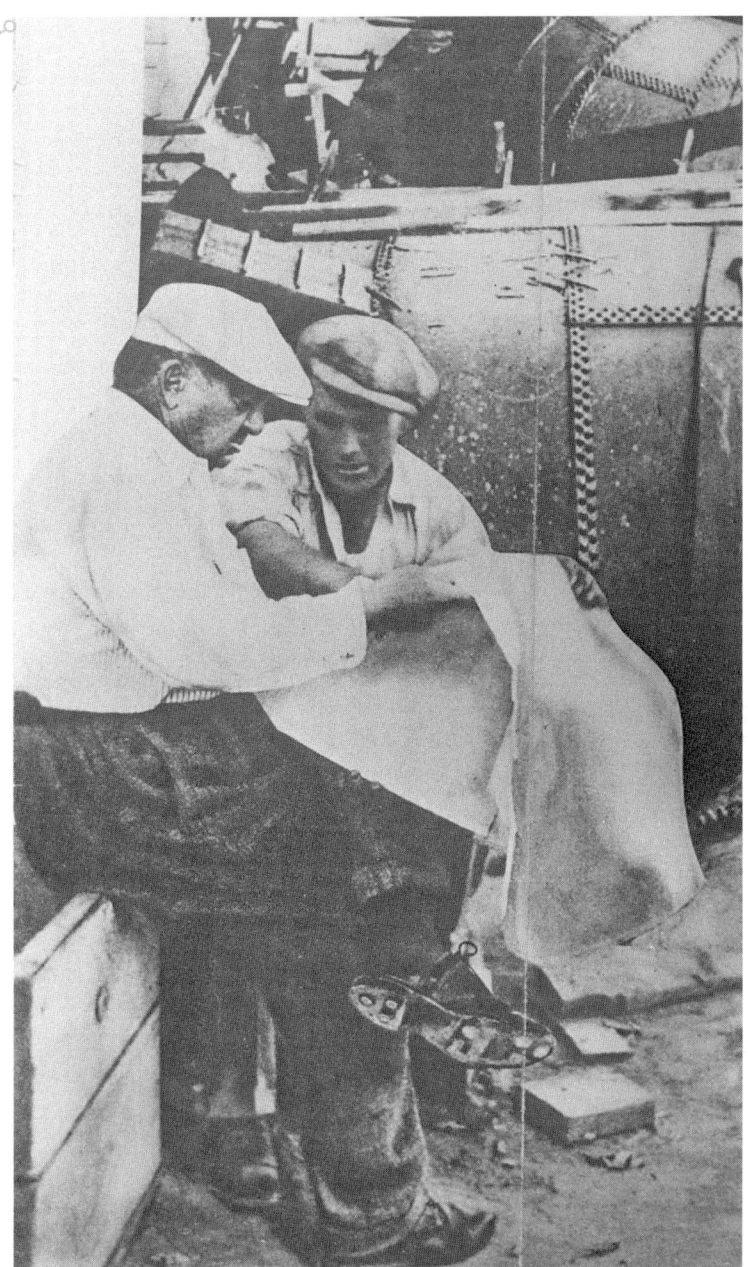

Fig. 6.5 Capitalism Meets Socialism: *The image shows the American engineer Hugh Cooper (left) advising a Soviet expert building the Dnieper power plant, by then the world's largest hydroelectric plant. In 1932, the Soviet government awarded Cooper the Order of the Red Banner of Labour. Cooper was hardly the only engineer from the capitalist West who advised the Soviet Union. At that time Soviet administrators were more willing to rely on foreign experts than was the case in later projects to foster industrialization.*

Forced Migration & New Knowledge Orders

The Second World War opened a Pandora's box that unleashed plagues of death, destruction, deprivation, and displacement upon

Europe. In the wake of the war, many of those who had survived the great slaughter were forced to leave. Millions of displaced persons and refugees moved in various directions in search of a safe place to start new lives; this mass relocation included thousands of scientists, architects, and technical experts. With the Second World War coming to a close, the great migration of experts that had started around 1930 also suddenly ended. Nonetheless, as the competing intellectual reparations programs of the former Allies showed, the circulation of expert knowledge continued well into the Cold War.

The great migration of knowledge was accelerated by the forced exodus of experts of Jewish origin from Germany immediately after the Nazi seizure of power. Over the next two decades, forced migration led to a fundamental reconfiguration of knowledge in Europe, a change that responded to the challenges of prewar, wartime, and postwar Europe.[37]

The transformative effects were particularly significant in Germany, which lost its leading position in many fields of science and technology, including physics, chemistry, and engineering. The nation that probably profited the most was the U.S. Although Jewish scholars were not always welcome everywhere, the U.S. provided fertile cultural ground for migrants. The rise of the American universities to global leadership was well under way in the interwar period; however, it received a significant influx of intellectual resources from experts fleeing Nazi Germany. U.S. industry benefited no less from the transatlantic transfer of knowledge. Many experts who fled fascist Europe got jobs in American industrial research labs. In biochemistry, refugees such as the Nobel Laureates Otto Loewi, Otto Meyerhof, and Otto Stern "propelled the United States to world leadership in the chemistry of life."[38] In polymer chemistry German researchers "revolutionized U.S. innovation" and helped to create the postwar boom of the synthetic fiber and plastics industry.[39]

The rise of nuclear technology is another telling case in point. The Hungarian refugees Leó Szilárd, Edward Teller, Eugene Wigner, and the Austrian-Swiss-German group based around Hans Bethe, Felix Bloch, James Franck, and Victor Weisskopf are just some of those who became indispensable for developing the atomic bomb within the Manhattan Project, which "changed the world."[40] Equally telling is the case of mathematics. A large group of German and Austrian émigrés, with Alfred T. Brauer, Richard Courant, Kurt Gödel, Emil J. Gumbel, Richard von Mises, John von Neumann, Emmy Noether, and Hermann Weyl at the top, marked

Fig. 6.6 Go East: *In the wake of the world economic crisis and its staggering unemployment the Soviet Union became the promised land for thousands of West European experts. Many were motivated by ideological reasons, others desperate to find employment. The journal* Das neue Frankfurt *was one of the major organs of modern architects in Germany. These visionaries hoped to find the opportunity to put through gigantic building schemes in Stalin's Soviet Union. After an impressive start, most of these architects left the Soviet Union.*

the intellectual shift of the discipline's center from Europe to the U.S. Richard von Mises exemplifies those individuals who often suffered from multiple displacements. In 1933, this aeronautics and applied mathematics expert was among those who emigrated to Turkey, where he assumed the chair for pure and applied mathematics at Istanbul University. After Kemal Atatürk died in 1938, von Mises lost confidence in the Kemalists' political ability to modernize the country; and a year later he accepted an invitation from Harvard University. With his move to Boston, von Mises also protested against the decision of Istanbul University not to extend the contract of Hilda Geiringer, with whom he had closely collaborated beginning in the 1920s in Berlin. Geiringer, who married von Mises in 1943 and helped edit his collected papers after his death in 1953, pointed to the "human side of the emancipation of applied mathematics" in the first half of the twentieth century.[41]

Her professional journey also uncovers the hardship of forced migration and the manifold problems that experts had adjusting to different cultural environments. It is fair to say, however, that experts were socially privileged compared to other refugees and migrants because of the intellectual capital they offered and the professional networks they enjoyed.

Historians have emphasized that "forced migration made possible careers that could not have happened in the smaller, more restrictive university and science systems of Central Europe," and that the pressure to respond to new circumstances led to innovations that might not have otherwise occurred.[42] In fact, for the recipient states, the unintended consequences of forced migration meant opening up new scholarly horizons, providing new technical opportunities, and developing new social networks, as the case of Turkey best exemplifies. Other receiving countries in Europe that profited from the great migration included Spain and Portugal in the south and Britain and France in the west. Indeed, the transformations resulting from expert migration also fundamentally affected the European center–periphery balance. Moreover, in the long run the migration strengthened the global bonds of knowledge. With European experts working in Argentina, Egypt, India, or Australia, European ways of knowing and innovating interacted with non-European knowledge traditions, often enabling cross-fertilization when numerous experts moved to Europe in the decades after the 1940s.

Forced migration of knowledge has long been framed solely in a discourse of loss and gain, Europe depicted as a "spender" transferring intellectual resources to "receivers" in other parts of the globe, with the U.S. in the fore. Such a view is shortsighted, for the great migration of knowledge was much more than a zero-sum game. In the long run, the cross-national transmission and transformation of scientific ideas, technical concepts, and innovation cultures that resulted from forced migration also accelerated denationalization of technoscientific expertise, both in Europe and beyond its borders.[43] Furthermore, one unusual aspect of European history is that forced migration leveled the inter-regional disparities of knowledge in Europe and strengthened its global linkages.

Finally, the great migration of knowledge in the second third of the twentieth century led to a growing social autonomy of experts. In the Introduction we argued that the authority of expertise is largely based on experts' relation to political and social institutions. However, the insatiable demand for technoscientific knowledge in

Fig. 6.7 Forced Orientalism: *In contrast to the emigrants to the Soviet Union, the German architects and other experts going to the young Turkish state had not chosen their new home because of political sympathy. Most had to leave Nazi Germany for racial or political reasons. Yet, Turkey's effort to modernize itself and the self-perception of experts such as Martin Wagner or Ernst Reuter went well together. The picture shows Reuter, who became the German mayor of West Berlin from 1948 to 1953, and his wife Hanna on their return journey to Germany in November 1946.*

interwar, wartime, and postwar Europe put the experts in the driver's seat. The case of the intellectual reparations from Nazi Germany documents experts' expanding bargaining power. The technoscientific specialists who had most actively supported National Socialism, such as Wernher von Braun and his Peenemünde team, managed to secure exceptional social and economic privileges for themselves and their families. Even the thousands of experts forced to leave Germany for the Soviet Union overnight were not only allowed to bring their families and belongings, but also led much more comfortable lives than their Russian colleagues, not to mention the average Soviet citizen—thanks to their Cold War role in brokering knowledge of strategic importance to the Soviet Union. Of course, technoscientific experts continued to have political patrons who often sought their own goals. However, the growing shortage of expertise and the rising political demand for it, along with the secular growth of relevant knowledge and the momentum of publicity, transforming people like von Braun, Ohain, or Taut into celebrities, all helped increase the autonomy of experts in and beyond Europe.

III
Cooperating Experts, Building Institutions

7
Geographies of Cooperation in Nuclear Europe

The European Super Machine

Some said it would become the most important day ever for Knowledge Europe. It would be a day of triumph, showing the world that Europe was able to pool its physical and intellectual resources to look for an answer to the biggest question of natural philosophy: the beginning of all beginnings. Some said it would become the darkest day for Europe, that the whole earth would be sucked up in a cataclysmic apocalypse, the end of all ends. The date was September 10, 2008, and the Large Hadron Collider (LHC), by far the largest machine the world had ever seen, was to be put into operation for the first time. This European super machine is a particle accelerator designed to smash atoms to create subatomic particles that enable physicists to inquire into the dynamics and structure of matter, space, and time. Built by the European Organization for Nuclear Research (CERN), the collider, a synchrotron, lies in a tunnel no less than 27 kilometers in circumference, 50–175 meters beneath the Franco-Swiss border at Meyrin near Geneva. The construction of the super machine engaged over 10,000 scientists and engineers from over one hundred countries. The budget was 7.5 billion euros, to be covered in the final instance by European taxpayers, all with the goal of testing various predictions in high-energy physics.

Why did so many people fear a machine developed to answer the most fundamental question of science? Were they scared by the fact that the LHC is not only the biggest but also the most complex artifact in the world of technoscience? The famous American author Dan Brown set the tone with his bestselling thriller *Angels & Demons*, published in 2000, in which antimatter created at the LHC is used in a weapon directed against the Vatican. The Hollywood version of the book shows footage filmed on-site during an experiment. Thanks to the book and the movie, many people associated the LHC with the forces of evil and danger long before it even went into operation. A far more serious source of concern came from the scientific community itself. The hypothesis of a number of physicists, namely that experiments at the LHC might create black holes, sparked fears of conspiracy and controversy among the public. The physicists were thinking of microscopic black holes that would only exist for a fraction of a millisecond. But that is not what a black hole meant to the general audience. Europeans and others imagined a black hole through which the entire earth would disappear into oblivion. People were unaware that only a black hole of astronomical proportions would be capable of swallowing the earth, the kind that can only result from a cosmic event like the inward collapse of a star much larger than our sun at the end of its life. Two safety reviews commissioned by CERN came to the conclusion that there was "no basis for any conceivable threat" and "nothing to fear from particles created at the LHC."[1] But the public statement of the CERN experts, endorsed by the prestigious American Physical Society, had little effect on public opinion.

In modern societies, public knowledge is shaped by the mass media, and it is often a hit-or-miss affair whether public and expert knowledge are in tune with each other. The two systems operate in fundamentally different modes, the media in the mode of seeking public attention, scientists in a truth-oriented mode. Moreover, in societies even scientific knowledge finds itself in competition with common-sense or religious knowledge, and constantly has to prove its definitional power anew. The physicists at CERN had to learn this lesson the hard way. Once the term "black hole" began circulating, they, as an expert community, lost control of their scientific use of the term.

Fig. 7.1 **Europe's Biggest Machine:** *The Large Hadron Collider allows the collision of opposing particle beams of either protons at the energy of 7 teraelectronvolts (TeV) per particle, or lead nuclei at the energy of 574 TeV per nucleus. Physicists at the CERN use the LHC to test various predictions of high-energy physics, including the existence of the hypothesized Higgs boson. Verifying the Higgs boson in the accelerator's four gigantic particle detectors would provide experimental proof for the so-called Standard Model of particle physics, a highly complex theory to which most physicists and cosmologists adhere. On July 4, 2012, CERN announced that experiments performed by the LHC had discovered a new particle that is consistent with the Higgs boson. The image shows part of the 27 km long LHC tunnel.*

Only nine days after CERN fired the first protons around the tunnel, a faulty electrical connection between two magnets led to a quench (that is, an abrupt shutdown). Six tons of liquid helium flooded in, causing a temperature rise of about 100 degrees kelvin in some of the affected magnets and damaging fifty-three of them. The big machine had to be shut down for examination and repair, leading to over a year's delay in its operational schedule. For CERN physicist Lucio Rossi, who was asked to examine the causes of the incident, the apparently trivial technical fault "revealed a lack of adequate risk analysis...and of understanding all consequences, as well as an incomplete global magnet circuit protection analysis and an inadequate detection of a dangerous situation." The main lesson he drew was that "experts on superconductivity and experts on system integration should work in closer collaboration all along the project."[2] In other words, Rossi noticed that within the LHC project different networks of experts worked in isolation from each other. This was a recipe for technical failure in the program, which, like any complex project of such large magnitude and experimental character, is inherently fragile, uncertain, and risky. The LHC's breakdown allows deep insights into the fundamental openness of technoscientific knowledge and the fragility of expert networks.[3]

Fig. 7.2 Art Meets Science: *The Large Hadron Collider has become an icon for Big Science in public discourse. The LHC has also sparked a multitude of projects in art and music. This picture shows an installation by the Russian artist Nikolay Polissky, displayed in 2009 in the Grand Hall of Mudam Luxembourg – Musée d 'Art Moderne Grand-Duc Jean. The installation underlines LHC's fragility as a technoscientific artifact.*

In this regard, European expert networks in nuclear physics and nuclear technology do not differ from expert networks in any other region of the world. However, certain aspects of European nuclear research are unique and make it a particularly suitable focus for this chapter. Here we will consider the complex interplay between global connectedness, national embedding, and Nuclear Europe, and will start by referring, once again, to CERN.

Europe's Model of Technoscientific Collaboration

It would be misleading to narrate the history of CERN as a series of misfortunes. On the contrary: apart from the LHC's start-up problems scientists, politicians, and the public alike view the European Center for Nuclear Resarch as a guiding light in knowledge production, epitomizing the best that Europe has to show. CERN materialized a European collaboration of global significance. It is seen to be motivated by pure curiosity, by the urge to produce fundamental knowledge about the Universe. The project is eminently scientific, detached from political interference, vested national interests or ostensible economic applications. At the same time it is paving the way for "future innovation and prosperity." This was most spectacularly demonstrated in 1989: the CERN employee Tim Berners-Lee invented the World Wide Web, when responding to the need to develop an automated information-sharing system for 8,000 CERN-related scientists working in different universities and institutes all over the globe.[4]

Since its official foundation in 1954, CERN has served as a point of reference for practically every effort to institutionalize large-scale technoscientific collaborations in Europe and elsewhere in the world. From an institution-oriented perspective, CERN thus presents itself as a role model for Big Science in Europe; its characteristic features have migrated, spurring the formation of related sites in Knowledge Europe. From a historical perspective, the organization's role appears rather different. Historians of CERN describe it in less idealistic terms. They characterize it as a unique institution that emerged from a particular set of circumstances in postwar Europe, a constellation seized on by a small group of actors in the scientific realm to colonize a particular space and fashion it to meet their specific ideas and needs.[5] This space—the governance of transnational science and technology—had been left largely untouched by political and economic actors after the Second World War. In a way, governments' transnational science and technology policy emerged only in response to CERN.

If CERN is truly unique, then it can hardly serve as a role model for other institutions. Yet it is considered both to be and do so. It is this ambivalence that makes the history of CERN so relevant for a better understanding of European technoscientific expertise. This history's main outline has been recounted by others at length and it will not be repeated here.[6] We will investigate CERN to test some crucial notions about building Europe on expertise in the postwar era, particularly a recently published claim that familiarity with the basic principles of science was part and parcel of European identity in the postwar years.[7]

The stage was set for creating CERN in Lausanne (December 1949) and in Florence the following June. At a Lausanne Cultural Conference, a small group of European physicists, led by the French science administrator Raoul Dautry, resolved to create a European center for atomic research. The prominent American nuclear physicist and Nobel laureate Isidor I. Rabi seconded this proposal half a year later, when UNESCO met in Florence for its annual conference. At that time UNESCO, the United Nations Educational, Scientific and Cultural Organization, was developing into a crucial institution not only for educational and cultural projects but also for initiatives to foster international scientific collaboration. Rabi had been a key figure in creating the Brookhaven National Laboratory on Long Island (near New York City) in 1947, which quickly developed into one of the leading centers in the world for particle accelerators and high-energy physics. As a member of the U.S. delegation, he proposed creating regional European laboratories equivalent to Brookhaven, and the participating nations adopted this approach. In December 1950, Dautry and Pierre Auger, a French expert in cosmic ray physics, who also served as director of UNESCO's Department of Exact and Natural Sciences, organized a meeting of scientists and state administrators in Geneva. Those attending proposed that the joint European initiative should be dedicated to the construction of the world's biggest accelerator. They decided against building a reactor, to avoid the complications that would ensue if the project had military or industrial potential. Further meetings refined the plan, leading to creation of a temporary organization in February 1952. Finally, eleven Western European states signed the convention for a permanent organization. Among them were Britain, which had long been dragging its heels, and the non-aligned Yugoslavia. On October 7, 1954, the now-permanent CERN Council met for the first time; construction of a laboratory in Meyrin, just outside Geneva, was already under way.

CERN came into being as a response to the Second World War, but was equally a product of the Cold War. The War had engendered a dramatically growing infrastructure in nuclear research. The Americans' gigantic Manhattan Project—the initiative which resulted in the two atomic bombs that destroyed Hiroshima and Nagasaki—required 10,000 experts to resolve numerous scientific and technical problems related to developing nuclear weapons. The researchers needed particle accelerators to answer basic questions about the dynamics and structure of matter, space and time. In Europe, too, the Second World War had triggered the development of accelerators at numerous laboratories for atomic research. Thus, the burgeoning field of high-energy physics was based largely on accelerator projects, a research domain that continued to grow exponentially in size and budget.

In another respect, as well, CERN can be considered a response to the Second World War. In the wake of that catastrophe, most scientists in Europe felt that fundamental research should be freed from political interference and military interests. After all, technoscientific experts in practically every war-waging country had actively engaged in efforts to ensure victory. As a reciprocal benefit of devoting their skills to their nations' military-industrial-scientific complexes, they had obtained elevated social status and a prominent profile for their research fields. In the postwar period, many experts came to regard the military entanglement of their fields as a trauma, and sought to strengthen the autonomy of science from political and military interference. Basic science, in particular, should be controlled by the expert communities themselves. Transnational cooperation perfectly suited this vision of a "civilized"—and civilian—scientific community, uncorrupted by military or industrial interests.

Notwithstanding the institution's ubiquitous peaceful rhetoric and its self-positioning in a fundamentally apolitical realm, CERN was in many ways also a product of the Cold War—a compound of European collaboration and American hegemony, a "coproduced instrument of European and American political interests."[8] We will extend this portrait of the intricate interlacing of CERN and the Cold War by focusing on the institution's relation to Germany, Switzerland, and the Soviet Union. This multidimensional perspective will show the ways in which the first major transnational collaboration of postwar Europe experts, which quickly developed into a role model for Knowledge Europe, was connected to the vested interests of particular nations and particular interest groups within these nations.

If there was one person in postwar West Germany who could claim to stand and speak in the name of science in political and public affairs, it was Werner Heisenberg. The Nobel Laureate in physics (1932) for the creation of quantum mechanics had been heavily engaged in Germany's uranium project during the Hitler years, but after the war "had no diffidence at all about speaking of expertise carried over from the Nazi period."[9] In the early 1950s he morphed into a science advisor to Chancellor Konrad Adenauer and an informal ambassador in scientific affairs. Despite Allied restrictions on nuclear science and technology, which officially came to an end only in 1955, Heisenberg the *éminence grise* chose atomic physics as a test bed to boost a forceful role for the West German government in research policy. A strong nuclear program would enable a leap into a bright scientific, technical, and economic future. For Heisenberg, CERN was part and parcel of this agenda. By taking on a leading role in CERN, German nuclear science and technology would shed its irksome Nazi heritage and profit from a European collaboration in basic science for peaceful ends. CERN helped Germany to become reaccepted internationally; Heisenberg helped CERN to get off the ground and to overcome both administrative and budgetary crises in its foundational period. CERN, the "new common European enterprise," would catalyze a "spirit of cooperation and mutual understanding" within and "outside of the field of physics," Heisenberg explained to an American audience in 1953.[10] His attitude towards European collaboration signified a broad consensus among German experts and policy-makers alike. Strengthening Western Europe was both a means to overcome the threatening techno-scientific dominance of the superpowers and to justify rapidly growing domestic research expenditures. Furthermore, CERN would support training a young generation of German atomic physicists who would be internationally competitive. European and national interests were inextricably interwoven.

While most researchers shared Heisenberg's enthusiasm for CERN, some challenged his self-proclaimed leadership. The Heidelberg experimental physicist Walther Bothe, who received the Nobel Prize in physics in 1954, and Wolfgang Gentner, who succeeded Bothe as director of the Max Planck Institute in Heidelberg in 1958, observed that Heisenberg as a theorist continued his usual game, speaking for all of physics, if not for all of science and technology. Decisions related to experimental physics, they argued, should not be taken "by the theorists at the green table."[11] Bothe and Gentner advocated that Germany should take on a major role in the construction of CERN accelerators; in fact, Gentner headed both the 600 MeV (megaelectronvolt) synchrocyclotron team and a much larger project, the 28 GeV (gigaelectronvolt) proton synchrotron. He even served as scientific director of CERN from 1955 to 1959.

When the big synchrotron went into operation in 1959 and CERN started to conceptualize the next generation of accelerators, the German consensus about the nature of their participation in future European physics programs collapsed. The hunger of high energy physics for ever-bigger machines and new centers forced experts to prioritize their plans and ideas. While Heisenberg opted for a small set of basically domestic projects and questioned the necessity of building the 300 GeV accelerator being planned at CERN, Gentner favored a double strategy of building national centers and fully cooperating in the future European "monster machine."[12] Gentner participated actively in the Scientific Policy Committee of CERN and fought for siting the new European accelerator in Germany. He was convinced that transnational techno-scientific collaboration would eventually lead to European integration and even to the end of the Cold War. He interpreted the collaboration between CERN and the Soviet nuclear research center in Dubna—which flourished in large part due to his efforts—as a clear indication of the transformative repercussions of expert cultures on political affairs. To him, Europe was not a mirage, but an actual entity that resulted from "true collaboration" in transnational centers such as CERN.[13] In the end, the 300 GeV accelerator was not built in Germany but located in Meyrin at CERN's main site. Germany, however, continued to drive European projects in nuclear science and technology.

From the start CERN provided West Germany with a welcome opportunity to reinforce its integration into the West by means of technoscientific cooperation. Less obvious was that CERN supplied Switzerland with the no-less-welcome means to strengthen its proclamations of neutrality, the central pillar of the Swiss national identity. Much like Sweden, the country's neutral stance came under severe attack from abroad after the war. The U.S. government, in particular, heavily criticized Swiss neutrality "as a cover-up for its sustained commercial relationship with Nazi Germany and as an excuse for engaging in trade with communist countries."[14] In order for Switzerland to participate actively in professedly apolitical technoscientific efforts, its neutrality had to be redefined in more positive terms. A counterbalance was required to compensate for the country's pointed refusal to join such intergovernmental bodies as the United Nations, the European Council, and the European Coal and Steel Community. When the majority of Swiss experts strongly advocated that CERN should exclude any and all military issues and be open to states from the Soviet bloc, they had the government fully on their side. Alfred Picot, a Geneva politician and member of the Swiss delegation to the 1951 UNESCO conference that paved the way for CERN, came up with a suggestive phrase: the new institution, he said, should be a "glasshouse." By avoiding classified research, CERN was to be politically open to every country and scientifically transparent.[15]

Others quickly adopted the glasshouse metaphor and it became a touchstone in the discourse about nuclear research at CERN, in Switzerland, and beyond. CERN manifests the co-construction of two nonpartisan arenas: the apolitically 'neutral' field of nuclear science and technology and the politically neutral country chosen to host its laboratories.

Neutrality was surely not the objective of the Joint Institute for Nuclear Research (JINR), the Soviet response to CERN, located in Dubna, north of Moscow. In March 1956, representatives of eleven "fraternal" countries accepted a Soviet invitation to participate in a nuclear research institution and signed an official agreement to that effect in Moscow. The Institute's member states included most nations from the Soviet bloc as well as China, North Korea, Vietnam, and Mongolia. The director of Dubna, Vladimir Veksler, was supported by the Chinese founding member Wang Ganchang, a brilliant expert in cosmic-ray research and particle physics. In 1958, Wang, who later masterminded the Chinese nuclear deterrent program, was elevated to the rank of vice-director, advertising its transnational character.

The creation of Dubna was neither sudden nor surprising. On the contrary, it sprouted from a long tradition of excellence in Russian and Soviet physics. Large-scale technoscientific building blocks, with institutes of physics taking center stage, were already in place during the First World War and the Russian Civil War, leading one historian of science to speak of an early socialist variant of Big Science.[16] In spite of Stalinist antipathy toward the intelligentsia and massive losses of expertise and intellect incurred through purges, Soviet research and innovation systems expanded in the interwar period and got a big push after 1941, when Soviet leaders mobilized the country's technoscientific capacities for the war effort. As the Cold War took shape, nuclear research became a high priority, with the aim of catching up with the U.S. in atomic fission and taking the lead in fusion research.[17]

Although the Joint Institute (JINR) was an integral part of the Soviet aspiration to hegemony in nuclear energy, the outcomes were mixed. On the one hand, it gave Soviet satellite states in Eastern Europe access to highly advanced research technologies. A 70 GeV synchrocyclotron became operational at Dubna as early as 1949.

Construction of a new synchrotron followed, in direct competition with the 28 GeV proton accelerator at CERN. The Joint Institute developed also into a giant dissertation factory. At any given moment, it had several hundred doctoral candidates from abroad in training. By the-mid 1970s, more than 3,000 East European nuclear experts had received their education in the Soviet Union, while about 1,000 Soviet specialists had been sent to work in Eastern Europe.[18] This pattern of knowledge transfer allowed Soviet leaders to oversee the nuclear research programs of their allies. In the Warsaw Pact, the flow of experts and technical knowledge was never open and uncontrolled. Quite the opposite: Soviet leaders were very reluctant to deliver nuclear material to their "allies." This attitude was reinforced after the political split with China in 1960, when Khrushchev withdrew all nuclear experts from the People's Republic. The conditions governing the transfer of immaterial knowledge and material substances were spelled out in bilateral agreements that left the satellite states with little room for bargaining.

Thus, the Eastern mode of nuclear collaboration shows patterns of hegemony, colonialism, and imperialism. In general, these characteristics have been regarded as a consequence of Soviet power politics. This may be too limited a view. As an in-depth comparison between the nuclear sectors of Czechoslovakia and East Germany has recently shown, Soviet politics oscillated between rigid control and offers for open collaboration. Changes in Soviet politics opened maneuvering space for Eastern European countries, so that different satellite states followed significantly different paths. The East German nuclear program "struggled almost from the outset, and eventually stalled completely in the 1980s." Czechoslovakia, by contrast, developed a unique nuclear reactor design and "became the monopolist producer of Soviet-designed pressurized water reactors for the entire Soviet bloc."[19]

After the Cold War, the Joint Institute gradually stripped away its imperial roots. The Russian government managed to reach agreements with Western European countries like Germany and Italy. Today, the Joint Institute presents itself as a hub in the global network of nuclear research, collaborating with some 700 institutions in about 60 countries. It is a prime partner of its former competitor, CERN. With its neutron reactor IBR-2, JINR is fully integrated into the European long-term program of neutron scattering research. At long last, the Russian center has become a cornerstone of the envisioned European Research Area.[20] Thus, even the evolution of the originally Stalinist Joint Institute affirms the self-constructed master narrative of the experts: it is once again possible to maintain in all seriousness that European collaboration in high-energy physics and particle research has always been oriented to purely scientific ends, that non-scientific actors have been kept at arm's length and that no political agendas have been followed. Although we have dismantled this narrative as a historical myth, we must emphasize nevertheless that this foundation tale of CERN was very influential. The ostensibly apolitical nature of the center in Meyrin made it available as a role model for varied institutionalized collaborations in Knowledge Europe.

While scientists and science policy have constantly stressed CERN's significance as a role model, science and technology studies have focused on the specifics of its knowledge practices. In this view, CERN has functioned as a "superorganism" that fuses experts, laboratories, and large-scale projects into community-like organized collaborations.[21] Individual experts sacrifice their identity for the good of joint knowledge production. The communitarian mode of knowledge production is encouraged by the fact that the laboratory stands on its own transnational, European, territory. The extreme degree to which experts at CERN can be divorced from ownership of their research was made evident when the first paper related to the Large Hadron Collider was published. This fundamental article, proving that a larger than expected number of mesons were produced during the first particle collisions, was signed by more than 2,200 authors, a new world record in multi-authorship.[22] Ironically, in this situation, "the only people qualified to truly review the work are within the collaboration."[23] In fact, CERN and its research culture are in some ways unique, in that the center has always worked at the cutting edge of Big Science. When it comes to the relationship between experts and society, the center is typical for complex knowledge societies. While knowledge production in fields such as high-energy physics has become ever more arcane, politicians and the public have continued to place unqualified trust in its validity and societal relevance.[24]

Ways of Collaborating in Nuclear Europe

Experts from all corners of the Soviet empire had feverishly prepared for the conference for over a month. They had come together to hold a trial before the best of the delegates. This was followed by rehearsals and dry runs. The Soviet Union was well prepared to launch an impressive propaganda coup at Geneva.

Home of the former League of Nations and now of a multitude of UN organizations, the picturesque—and terribly expensive—city on Lake Geneva in neutral Switzerland, with CERN close by, hosted the first International Conference on the Peaceful Uses of Atomic Energy. The conference took place August 8–20, 1955 under the auspices of the United Nations. Some 1,500 nuclear experts from all over the world participated. Pictures of the nuclear reactors on display quickly became icons both of the Atomic Age and the possibility of global collaboration despite the Cold War.

Stalin's death in March 1953 paved the way for this landmark gathering. In Stalinist Russia, scientists and engineers had been treated badly, but under Nikita Khrushchev they underwent rehabilitation, achieving special prominence when representing the glory of the regime abroad. Vladimir Veksler, whom we have already met as co-founder of Dubna, was quoted saying that the meeting was "not only the first international conference in the field of physics," but also "a conference of scientists unique in history." The Soviet nuclear specialist predicted that the Geneva meeting's academic openness would strengthen "the atmosphere of mutual understanding and good will."[25]

Having prepared so intensively, the Soviets indeed performed very well. Their delegation of seventy-eight persons—including a substantial number of KGB staffers with a different expertise—delivered no fewer than 102 papers out of the total of 1,067 presentations. Western scientists were quite impressed both by the quality and the candidness of the Soviet contributions. So were the politicians. The British government invited the Soviet delegation to its nuclear research center at Harwell immediately after Geneva. At Harwell, Nobel laureate John Cockroft welcomed a "surprise guest," Nikita Khrushchev.[26] In return, the Soviets allowed Western experts glimpses of their once hermetically sealed nuclear research complex, including the famed Institute of Atomic Energy in Moscow, led by renowned physicist Igor V. Kurchatov.

At the Second Conference on the Peaceful Uses of Atomic Energy in Geneva (September 1958), the Soviets again surprised their Western colleagues with the latest news on their achievements, including a report on the *Lenin*, the world's first nuclear-propelled icebreaker. They even honored a promise made by Kurchatov at Harwell to declassify research on nuclear fusion. The head of the Soviet delegation, Vasilii Semenovich Emelianov, presented a collection of 100 articles in four volumes on the theoretical and experimental physics of fusion. Khrushchev's visit to Washington one year later expanded these contacts by enabling intensive mutual exchange of nuclear knowledge between American and Soviet specialists. What until then had been the most secret item of national security now traveled openly through the Iron Curtain: the knowledge basic to the development of nuclear weapons, even hydrogen bombs.

This global political climate of nuclear détente allowed the emergence of European collaboration in the atomic sector. The project could not have come into being without the acquiescence, indeed the active support, of both superpowers. This was forthcoming, and Europe launched a transnational nuclear program: the European Atomic Energy Community, in short Euratom, established on March 25, 1957, simultaneously with the European Economic Community (EEC), by the Treaty of Rome. Like the latter, Euratom had the "inner six" states of European integration as founding members: Belgium, France, West Germany, Italy, Luxembourg, and the Netherlands. Jean Monnet, the mover and shaker of European integration, had dreamt of fostering integration by ensuring transnational collaboration. Now his dream came true for the strategic area of nuclear technology. Both "parents" of Euratom, the U.S. and the Soviet Union, shared an interest in stimulating technoscientific internationalism. For both, "a demonstration of scientific and technical generosity and prowess on the international stage was intended to win the hearts and minds and to confirm the legitimacy or even the superiority of rival politico-economic systems."[27]

While the Soviet Union merely tolerated the emergence of Euratom, the U.S. enthusiastically supported the new institution. President Eisenhower had set the tone with his famous speech before the UN General Assembly on 8 December 1953, kicking off the Atoms for Peace program highlighted in the two Geneva conferences. Atoms for Peace had more than one meaning. By enabling access to formerly restricted nuclear knowledge, it undeniably advanced reactor technology. At the same time, it was part and parcel of a policy to solidify a U.S. global hegemony by fostering scientific and technical collaboration in Europe. Euratom represented an extension of this policy and quickly developed into a major "cornerstone of Eisenhower's grand design for a United States of Europe."[28] The

Fig 7.3 Displaying Nuclear Modernity: *The first International Conference on the Peaceful Uses of Atomic Energy in Geneva in August 1955 was a scientific gathering of nuclear experts from all over the world. For the U.S. Government the meeting was also an opportunity to promote both its "Atoms for Peace" ideology and its industrial technology. The Geneva public show of nuclear information and equipment became a symbol of nuclear modernity. Visitors such as those shown by this image were particularly fascinated by the characteristic blue glow of nuclear reactors due to the Cherenkov radiation that they could observe in the reactor core.*

State Department supported the idea of Euratom enthusiastically, hoping that a transnational program would knit the European allies closely together, distract them from pursuing autonomous national programs, and secure a technical and economic lead for the U.S. in the nuclear field. That was the idea behind the so-called Joint Program between the U.S. and Euratom that Congress ratified in August 1958. The first aim of the program was that by 1962/63, Euratom would construct several nuclear power plants designed

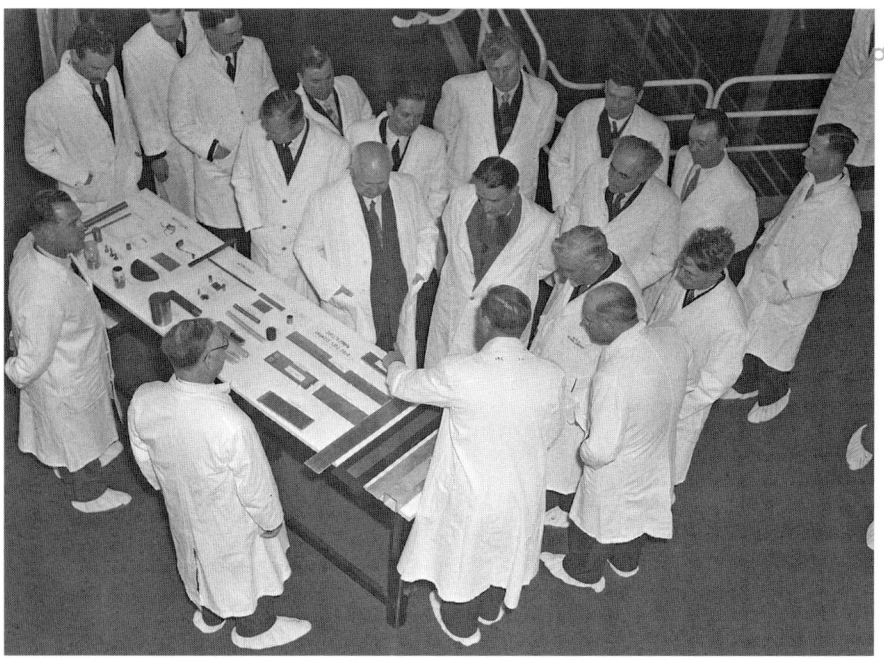

Fig. 7.4 **Nuclear Experts as Peacemakers:** *The picture shows Igor V. Kurchatov (in the middle, with beard), director of the Soviet atomic bomb project, during his visit to Harwell, Britain's Atomic Energy Research Establishment, on April 25, 1956; on his right is Nikita S. Khrushchev, to his left is Prime Minister Nikolai A. Bulganin, opposite is Sir John D. Cockcroft (gesturing), director of Harwell. Kurchatov's openness and deep insight into the problems of controlled thermonuclear fusion paved the way for a technoscientific détente by declassifying fusion research in the midst of the Cold War.*

after U.S. models and together capable of generating 1,000 megawatts of electricity. The second objective was to launch a ten-year collaborative research and development program to improve knowledge of reactor technology and to lower "fuel cycle costs."[29]

The American government had good reason to keep a sharp eye on the European nuclear scene. Across most of Western Europe, national programs to build autonomous nuclear capacity developed during the postwar years. Drawing on the expertise of the British-Canadian team from the Manhattan Project, Britain first challenged U.S. nuclear hegemony. In the mid-1950s, the U.K. nuclear industry was still technically and commercially in the lead, outperforming U.S. on the world market. France followed suit. The first French nuclear reactor went critical—that is, started operation—in 1948, allowing the extraction of small amounts of plutonium. Plans to build reactors for the large-scale production of plutonium followed, and in 1954 the French government formally launched a program to build atomic bombs. Eventually, after Charles de Gaulle returned to power in 1958, the program solidified into the concept of a fully independent *Force de Frappe* capable of protecting the nation from foreign attacks. For postwar France, nuclear technology and national identity became mutually shaping forces.[30]

Interestingly enough, it was not only France, Britain, and Germany, with their advanced knowledge base in nuclear science and technology that embarked on national nuclear programs. Smaller countries too, like Norway, Switzerland, Sweden, Finland, Austria, and the Netherlands, not to mention Italy and Spain, developed autonomous nuclear programs. In early postwar Europe, the intricate connectivity of industrial and security issues, civilian purposes and military interests favored the pursuit of national programs. Nation-building thus was very much shaped by nuclear technopolitics. The nation state was the focus of discourse on nuclear projects, after all.

Across the 1950s, European actors gradually realized that building nuclear facilities tended to overstretch national capacities. One way out of the dilemma was transnational collaboration and Europeanizing formerly national programs. A case in point is Belgium. This interesting story reveals the great significance of the colonial and postcolonial dimension of technopolitics in the nuclear realm, with roots going back to the Second World War. The uranium for the Manhattan Project and its devastating atomic bombs had come from the Belgian-Congolese company *Union Minière du Haut Katanga*. In 1944, the Belgian government in exile officially backed the deliveries from its Congolese colony in a tripartite pact with the U.S. and U.K., hoping to get in return privileged access to scientific information. The one-sided agreement, which was renewed in 1955 under conditions only slightly more favorable to the Belgian partner, was negotiated and signed by Paul-Henri Spaak. Two years later Spaak, now Belgian Foreign Minister, informed the U.S. government that starting in 1960 Belgium would send all its uranium ore to Euratom. The Benelux country was free to do so according to the agreement of 1955. This political turnabout reflects Belgian disappointment with the U.S.: America had denied it a privileged nuclear partnership. The pioneering role taken on by Belgium—in particular by Spaak—to foster Euratom reflected his nation's aim to build a nuclear power plant. The reactor that supplied power to the World's Fair in Brussels in 1958 sought to demonstrate that Belgium was capable of leapfrogging into a high-tech future.

Belgium founded its national Research Centre for the Applications of Nuclear Energy at Mol, east of Antwerp, quite early on, in 1952. The Centre quickly developed into a large-scale laboratory for reactor research. The first two experimental projects carried out there—BR1 and BR2—were performed on a national basis. BR3 started as a national project too, but later was Europeanized. BR3 was the first pressurized water reactor in the world, which added to its significance. The turn to Europe in the Belgian nuclear narrative converted the uranium agreements of 1944 and 1955 from a "fool's bargain" into a "diplomatic masterpiece."[31]

Euratom raised high expectations beyond Belgium. It had proponents on both sides of the Atlantic, who expected it to ensure the Continent's energy future and to foster European integration. They looked to technoscientific collaboration as a building block for Knowledge Europe. Euratom's foremost research site was located in Ispra on the banks of Lake Maggiore, northwest of Milan. In 1960 the European Commission proudly presented Ispra as "one of the largest building sites in Europe" and a center that eventually would employ "several thousand technicians and specialists." Indeed, in

Fig. 7.5 Visualizing the Atomic Age: *The 102 meter high Atomium became an icon for the World's Fair in Brussels in 1958 and a symbol both of modern architecture and the atomic age. Half a century later, it even lent its name to a joint initiative of Europe's leading universities, newspapers, and businesses, when in November 2009, the former President of France Valéry Giscard d'Estaing launched "Atomium Culture" to promote transnational knowledge sharing, to increase public interest in science and technology, and to prevent a brain drain from Europe.*

1967, at the end of its first decade, Euratom had about 3,200 scientific and technical personnel.[32] With four Joint Research Centers in full swing, of which the vibrant Ispra was only one, Euratom seemed well on the way to fulfilling its challenging tasks. The euphoria did not last long. Only a few years later, great expectations gave way to great disappointment, again on both sides of the Atlantic. From the U.S. perspective, Euratom had failed with respect to both of its anticipated functions. Because Euratom did not knit together the nuclear capacities of America's European allies, it neither blocked the nuclear projects of individual nations nor controlled the proliferation of radioactive material that could be used to build nuclear weapons. The U.S. government concluded that Euratom was "a pale shadow of its original grand design."[33]

The European Commission was frustrated for another reason. Europe had spent even more than the U.S. as a proportion of the respective countries' gross national products. But most of the large-scale investments in research and development for civilian nuclear research had gone to national programs, not coordinated at the European level. In the first decade, Euratom's six member states had allocated $650 million to two five-year research programs. That grand total was no larger than those countries' aggregate spending on their national programs *in the year 1967 alone*. In particular, France, which provided no less than two-thirds of Euratom members' overall expenditures, maintained a strong national program, channeling only about 8 percent of its nuclear budget to Europe. While the joint efforts had led to positive research results, they had been "disappointing in sectors with more immediate industrial applications." The prominent French writer Francis Gérard seconded this judgement when he concluded in 1969 that European nuclear cooperation was intergovernmental rather than supranational and continued to depend "on the good faith of the national governments."[34]

Even in Euratom's initial period, the idea of a permanent joint organization for nuclear research and technology met with more criticism, heel-dragging, and outright opposition than enthusiasm. There was a vast gap between the rhetoric propagated by Euratom's founders and the actual situation. Its father, Jean Monnet, depicted the nuclear sector as virgin territory, "uncontaminated by the protectionist spirit of established industries," a field in which the participating countries would necessarily be dependent on international collaboration.³⁵ But the opposite was closer to the truth. Euratom found itself in competition from the start with powerful vested national interests. Britain, in particular, as the leading nuclear nation in Europe, was never willing to share its technology with the Continent. Most of the six member states ratified the Euratom agreement only after fierce political debates. In the Netherlands, for example, some members of the government favored a continuation of bilateral collaboration with Norway. This arrangement dated from March 1951, when the Institute of Atomic Energy of Norway and the Dutch and Norwegian Research Councils had agreed to establish Jener, the Joint Establishment for Nuclear Energy Research. The aim of the project was to build a research reactor based on uranium oxide and heavy water in Kjellen, Norway. The Dutch cabinet decided to join Euratom only after a long and heated debate. The argument that finally brought the reluctant Dutch Minister of Foreign Affairs, Johan Willem Beyen, to accept Euratom was that the success of European integration in general depended on this program. Nuclear energy, proponents maintained, would serve as the "carrot required to keep the donkey moving."³⁶

Time did not cure Euratom's ills. The longer the organization was in business, the more pronounced the conflict of interests between the "big two"—France and West Germany—became. They were simply not prepared to reduce their strong national programs to accommodate Euratom and the other member states. The less-advanced countries responded by inaugurating the principle of *juste retour* (fair return), demanding that their financial investments in the supranational program should be equaled by contracts for their national research laboratories and industries. Having no major national programs of their own, they were no longer willing to subsidize the activities of France and West Germany, who unsurprisingly argued that all members profited equally from the shared expert knowledge produced in the Joint Research Centers.

Euratom was by no means the only avenue of transnational European collaboration in atomic research and industry. The map of nuclear Europe was a dense network of border-crossing connections, most often bilateral or multilateral in nature. Some of these ties were transient, like the secret Franco–Italian–German project to build an isotope separation plant. The plan emerged from the Entente Cordiale between the French Premier Guy Mollet and the

German Chancellor Konrad Adenauer.[37] Its main aim was to counter the vaunted British supremacy in nuclear Europe; a secondary goal was to mutually reinforce the three partners' national nuclear programs. The short-lived trilateral project illustrates an important point: European technoscientific cooperation could serve seemingly contradictory ends at the same time. More often than not, the national and transnational dimensions of collaboration were inextricably entangled. When General de Gaulle returned to the Élysée in 1958, he immediately canceled the tripartite scheme in favor of collaborating with Britain and the U.S.

Another transient project of crucial importance to one of the partners was a Spanish–German collaboration. In the first postwar decade, when Fascist Spain was internationally isolated and West Germany was bound by an Allied ban on nuclear technology, the two outsiders entered into secret collaboration. The Franco regime managed to mobilize its old boys' network to gain technoscientific advice from Germany, while a number of Nazi atomic scientists went into exile in Spain to work for the Junta de Energía Nuclear. Werner Heisenberg and Carl Wirtz trained a whole cohort of nuclear experts at the famous Max Planck Institute of Physics in Göttingen. Moreover, Spanish military intelligence was able to recruit German experts to work out long-term plans for a nuclear research center and a nuclear power plant. The German firms *Degussa* and NUKEM even illegally exported highly sensitive uranium material to Spain.[38] The collaboration ended in the mid-1950s, when the Allies lifted restrictions on German nuclear research and development and West Germany sought to become a reliable partner of non-communist nations.

Other connections, more enduring and supranational in character, commenced around the same time as Euratom, if not earlier. Two of these collaborations, initiated by scientists and developed according to technoscientific principles, are of particular importance here. The European Atomic Energy Society (EAES) was launched during a conference at Kjeller's joint Norwegian–Dutch nuclear research center in August 1953. Eighty nuclear experts from nearly 20 countries met not only to discuss the technical problems involved in designing heavy-water reactors but also to plan an organization for nuclear research and technology. After the meeting, the statutes of the Society were worked out by Gunnar Randers, the center's director, and John Cockroft, founder of the

Fig. 7.6 **Nuclear Experts at Work:** *The Norwegian–Dutch Nuclear Establishment for Energy Research (Jener) was founded in 1951 at Kjeller, Norway, to build a joint reactor, based on Norwegian heavy water and Dutch uranium. The image shows nuclear experts supervising the reactor, which went critical in June 1951 and became the first reactor outside the major five atomic nations of the time (Britain, Canada, France, the Soviet Union, and the U.S.). In 1953, Jener was the birthplace of the European Atomic Energy Society, when eighty experts gathered at Kjeller for the first ever open international nuclear conference.*

British Atomic Energy Research Establishment. The EAES, founded in June 1954 in London, with Cockroft as its first president, deliberately avoided being an intergovernmental agency. Instead, it was constituted as a forum for the atomic energy organizations of the eight founding member states—Belgium, France, Italy, the Netherlands, Norway, Sweden, Switzerland, and Britain. Associations in Austria, Denmark, West Germany, Portugal, and Spain joined later in the 1950s. The Society developed into a European clearing house, openly to discuss and freely to disseminate nuclear knowledge, to standardize nuclear nomenclature and technical symbols, and to advise governments on technical matters. These were, in Cockroft's view, "precious components which other international organizations could not possibly provide."[39] With glasnost and perestroika, the Society expanded both its membership and its activities into Eastern Europe. An (inadvertent) Eastern European contribution of another kind was the Chernobyl disaster of 1986, which gave European expert networks much food for thought.

Addressing the consequences of fatal nuclear accidents came to be a task of the European Nuclear Energy Agency (ENEA), a challenge unforeseen by its founders. Although it was set up by the political arm of European cooperation, the Agency was basically an expert network. In December 1957, after years of deliberation, the Organisation for European Economic Cooperation (OEEC) decided to create an expert forum for coordinating the member countries' national nuclear programs, particularly in health, safety, and regulation. The Agency went through a number of fundamental changes in aim and function and finally also in name, when in 1972, acknowledging its growing non-European membership, it dropped "Europe" from its title to become simply the Nuclear Energy Agency. The Agency's ambitious research and development program was based on three pillars. A joint project with Norway aimed at operating an experimental boiling heavy-water reactor in Halden, Norway. The project achieved remarkable momentum: in 2008, after 50 years of nuclear knowledge production, it was again extended and is still up and running. While Halden involved seven European states, the Dragon project managed to recruit twelve countries. Dragon was an experimental 20 megawatt high-temperature, gas-cooled reactor. It went critical in 1964 and became a resource for research into high-temperature reactors both for technical purposes and industrial applications. The Agency's third and probably most distinctive project was the European Company for the Chemical Processing of Irradiated Fuels, in short Eurochemic, established in July 1959 by thirteen European countries as an international shareholding enterprise. With Eurochemic, nuclear Europe started to investigate the nuclear waste cycle. After performing a wide range of research and training activities in nuclear chemistry in its first phase, Eurochemic next focused on radioactive waste management programs. Significantly, the American government kept close watch on its activities and "provided assistance" to the company in reprocessing spent nuclear fuel.[40] For the U.S., nuclear waste management was too sensitive an area to let the European allies go their own way unobserved. Consequently, the U.S. joined the Agency in 1976 as its twenty-third member, right after Canada. The enterprise came to an end in 1990, when the company was liquidated.

In the course of the 1970s, the Agency's initial research functions were absorbed by national programs and commercial activities. But the Agency was not left empty-handed. Its network of experts found new territories of knowledge to research. The nuclear catastrophes at Three Mile Island and Chernobyl created intense challenges. This involved not only laboratory research but also improved techniques for communicating with the public about nuclear energy. Following the fall of the Iron Curtain, new possibilities opened up for collaboration with colleagues in Eastern Europe. In the later 1990s the relation of Nuclear technology to sustainability became the hot issue it remains today. The constantly changing institutional setting of Nuclear Europe demanded an attitude of "pragmatism and flexibility." Today, the Agency understands itself as a transnational center of excellence which, through its network of some 3,000 national experts, is "capable of pooling and maintaining expertise."[41]

Euratom has been described as an institution run by specialists who combined technoscientific "expertise with a commitment for Europe."[42] Indeed, Euratom, as well as its two sister organizations, the Society and the Agency, followed the European model of technocratic internationalism. In this approach, pride of place was given to experts who defined the institutional agenda in terms of their technoscientific orientation. This is why Euratom worked rather well when it came to fostering transnational nuclear research and engineering, but was basically unable to coordinate European nuclear policy and to foster industrial development. This has led some historians to conclude that Euratom was a "debacle," an institution "in crisis," an organization that failed "in virtually every respect," and in the final analysis a *gescheiterte Gemeinschaft* (failed community).[43] But such a linear view, which views Euratom exclusively in the light of the grandiose expectations of enthusiasts like Jean Monnet in the 1950s, misses a number of basic points. It overlooks Euratom's dynamic adaptation to the changing face of Nuclear Europe. It also fails to recognize that for EU member states, Euratom, not the International Atomic Energy Agency, controls the global regime of non-proliferation. This "European exception" is a phenomenon of no little importance.[44] In reality, the organization is a body of experts who, in following a technoscientific rationale for their work, produced knowledge that had an impact of its own on politics, even if political goals were not on their agenda.

Euratom shows the close entwinement of actors from state, science, and technology that characterizes modern cultures of knowledge production. In its period of strongest expansion, Nuclear Europe followed the lead of technoscientific experts and politicians, while capitalist industry was reluctant to enter the scene. Such a constellation of actors is inconceivable today, when industry is the driving force in nuclear energy. Technical knowledge and political power are instrumental for attaining economic results. In the 1950s, industry was put off by the many "unknown unknowns" and incalculable risks in the nuclear sector. Not until the 1960s were utilities willing to make large-scale investments in nuclear power. Even then, they left it to politics to clear the way, expecting governments to fund the science and engineering necessary to build the knowledge base for major implementation of nuclear technology. Out of a mistaken belief that "nuclear power = modernity," Europe's political actors saw nuclear energy as a necessary condition to revitalize the postwar economy. They were ready to think and act along industrial lines. If their vision did not take shape immediately in policy, this was due to the success of technoscientific experts in framing transnational nuclear collaborations as scholarly undertakings, seemingly above vested political and industrial interests.[45] This is why highly self-interested and opportunistic technoscientific experts and politicians marched in the front lines of Nuclear Europe for so long, with business bringing up the rear.

Breeding & Fusing Europe

The oil price crises of the 1970s prioritized two cases of big technology in Nuclear Europe: the efforts to build a European fast breeder reactor and the project to develop a fusion reactor capable of delivering energy. Both technologies exemplify, in differing constellations, the ambivalences of Knowledge Europe, such as tensions between nationalism and transnationalism, the interrelatedness of military and civilian interests in the nuclear sector, and the tension between Europeanization and globalization.

History sometimes has a sense of humor. An unpredictable turn of events in the early 1990s allowed a rare experiment to take place. A vast white elephant of European engineering expertise, a failed project of big technology—the pilot-scale fast breeder reactor SNR 300, built to produce 327 megawatts of electricity for the Rhineland—was appropriated by civil society and turned into an amusement space for rapidly growing leisure society.[46] The experiment took place near the small city of Kalkar on the lower Rhine. A prior condition was fulfilled when the German Federal Minister of Research and Technology, Heinz Riesenhuber, met with Siemens and representatives of the utilities RWE, *Preussen/Elektra* and *Bayernwerke* on March 21, 1991 to decide on the reactor's future. The next day, a ministry press release announced that the project would be stopped, attributing full responsibility for closing the Kalkar reactor to the refusal of the state government of North Rhine-Westphalia to issue the legally required operating permit.[47] The project had started in 1972 as a collaboration between Germany, the Netherlands, and Belgium. An agreement covered its first phase, building the SNR 300. A second phase would feature an even bigger, natrium-cooled reactor. Planned to produce 2,000 megawatts, it would have been the world's largest breeder. Completed in 1985, SNR 300 never went operational. Rather than take upon itself the cost of dismantling the nuclear facilities at Kalkar, the German government sold the complex at a public auction in 1995. The venturesome Dutch entrepreneur Hennie van der Most acquired it for 2.5 million euros—cheap, one might say, for an object that cost some 3,600 million euros—and converted it into a leisure park. At first it was labeled, almost sarcastically, "Corewater Wonderland." In 2005 it was renamed "Wonderland Kalkar." The space, originally intended to become a landmark project of Nuclear Europe, is now open to the general public, who experience it by climbing the giant cooling tower, swimming, playing soccer, tennis, and beach volleyball or just relaxing in a "radiation-free" space, as the marketing experts of the park are eager to reassure visitors.

When Riesenhuber and the utility companies decided to terminate the breeder at Kalkar, they hastened to make it clear that they had not entirely abandoned the project of building breeder reactors. The expert knowledge behind the project was preserved in a European-wide collaboration that continued, namely the European Fast Reactor (EFR). Belgium, France, Germany, Italy, the Netherlands, and Britain launched EFR in 1970. The British government did not agree to join the project until 1984, and then only under certain conditions, an attitude typical of Britain's reluctance to enter European collaborative efforts.

Why did EFR enter the scene so late? By 1970 multiple national breeder programs already existed, some of them already in operation for ten years or longer. To answer this question, we must look first at the technical specifications of breeders and second at their history in the 1950s and 1960s, when Europe was in the throes of a

Fig. 7.7 From Nuclear Site to Leisure Park: *What used to be the cooling tower of the fast breeder reactor SNR 300 at Kalkar, Germany, now serves as a climbing wall at an amusement park. After the reactor was shut down in 1991, the site was sold at a public auction. The Dutch investor Hennie van der Most acquired the site and transformed it into a leisure park.*

virtual breeder fever. A breeder reactor generates more fissile material in the core's fuel than it consumes. It has a core of plutonium and uranium-238 and a mantel of uranium-238. The neutrons emitted in the core are not slowed down as in conventional reactors—that is why the machine is called fast reactor—but continue at high speed, transforming the mantel's uranium into plutonium. The resulting plutonium can be reprocessed and used to fuel other types of reactors. It can also be used for atomic bombs, making the technology highly relevant for military purposes. Breeders offer spectacularly efficient resource consumption. In theory, breeders can exploit 80 percent of the potential thermal energy in uranium, as against the mere 1–2 percent attainable by conventional nuclear reactors.

The military value and superior fuel economy of breeding make the concept attractive to many technoscientific experts. For Europe during postwar recovery, with its rapidly growing demand for energy, breeders seemed to offer an ideal solution. While the U.S. had started breeder programs immediately after the war, with the Soviet Union following suit, breeder projects in Europe first took off in the 1950s and 1960s and became a veritable craze. Even smaller countries like Sweden launched autonomous national programs to develop expertise in breeder technology; Sweden managed to build a prototype named FR-0 by 1964.[48]

With its advanced nuclear expertise, Britain had a head start in breeder development, launching the fast reactors Zephyr in 1954 and Zeus in 1955. Its larger Dounreay reactor went online half a decade later. France pursued breeder reactor development the most intensively, however. The climax of the ambitious French program was the breeder power plant Superphénix at Creys-Malville on the river Rhône. The construction of the breeder, designed to deliver an output of 1,200 megawatts, was approved in 1972 and accomplished by 1981. The planning called for another four plants of this type. In time this large-scale "technopolitical" artifact developed into a symbol for French national competitiveness and even into an icon for a French-led socialist future, over and against the world vision projected by American imperialism.[49]

We now return to the question of why the collaborative European breeder project took off so late, a question that leads to another: why did it fail? The national breeder craze came to an abrupt end in 1993, two years after Kalkar was killed and four years before Superphénix was completed. Only when this breeder lust cooled down with the 1970s' end of the postwar boom did European governments show themselves ready to join forces in a collaborative project. By that time, governments and experts had learned that developing a large-scale breeder was an enormously costly and time-consuming undertaking. Pooling efforts on a European scale promised many advantages: to share resources, to save time, to increase the totality of knowledge to be gained, and to protect domestic projects in the face of growing social protest. For a decade or so, the breeder was seen by experts and politicians alike as an opportunity to develop a distinctive European nuclear-technological profile in a global perspective. This seemed all the more feasible since the U.S. had more or less left the field. Soon after the former naval nuclear engineer Jimmy Carter moved into White House in 1977, the U.S. government began opposing further development of breeder reactors and plutonium fuel cycles, owing not only to major concerns over nuclear proliferation but also to safety issues.

Protest against breeders grew rapidly in Europe as well as the U.S., not only in the general public but even in the expert community. Klaus Traube, who as a senior manager of the Siemens daughter company Interatom had been responsible for the Kalkar breeder, did a complete about-face. After he left the company in 1976, he became one of nuclear energy's most prominent and outspoken opponents, with a particular animosity to breeder reactors. In the same year Pugwash, the international movement of critical scientists, held a public debate at the Royal Society in London on the question "to breed or not to breed," concluding that breeders constituted a threat to international peace.[50] The eagerness of European governments to take part in a large breeder project faded away. A new sentiment now prevailed: the technological risks of the breeder were too high, the economic "uncertainties" too big.[51] These worries were not the only problems the breeder brought with it. There were a number of other pressing issues: controlling the international flow of weapons-grade plutonium; securing adequate cooling and safety; and the stagnant price of uranium, which undermined business plans for breeders. These were perhaps the most important reasons why the European breeder reactor finally failed to go operational and was discontinued in 1993.[52] Euratom, too, was committed to the European fast breeder program, and it too failed, as we have seen, to "unify all research work on fast breeders" in the European Community.[53] The European nuclear agency continued to invest in breeder research, but it gradually redirected its priorities and budgets toward fusion research and technology. Since the 1980s, it has invested about two-thirds of its annual budget in the development, at some unspecified time in the future, of a fusion reactor capable of delivering energy.[54] In a way, Euratom has mutated into an agency with the almost exclusive task of administering international fusion programs on behalf of the European Union.

As Euratom sees it, the history of fusion research is quite straightforward, and Europe has always been in the lead. With his famous equation $E = mc^2$, Albert Einstein provided the theoretical basis for converting a small mass (m) into a very large amount of energy (E), the conversion factor being no smaller than the square of the speed of light (c). The interwar period saw some key experimental advances in Britain and Germany, but fusion research only really gained momentum after the Second World War, when nuclear fission weapons were being built on a grand scale. The first large experimental fusion device was built in the late 1940s and early 1950s at Harwell, the British nuclear research center. Experimental programs in France, Germany, the Soviet Union, and the U.S. followed suit. According to Euratom's linear narrative, the Atoms for Peace conference in Geneva in 1958 "formally sealed the start of truly international collaboration that would in time lead to today's ITER experiment in southern France."[55] ITER stands for International Thermonuclear Experimental Reactor, a global megaproject under European leadership, currently under construction at Caderache near Aix-en-Provence in southern France.

This brief "official" history of fusion in Europe overrates Europe's role in the global setting. It also fails to address the complex nexus of national and transnational forces and to pay proper due to the fact that the "contingencies and rationales, purposes and valences" for technoscientific collaboration change over time, often "in ways not entirely within researcher's control."[56] The European fusion program exemplifies the epic struggle of experts to keep control over "their" field and to limit political interference into technoscience, even though their costly big machines solely depend on government sponsorship.

With their fusion research program, which started in the late 1960s, European experts pursued a multitude of goals. These included "long-term energy independence, international standing in the physics community, and the continued survival of research programs dependent on building the 'next big machine.'"[57] It was also designed to ensure the survival of an independent European program should the effort to globalize fusion reactor development fail. European researchers shared most of these aims with colleagues in other parts of the globe. This raises a sensitive question: what was then specifically European about the European fusion program?

In many ways the initial phase of fusion research can be called typically European. To begin with, the first option for the housing of a transnational collaboration for the development of nuclear fusion was once again CERN. Technoscientific experts and policy makers planned to form a joint study group of CERN and Euratom, with the former providing expertise and the latter the necessary budget. The idea ran aground on the strong opposition of some CERN and Euratom member states. Britain feared that its strong national knowledge base would be weakened. Perhaps it was typically European that the project positioned Europe as a junior partner to the two technoscientific superpowers. As European experts expanded their fusion programs, they also directed their efforts towards beating the U.S. and the Soviet Union. In addition, the first efforts of Euratom to foster fusion research followed the "subsidiarity principle" of European politics, through which national programs were given priority over communitarian ones. European agencies were only allowed to support and to coordinate national initiatives, not replace them. The Italian expert Donato Palumbo, who was in charge of Euratom's fusion program, honored that principle by supporting national laboratories in Europe with a series of contracts. Pursuant to the technocratic terminology of the European Community, these contracts were "indirect actions" to enable work to be done which might not otherwise be feasible in the member states.

During the 1960s, a number of Western European countries launched elaborate fusion research programs. After Britain, West Germany was particularly vigorous, with Werner Heisenberg as the lead actor. The Munich-born Nobel laureate managed to found a Max Planck Institute for Plasma Physics in Garching, near his home town, that quickly evolved into a worldwide node of fusion research.[58] Results on the research front were, nevertheless, "rather disappointing," as one eyewitness later recalled. That was before a big push from outside reinvigorated the flagging European activities.[59] At the 1968 International Plasma Physics Conference in Novosibirsk, Soviet physicists of the famous Kurchatov Institute reported sensational new results from their latest Tokamak machine, T3. A fusion reactor is designed to confine the plasma created inside a vacuum chamber by using magnetic fields. The Soviet Tokamak—an acronym of *toroidal'naya kamera s magnitnymi katushkami*—was a device in the shape of a torus, or doughnut, in which the plasma was corralled by strong poloidal magnetic fields and an even stronger toroidal field. It allowed the plasma to be heated to 10 million degrees kelvin, without melting down the surrounding apparatus.

The Soviet breakthrough in reactor technology caused a Tokamak craze, spurring similar efforts all over Europe and encouraging a paradigm shift in European technoscientific politics. To make a long story short: in order to shoulder the high cost of building a large

Fig. 7.8 **Crowning Nuclear Technology:** *Only the highest representative of Britain, Her Majesty Queen Elizabeth II, was dignified enough to officially open the Joint European Torus on April 9, 1984. On this seminal day for Nuclear Europe, the Queen showcased fusion research as the new flagship of European technoscientific collaboration. The image shows Elisabeth II, JET's Council chairman Jacques Teillac (middle), and French President François Mitterand (right).*

machine to compete with and to finally outdo the Soviet reactors, Western European countries joined forces under the umbrella of the European Commission. Their aim was to collaborate in building a substantially improved model that would be a landmark on the long road to commercial fusion. Palumbo explained in the name of the Commission that the final goal of long-term cooperation was not to foster science but to construct prototypes "with a view toward their industrial production and marketing."[60] Only after a five-year struggle among France, Germany, Britain, and Italy could Nuclear Europe agree on a site: Culham near Oxford. Construction work for the Joint European Torus (JET) began in 1977 and on April 9, 1984 Her Majesty Queen Elizabeth II officially opened the new center. By then, JET was already praised as a flagship of European technoscientific collaboration; fusion research had developed into the largest single research program of the European Commission.

Experts had differing interpretations of the controversial process by which JET was established. A case in point is Harold Montague "Monty" Finniston, lead author of of a late 1970s report that deplored the low status of engineering in Britain. To him, the story of JET combined "elements of a farce and near-tragedy presented in the form of a political ballet, to a gavotte played by politicians"; as a fiction he would have ascribed it "to Kingsley Amis or to George Bernard Shaw in the more cynical passages of his political plays." From Finniston's technocratic perspective, "chauvinistic pressures" had hindered JET from building the most advanced experimental reactor in the world, thus putting Europe into the lead in fusion technology.[61]

It was not only the politicians who displayed chauvinism. European expert communities, too, were fragmented along national and even regional lines. In Germany, for example, expert groups in Munich, Karlsruhe, and Jülich would not unite under the national flag when it came to the crucial point of getting priority to apply for hosting the next big European experiment. At issue were not just matters of administrative competence and national pride but also disagreements about design and location. The British fusion authority Denis Willson observed from inside the expert community that the "perpetual conflict between self-interest and the need to collaborate" extended to scientific bodies as well.[62] Local cultures of experimentation clashed with differing institutional agendas, bringing forth not just one, but a multitude of knowledge societies and expert networks in fusion Europe.[63]

Long before JET reached out to set new world records in heating plasma to temperatures of over 100 million degrees kelvin, experts had already begun to plan the next generation of fusion reactors. In 1978 a team of scientists and engineers started to draft design specifications for what was now called the Next European Torus. They were quickly reminded that Europe was not the only player in international fusion technology. The Soviet Union proposed building a large-scale Tokamak not on its own but under the auspices of the International Atomic Energy Agency (IAEA). The latter had installed an advisory group on fusion research as early as 1972, which now discussed the Soviet proposal in a series of international workshops. The European Commission responded by adopting a wait-and-see strategy, keeping the doors open for both alternatives: a European-centered and a broader international cooperation. European experts could hope that their leading position in fusion technology would qualify them to head an international planning team and even to host the future experimental reactor station.

That Europe was just one more player in the global fusion theater became clear once again when Ronald Reagan and Mikhail Gorbachev met for the first time at the famous "fireside summit" in Geneva in November 1985. Thirty years earlier, at the first Geneva summit, U.S. President Dwight D. Eisenhower, British Prime Minister Anthony Eden, Soviet Premier Nikolai A. Bulganin, and French Prime Minister Edgar Faure had discussed nuclear disarmament and Eisenhower's Atoms for Peace initiative. Reagan and Gorbachev resumed that discussion in a highly symbolic act of technopolitics. They agreed on an international collaboration to utilize "controlled thermonuclear fusion for peaceful purposes and [...] for the benefit for all mankind."[64] Nuclear fusion was as much shaped by the Cold War as it contributed to the end of that era, paving the way for global cooperation. When all was said and done, Europe had to redefine its place in the changing geography of nuclear collaboration.

Two full decades passed between the fireside summit's big push for ITER and the final decision taken in 2005 to build the multi-billion-euro machine in France. It was a global effort, but the EU and France paid half of the construction costs. These two decades saw the European fusion community struggling to balance national, European, and global efforts without losing sight of ITER's two main aims: to prove that an energy-delivering reactor was technically feasible and to display the technoscientific capabilities of Europe's fusion experts. The struggle was fierce, all the more so as fusion evolved into a technopolitical showcase of the first order. Europe faced global competition in the siting question from Canada and Japan, while at home an intense duel between France and Spain had to be fought out before the European Union was able to speak with one voice.

The Multiple Geographies of Europe

Nuclear science and technology in postwar Europe developed into a dialectic force that both enabled European integration and divided the Continent along Cold War lines. Both superpowers, the U.S. and the Soviet Union, tried to secure control over the nuclear knowledge arsenals of their respective European allies in order to gain global hegemony in the long run. Supporting expansion of the nuclear knowledge base, as the U.S. government did most visibly with the Atoms for Peace program, was not motivated by technoscientific philanthropy, but by international power politics. Western Europe responded to the Cold War power play by pooling its domestic resources in transnational programs. CERN, in particular, became the institutional role model for a European collaboration built on expertise and a role of politics confined to secure an adequate institutional and financial environment.

CERN is, however, just one node in the seamless net of transnational cooperation. These overlapping collaboration spaces illustrate a Nuclear Europe of multiple dimensions which can be traced from the local to the global.

The *local dimension* of transnational collaboration is, again, best illustrated by CERN. The institution is housed at one site, Meyrin, Switzerland, near Geneva. In the 1970s, when CERN's first generation of accelerators was phased out, the member states became involved in a fierce competition regarding who would get to host the center's next big machine. West Germany, in particular, campaigned for the 300 GeV proton synchrotron that was then being considered. From this conflict, CERN drew the lesson not to decentralize but to concentrate its facilities. CERN's gigantic subterranean particle storage rings are transnational even in a geographical sense: they cross the border to France. However, CERN's location in Switzerland (and France) is unimportant for the thousands of scientists and technicians working at the center for months or years—except for the need to cover the high costs of living in the area. For them, what counts is performing experiments with the most advanced machinery world-wide and thus being able to compete with their colleagues in the U.S. and other regions of the world. For them, it is Europe that provides this unique opportunity, and CERN is the place to be in order to advance both their individual careers and their collective sciences.

Examples of *regional* and *bilateral dimensions* of transnational collaboration are manifold. Often, both kinds of cooperation were born in experts' minds and later mingled with political interests to gain momentum. A case in point is the Joint Establishment for Nuclear Energy Research (Jener), founded, as noted earlier, in the early 1950s by Norwegian and Dutch physicists to continue a collaboration that dated back to the immediate postwar period. Later the bilateral project developed into a key component of Euratom, Western Europe's flagship institution for nuclear technology. A secret Spanish–German collaboration in nuclear technology also grew out of the specific context of postwar Europe, when both nations were internationally banned from research having military potential. Spanish experts were trained in Germany, and German experts supported the isolated Franco regime in developing nuclear research centers and power plants.

Nuclear technology is based on experts' knowledge—and on uranium ore. Without uranium there are no reactors, no nuclear energy, and no nuclear deterrents. During the Cold War, access to uranium mines developed into a strategic issue. The case of Belgium points to the *colonial and post-colonial dimension* of transnational collaboration here. As previously discussed, in the late 1950s the Belgian government decided to ship all the uranium ore mined in its Congolese colonies to Euratom. This decision resulted from frustration about the one-sided collaboration with the U.S. that never allowed Belgium to cooperate on an equal footing. The Europeanization of its uranium ore, however, did not prevent Belgium from finally losing its colonial access to nuclear raw material, when the Democratic Republic of Congo gained independence in 1960. By turning to Europe, Belgium tried to adapt its nuclear policy to the changing environment of decolonization.

The U.S. was a major point of reference in the nuclear realm for other countries as well. For most nations of Western Europe the *transatlantic dimension* of transnational collaboration was always present, whether as an opportunity to access the latest technical knowledge, or as a threat of being dominated as the U.S. sought technoscientific hegemony. Collaboration between experts in Nuclear Europe often meant trying to find answers to the challenges presented by Europe's transatlantic partner during the Cold War. Transatlantic cooperation was also an option for Nuclear Europe in the 1970s, when the oil price crisis spurred efforts to develop both fast breeders and fusion reactors capable of delivering energy. When the Carter administration in the U.S. opposed breeder development due to concerns about nuclear proliferation and safety, experts and politicians in Western Europe saw breeders as an opportunity to develop a distinctive European technological profile and pooled their resources in cross-national projects. However, this did not prevent Europe from finally ending

Fig. 7.9 Closing the Cold War with Collaborative Fusion Research: *The image shows Ronald Reagan and Mikhail Gorbachev at the "fireside summit" in Geneva, November 19–21, 1985. The initiative to put international cooperation in fusion research on the summit's agenda came from the Soviet delegation. Gorbachev's science advisor Evgeny Velikhov suggested it to Foreign Minister Eduard Shevardnadze, who proposed it to the American Secretary of State George Schultz. Schultz was able to block opposition from the Pentagon and to get approval from the White House.*

the fast breeder programs in the early 1990s because of public protest and safety issues.

Cross-national collaboration in fusion technology started as early as the 1950s, in both Western and Eastern Europe. When the big four of Western Europe—France, Germany, Britain, and Italy—agreed to a cooperative fusion reactor project with the telling name Joint European Torus in the early 1970s, they were also responding to the technically advanced Tokamak program of the Soviet Union. When the Cold War ended in the late 1980s, *the global dimension* of transnational collaboration came to the fore. In Geneva in 1985 Reagan and Gorbachev paved the way for a global project in thermonuclear fusion. Western Europe discovered that it was just one player in the new arena of globalized technoscientific collaboration. Only after a long struggle of national, European, and global dimensions was the European Union able to secure its leading position by constructing ITER and locating it in France, its most pro-nuclear member state.

8

Contesting Europe in Space

Space Europe Eastern Mode: Interkosmos

Space has no borders. Space is almost endless, truly universal. When the lunar module *Eagle* landed at Mare Tranquilitatis (Sea of Tranquility) on July 21, 1969, and Neil Armstrong planted the U.S. flag on the moon, it became clear, however, that outer space was neither beyond the anthropological struggle over power, nor beyond human efforts to delineate spatial dimensions according to the principle of the nation state. Before the epochal launch of the Soviet satellite *Sputnik* in 1957, which triggered the race to the moon, the global conflict between the two superpowers had already expanded from earth to the heavens. The Cold War spurred multiple efforts to attack and counter-attack from space, long before U.S. President Ronald Reagan launched his infamous Strategic Defense Initiative (SDI), which soon became known as "Star Wars." *Sputnik*, then, started the race to the moon. For more than a decade, the space race absorbed vast resources of the two superpowers, financially and intellectually. Other nations, concerned that the U.S. and the Soviet Union would acquire a hegemony in space, joined in and developed autonomous space programs, in particular countries in the Global South. China and India, for example, were keen to enter the club of the spacefaring nations as quickly as possible. Space capacities and launch technologies grew into indicators of international power, economic competitiveness, and technoscientific capabilities.

The space race produced numerous personality cults, although the first-ever creature in space was the famous Russian dog Laika. Astronauts, or cosmonauts as they were called in the Soviet bloc, evolved into icons of popular culture, public figures of crucial importance to the identity of a nation. Yuri Gagarin is a case in point. Soviet Russia cast its first cosmonaut as a hero, leaving the nation no escape from the Gagarin cult. The sustained cultural effects of this cosmonaut-heroism are evidenced by recent sociological studies in Russia, in which respondents rank Gagarin's space flight second (91 percent) in their list of Russia's historical achievements—closely behind victory in the Second World War (93 percent), and well ahead of the launch of *Sputnik* (84 percent).[1] Astronauts and cosmonauts are virtually omnipresent in their home countries, in particular in the smaller spacefaring nations. Even the fragile nation of Afghanistan has its nationally renowned space champion: Abdul Ahad Momand, who in 1988 spent nine days aboard the Russian *Mir* space station.

The Afghan Momand was offered a space flight within the framework of the Soviet Interkosmos program. The Council for International Cooperation in the Exploration and Use of Outer Space (in short, Interkosmos) was created in 1966 as part of the Soviet Academy of Sciences, in order to coordinate the work of numerous Soviet ministries in international cooperative space programs between the Soviet bloc and other communist states. In April 1967, the Soviet government formally initiated the Interkosmos program, which over the next decade pursued numerous space exploration missions in the five fields of physics, communication, biology and medicine, meteorology, and remote sensing. In 1976, the program expanded to include manned missions. Starting with Vladimír Remek of Czechoslovakia in February 1978, Interkosmos enabled fourteen non-Soviet cosmonauts to participate in flights with the Soyuz space capsule, including Momand. France's Jean-Loup Chrétien was the last to fly on the Soyuz before the program ended in November 1988, just as the Cold War collapsed. During the Cold War, however, Interkosmos was a powerful tool that enabled the

Soviet Union to exert political hegemony by means of technoscientific collaborations. In Interkosmos, the exchange of knowledge and resources was asymmetric because of the Soviet Union's political dominance, and collaboration on equal footing was never intended. The participation of states outside the Soviet bloc, such as India, Syria, and even France, reflects the Soviet Union's politics of cooperation with non-aligned states to win them over during the Cold War.[2]

Interkosmos offered opportunities for the Warsaw pact allies to obtain limited access to the Soviet space sector's gigantic knowledge machine and to develop autonomous expertise, even if the program always stood under unquestioned Soviet supremacy. In East Germany, for example, the Institute for Cosmic Research of the Academy of Sciences used the Interkosmos program to develop skills in designing cameras for space missions. Based on their longstanding regional expertise in optics, scientific instrument making, and precision mechanics, the experts in Berlin-Adlershof built the multi-spectral camera MKF-6M key to many Soviet space missions, both inside and outside the Interkosmos program. After the fall of the Berlin Wall, its expertise in high-precision cameras was the institute's most important asset in the contested process of scientific evaluation and integration into the new German system of research and innovation.[3]

Today, for the small expert community of former East German space scientists, the MKF-6M signifies their greatest national achievement. In their cultural memory, the camera is only partially linked to Interkosmos and the Soviet-led space program, and predominantly perceived as national icon, now displayed at the *Deutsches Museum*, Germany's national museum dedicated to science and technology.[4] The same holds true for the cosmonaut Sigmund Jähn, who went into orbit with the Interkosmos mission Soyuz 31 in August 1978. In Erich Honecker's East Germany, Jähn was cast as a national champion, signifying "eternal friendship" between the socialist brother states, as well as, and more importantly, the country's technoscientific capabilities. In today's popular culture, Jähn continues to be celebrated in nostalgia for the former East German state. The museum devoted to his memory in his home town Morgenröthe-Rautenkranz in Thuringia proudly displays his space suit as an iconic object.[5]

Fig. 8.1 A Space Camera for the Soviet Bloc: *Built by Carl-Zeiss in Jena, the MKF-6M is a multispectral camera with six lenses for photographing the Earth's surface from space. The camera consists of the body, six film cassettes, and a control panel. Used for the first time in 1976 as part of the Soviet Soyuz 22 mission, the robust camera developed into a standard element of space technology for the Soviet-led Interkosmos program. The camera was East Germany's most important technological contributions to manned space flight. Today, it is remembered as a national icon, and is displayed at the* Deutsches Museum *in Munich.*

Interkosmos, after all, highlights a fundamental dichotomy of the space age: the national–transnational dichotomy. On the one hand, space activities are inherently nationally framed and coded. The U.S. Apollo program was, at first, an effort to identify outer space as the new frontier of the twentieth century. It was aimed at stabilizing the stunned American nation after the *Sputnik* shock by jointly realizing one of mankind's most potent dreams through a gigantic technical project, much as closing the Western frontier had helped to unify the nation in the nineteenth century. Many other countries followed the American model. Postcolonial India started its own space program in order both to knit together its multi-ethnic society and to leap into a Western-like modernity based on technoscientific knowledge, a process full of complexities, ambiguities, oscillations, and contradictions. [6]

On the other hand, space activities show evidence of transnationalism, if they are viewed, not so much from a political, but from a knowledge perspective. A historian of technology, who studied the Soviet, the U.S., and the Indian space programs, has emphasized that the "history of spaceflight has been part of a consistent flow of knowledge and technology across (geographical) space and time." He advocates replacing the "national" with the "global," when it comes to space history.[7] In space, as in the nuclear realm, "practically no state travelled alone."[8] A constant flow of knowledge and expertise weaves together the U.S., Soviet, British, French, German, Italian, Japanese, Indian, Chinese, Brazilian, Australian, and other space efforts into a global network. However, this global net is not seamless. On the contrary, it shows political ruptures and national boundaries. Space history oscillates between the poles of the national and the transnational. And this holds particularly true for European space history, as will now be illustrated from a knowledge perspective.

A Fragmented Europe: ELDO

November 5, 1971 was a bright day at Kourou in French Guiana. The sky was clear, allowing an attentive audience to watch an attempt to launch the European flag into space, as the third "nation" after the two superpowers, the U.S. and the Soviet Union. What the assembled spectators saw was an impressive four-stage rocket, including a new "perigee-apogee" stage, with the telling name Europa 2, mounted at Europe's remote spaceport in equatorial South America. The officials of the European Space Vehicle Launcher Development Organization (ELDO), founded in 1962 as an institution with the aim of designing and building European spacecraft, were quite nervous. Between 1964 and 1970, the patient European public had witnessed no fewer than ten test flights. In a few cases, they observed limited operational success, but mostly they saw only technical failures. Flight F 11 was intended to prove that Europeans could collaborate effectively in space technology, the ultimate technoscientific challenge.

The countdown went well. After ignition the vehicle left the launch pad and disappeared into the skies. The crowd applauded enthusiastically and, in fact, for a short while the flight seemed to be successful. But after two and a half minutes the first stage disconnected, immediately followed by the explosion of the second and third stages. Europa 2 perished, and the pride and hopes of Knowledge Europe with it.

What had gone wrong? Why had Europe experienced such a disastrous series of failed test flights, which ultimately led European officials to rethink the entire collaborative space program? Answers to these questions can be found on the technical level but also, more importantly, in the political culture of European technoscientific collaboration. ELDO's failure allows us to identify the key factors of European collaborative efforts in science and technology, to trace the complex networks of experts, and to explore the tensions between transnational orientations and vested national interests. As in the case of CERN's Large Hadron Collider, it is the failures and faults which provide the greatest insights into the societal conditions of innovation processes and the nature of socio-technical change.[9]

The technical factors that led to the Europa 2 disaster were relatively easy to identify. Already in flights F7 (December 1968) and F8 (July 1969), which aimed to test the operability of Astris, the third stage, a West German responsibility, the spacecraft had exploded shortly after separation. In contrast to design expectations, the investigators found that the explosions resulted not from the propulsion system of Astris but rather from an electrical failure between the third stage and the test satellite created under Italian responsibility. Fixing the electrical problems did not prevent Astris from malfunctioning on the next test flight in June 1970. The latter disaster convinced the ELDO Council of the necessity to create a Quality Assurance Association, "but due to a lack of staff, it could not cover all sites and processes."[10]

When the ELDO Council met some weeks after the devastating test flight of November 1971, the participants set up an investigation committee of senior experts and executives from government and industry in Europe and the U.S. The committee found that one of the key problems was the poor communication between the British contractor Marconi and its West German corporate counterparts. Fixed at the upper end of the inner side of the main bulkhead of Astris, some small black units carried the devices for guidance, control, and the telemetry technology that allowed data measurements to be made at a distance. For the German experts, these units were literally "black boxes." They also contained the computers to guide and control the first stage of the launcher. Marconi had built an impenetrable information barrier around the modules it had produced. The West German engineers accepted this boundary and responded by refusing any responsibility for these parts of the system. Lack of coordination led to a technical design that, according to the experts' ruthless report, obeyed "none of the elementary rules concerning separation of high and low level signals, separation of signals and electrical power supply, screening, earthing, bonding, etc."[11]

Europa failed for these technical reasons, but, more important, it also failed because of ELDO's poor organization and its massive management and communication problems. The ELDO secretariat had no say in the central task of contracting. How industrial contracts were distributed was the *arcanum imperii*, the prerogative of politics. National governments jealously controlled their financial investments in ELDO so that these were returned as contracts for their national research laboratories and industries, whatever their capabilities. This policy of *juste retour* (fair return), which we observed in Nuclear Europe, was identified early on as a key misconception of European "cooperation" in space.[12] Rather than fostering transnational collaboration from the bottom up, the member states sought to acquire as much knowledge and resources from the joint undertakings as possible, in order to strengthen their economic positions in international markets. The individual countries used the cooperation not as a way of combining forces, but in order to enhance their own status and prestige. In other words, the supranational body, ELDO, continued to organize its institutional structure in a way that reflected the concept of the nation state which had dominated European history for many centuries.

In order better to understand the persistence of national interests in ELDO as a transnational body, we now leap back in time to the early origins of European collaboration in space.

In the beginning there was *Sputnik*. The beeps of the Soviet space probe, launched into space on October 4, 1957, came as a shock to the American public.[13] The U.S. media pushed the panic button and called for an immediate response. The race to space by the two superpowers began to accelerate, again stimulating physics and the related sciences, which had already expanded enormously during the first phases of the Cold War. Leading scientists in Europe were anxious not to be left behind and launched the idea of pooling their technoscientific resources in the space sciences and rocketry to counterbalance the threatened superiority of the superpowers. Taking CERN as a model, Pierre Auger and Edoardo Amaldi, the two veterans of European cooperation in the physical sciences, proposed a joint organization for space research. While walking in the Jardin du Luxembourg on a sunny Paris spring day, so the story goes, Amaldi informed his close friend about his idea to create "euroluna."[14] Soon afterwards, the Italian physicist formulated a famous memo, "Space research in Europe," the foundational document of Space Europe. Auger and Amaldi quickly gained support from other eminent European scientists, such as the atomic and atmospheric physicist Harrie Massey from University College London. They set up a commission of representatives from ten European governments to determine the possibilities for European cooperation in space. This organization, the *Commission préparatoire européenne de recherches spatiales* (COPERS), and the initiative to ensure freedom for the civil exploration of space, which the U.S. government had launched after *Sputnik*, provided the arena within which to investigate the political and diplomatic feasibility of the scientists' ideas.

The push from the scientific community resonated well with shifting concepts of nuclear armament, security policy, and European political collaboration in Britain and France to counter U.S. dominance. In a series of international meetings, the two nations took the lead in ensuring support for a European-wide space collaboration based on their respective programs to build nuclear ballistic missiles. West Germany was at first reluctant because of internal problems within government and reservations on the part of its technoscientific experts, but it joined in after receiving signals from Paris and London that the technical concept of the planned spacecraft would be improved as the West German experts requested. Within a remarkably short period, an initiative by a few scientists gained political traction and manifested itself in two complementary transnational agreements, both reached in 1962. ELDO was the technical arm of European space collaboration, while the European Space Research Organization (ESRO) was designed to conceptualize and implement planetary and deep space missions.

ESRO, and in particular ELDO, embodied the political will to keep Europe independent from the super powers in general

Fig. 8.2 A Europeanized Rocket: *The medium-range ballistic missile Blue Streak was originally intended to maintain an independent British nuclear deterrent. When design was completed by 1957, it became clear that the missile system was too expensive and too vulnerable to a pre-emptive strike. The British government cancelled the disastrous project and offered it to France and Germany as the first stage of a joint European civilian launcher. The picture shows Blue Streak dominating the approach to the main exhibition hall at the famous Farnborough Air Show in September 1961.*

and from the U.S. in particular. From one perspective, European collaboration in space was the extension of national interests into the European arena, essentially "Euro-Gaullism," that is, to strive for national independence on a European level. In this view, ESRO and ELDO resulted primarily from the extension of global power struggles into the realm of space.[15] Yet, even though the European nation states channeled national technoscientific knowledge and resources into transnational bodies in order to place Europe on an equal footing with the two superpowers, they continued to act in their own national interests. The motivations of the European nations involved in ELDO were highly diverse. For Britain, ELDO meant the opportunity to "europeanize" the development costs of the canceled British military rocket Blue Streak, which was chosen as the first stage of the joint European launcher. It also meant the opportunity to preserve its advanced knowledge base in aerospace engineering and its industrial skills and facilities. Furthermore, ELDO resonated with the British strategy of moving closer to continental Europe, which eventually led to entering the Common Market.[16] For France, ELDO provided the opportunity to share the development cost of its Coralie missile, which would become the second stage of the joint launcher. It also allowed access to know-how on launcher technology with military implications for France's *force de frappe* program, which aimed at guaranteeing nuclear deterrence independent of NATO and the U.S. In a way, ELDO was the result of the Franco-British attempts to identify a "third way between science's universalism and US hegemony."[17]

Britain and France asked West Germany to join and take over the responsibility for the third stage. For German scientists, engineers, and politicians alike, the Anglo–French offer was an attractive possibility to re-enter the arena of rocketry, to underline its leading role in European unification, and to show the competitiveness of its industry in the most advanced field of science and technology. It also entailed stepping out of the long shadow of Peenemünde, where Nazi Germany had located its notorious research and testing facilities for the destructive V-2 missile. Due to the historical burden of National Socialism, German governments tended to favor transnational projects in general, and collaborative efforts with its European partners in particular.[18]

Italy joined in to build a test satellite. From an early stage on, however, Italy was a reluctant partner, complaining about the "unfair" return in terms of contracts assigned to Italian industry. In contrast, Belgium and the Netherlands, who assumed responsibility for ground control and telemetry, became devoted supporters of ELDO as a means for small countries to participate in large-scale international projects on an equal footing.

Parallel with ELDO, European industry mobilized its resources and founded the supranational organization Eurospace in 1961, which soon comprised about 1,000 member companies. It lobbied openly for the establishment of ELDO and ESRO and offered its services on different levels. With governments investing in space, industries increased their activities in space research and development. The European dimension also made investments more secure, since European programs were safeguards against national governments' cancellation of projects. By pooling resources at the European level, the industry could invest in space on a long-term basis while simultaneously improving its global competitiveness.

Initially, a number of smaller countries also participated in the negotiations to create ELDO, including Austria, Denmark, Greece, Norway, Spain, Sweden, Switzerland, and Turkey. Ultimately, however, these countries did not sign the convention. The motives behind the refusal to join differed. Norway declared that it did not have the technical capacities to benefit from the program, and deliberated about cooperating with the U.S. instead.[19] Denmark and Spain referred to high costs. Other countries stressed the potential military implications of launchers and rejected compromising their policy of neutrality. Sweden even insisted that plans for a European organization be officially communicated to the Soviet Union. For Switzerland, any participation of either the U.S. or the Soviet Union in Space Europe was unacceptable, utlimately withdrawing because becoming involved in "international cooperation in the field of launchers as a dual-use technology" would risk its neutral position.[20] While ELDO and ESRO were both created to improve Europe's role in space sciences and technology, a comparison of the two allows us to identify different cultures of governance in different networks of experts. Most member states of ELDO understood the joint development of launchers as a means for boosting their respective industries. Investments in European collaborations were intended to foster competitiveness and to reward national companies through industrial contracts. Similarly, technical experts and industrialists also regarded Europe mainly as a political vehicle to justify their demands for national-level financial support. Most academic scientists opposed ELDO precisely because of its obvious political character. ESRO, in contrast, was understood as a primarily scientific endeavor. Consequently, it was granted much more autonomy to choose its organizational structures and scientific programs according to the needs and aims of international knowledge cooperation. Government interference, the fair-return principle, and vested national interests played much smaller roles in ESRO than in ELDO.

Consequently, prior to the official debut of ESRO, an expert group headed by Lamek Hulthén from the Royal Institute of Technology in Stockholm and Reimar Lüst from the Max Planck Institute for Physics and Astrophysics in Garching near Munich worked out a long-term plan for space missions, referred to as the "Blue Book." Lacking experience in complex space programs, the European experts radically underestimated the costs for their projected programs. As early as 1966, ESRO went through a severe budget crisis, resulting particularly from the high running costs of ESRANGE, the launching site for ESRO's sounding rockets, located in Kiruna, Sweden; of ESLAR (later renamed ESRIN), a laboratory for advanced research based in Frascati, Italy; and of the European Space Research and Technology Centre (ESTEC), which was first located in Delft and then moved to Noordwijk on the Dutch coastline. A group of high-profile experts, led by former CERN Council president J.H. Bannier, proposed substantial organizational reforms that allowed ESRO to solve most of its problems and to implement a number of space missions regarded as highly successful scientific endeavors, not only in Europe but in the U.S. as well.

By the time Reimar Lüst left ESRO in 1970 to return to the Max Planck Society in West Germany (where he later became president), ESRO had established not only space programs but also physical spaces of European collaboration. Its laboratories in Noordwijk and Frascati, and its launching facilities in Kiruna were sites where a European knowledge society materialized in quotidian technoscientific routines. Furthermore, collaboration in space science covered a much larger territory than the collaboration in space technology. Besides the six ELDO members, ESRO included Denmark, Spain, Sweden, Switzerland, and (with observer status) Austria and Norway.

A closer look shows that the smaller countries had different motives to participate in European space efforts. The Belgian case is interesting since it reveals how the Europeanization of space substituted for creating a national structure. Belgium never developed a national space program. It chose the European option, hoping that contribution to ESRO and ELDO would pay off scientifically and technically in the long run.[21]

Fig. 8.3 A Division of Labor for a European Launcher: *This image shows the ELDO A-rocket, later renamed Europa 1, mounted at the spaceport Woomera in Australia. The national flags illustrate the division of responsibilities: Britain contributed the first stage, France the second stage, West Germany the third stage, and Italy the research satellite on top, while Belgium and the Netherlands bore responsibility for telemetry. Breaking the project down into national subprojects was one of the reasons why Europa 1 ultimately failed. Experts' work was fragmented and largely uncoordinated.*

The Nordic countries acted differently. Whereas the efforts of Denmark and Finland were moderate, both Sweden and Norway displayed great interest in space activities early on. Their geographical position close to the North Pole put them center-stage for sounding rocket programs in the auroral zone, allowing the study of phenomena such as the northern lights, which were of high interest to astrophysicists. However, Norway and Sweden chose different routes. They both developed national space programs. While Norway adopted an observer status and formally entered

Space Europe only in 1985, Sweden was part of ESRO from the start. In Sweden, the international involvement served to justify national expenditures for space programs.[22]

The geography of Space Europe also includes a regional dimension. In addition to performing space missions in either a national or a European setting, the Scandinavian nations developed a third option. As in other fields of policy, economy, and technology, they collaborated in transnational "Nordic" programs. Their plans for a joint Nordic space project materialized in the communication satellite *Tele-X*, launched by an Ariane 2 rocket in April 1989.[23]

Finally, European space collaboration also yielded bilateral projects. This option was most often chosen by the two front-runner nations of European integration, France and West Germany. The Franco-German project to build the experimental communication satellite system Symphonie is a telling case, a story that exemplifies the politicized character of big technologies "whose costs often spiral dramatically and whose benefits become increasingly difficult to see."[24] When President Charles de Gaulle and Chancellor Konrad Adenauer signed the Élysée Treaty in January 1963, they paved the way for Franco-German cooperation in science and technology.[25] After some deliberations, on June 6, 1967 France and West Germany signed an intergovernmental agreement to jointly build Symphonie and two ground stations.[26] The project, the name of which connoted unity, harmony, and beauty, sought to become programmatic for Space Europe.

The ground segment comprised a series of earth stations which allowed for a broad spectrum of utilization, including transmission of television and radio programs, telephone calls, telexes, and data. The space segment consisted of two three-axis stabilized geostationary satellites, *Symphonie 1* (launched on December 19, 1974) and *Symphonie 2* (launched on August 27, 1975). They enabled simultaneous communication between several earth stations in the 4 and 6 GHz (gigahertz) frequency ranges. During their lifetime, joint German–French centers controlled both satellites on a time-shared basis. After ten years, the satellites were switched off and removed from the geostationary orbit into the so-called graveyard orbit, an orbit outside the path of other satellite traffic.

Symphonie resulted from a desire to instantiate the Élysée Treaty, but it also responded to a major setback for European integration.

In 1964, the U.S. government had succeeded in establishing an international organization for satellite-based telecommunication (INTELSAT). Western Europe feared an American monopoly in this crucial field of technology, economy, and military security and responded by initiating the European Conference on Satellite Telecommunication (CETS). As early as 1967, Britain opted out of the European program, and developed an autonomous British satellite within the framework of an Anglo–American military space communications program, thereby torpedoing European efforts to counter the U.S. initiative. Furthermore, Symphonie resulted from both France's and West Germany's acute awareness that national projects in satellite technology tended to exceed national capacities. Both the French SARGOS project and the West German OLYMPIA project to develop communication satellites (in the French case with the aim of ensuring transmission to the former colonies in North Africa, and in the West German case to allow TV transmission of the Munich 1972 Summer Olympics) quickly ran into national limits, necessitating "internationalization." For France and West Germany, bilateral cooperation made sense from a technological, economic, and political perspective.

Symphonie caused resentment and concerns that a Franco-German solo attempt would create an enduring political and technoscientific axis in Europe's space activities marginalizing the smaller European nations. To overcome these tensions, others were invited to join, and finally Belgium participated with a share of 4 per cent, while Italy declined the offer to take over responsibility for the apogee engine of the satellite.

In many ways, the program was a favorable one. The two spacecraft, *Symphonie 1* and *Symphonie 2*, as the first triaxially stabilized civil communication satellites, provided 600 telephone channels or two color TV channels. The innovative apogee engine could effectively launch the satellites into a geostationary orbit. Most of its technical components and two-thirds of its electronic devices were developed by European enterprises. A Franco-German industrial consortium that tied together *Aérospatiale*, SAT, and Thompson-CSF on the French side with AEG-*Telefunken*, MBB, and *Siemens* on the German side produced the functional model, the prototype, and the two spacecraft. While creating cutting-edge technical solutions for hitherto unknown technoscientific problems, such as hot gas attitude control systems, the Franco-German experts gained

Fig. 8.4 **A Symphonie for Europe:** *The Franco-German Symphonie satellite program made it into a stamp series of the West German Postal Office (Deutsche Bundespost) on industry and technology. The 5 Pfennig stamp was first issued on November 14, 1975, shortly after Symphonie 2 was launched into orbit. The project name, which connoted unity, harmony, and beauty, sought to become programmatic for Space Europe. In the end, however, the Symphonie project raised concerns of a Franco-German solo attempt in space and was hindered as a commercial undertaking by the U.S. government.*

experience—independently, for the first time, from the U.S.—in complex attitude maneuvers with geostationary satellites.

But independence from the U.S. had its limitations in high-tech sectors such as communication and space technologies, because the Franco-German program had to learn the hard way: the European space community bitterly remembers this lesson about the asymmetries of space power even today. Thus, a closer look at the implementation of the Symphonie program, which was only possible with support from the U.S., enables us to gain insight into the fragile nature of transatlantic space collaboration in the post-Apollo era, when NASA tried to adjust to drastic budget cuts and the U.S. government was desperately searching for a new overall space strategy.

What happened? Why has this experience been engraved so deeply in the collective memory of the European space community? Initially, the two Symphonie satellites should have been launched by the Europa 2 rocket, which was designed to carry spacecraft into geostationary orbits. Once Europa 2 failed with the unsuccessful F11 test flight in November 1971, and the overall ELDO program was canceled in April 1973, Europe had to purchase launching capacities from the U.S. Finally, a Thor-Delta launched *Symphonie 1* from Cape Canaveral on December 19, 1974, two years after the Olympic Summer games of 1972, which had been broadcast via INTELSAT IV satellites. *Symphonie 2* reached orbit on August 27, 1975. The satellites enabled a series of effective radio and television transmission tests, telexes and all kinds of data, which proved the great economic potential of satellite-based communication technologies. The reported conversation between French President Giscard d'Estaing and West German Chancellor Helmut Schmidt in January 1975 was certainly a symbolic act of European technological collaboration. The TV broadcast of the launch of a Soviet cosmonaut crew as part of the American–Soviet Apollo–Soyuz test project in July 1975 was an outstanding European media event.

The resulting financial and technopolitical costs were high, however, and the frustration among the European actors was even greater. The U.S. administration only agreed to the tripartite contract between NASA and their respective French and German partner institutions on the condition that Symphonie was used solely for experimental purposes and that no commercial activity would result from it. Washington attempted to control the market for satellite-based communication technologies by sheltering the U.S.-led INTELSAT consortium from European competition. European experts had to learn that a large and costly program, which was designed to open up a promising high-tech market for European business and which proved highly successfully technically, could be thwarted by U.S. interference. Furthermore, Washington taught European space actors another lesson: even when NASA had given its approval, a project could be stopped either by the U.S. administration or by Congress. This happened when the Department of Defense canceled the delivery of attitude control rockets and of infrared horizon sensors, needed to deploy the satellite antennas, just four weeks before the agreed date. A number of similar cases in the post-Apollo era solidified the European perception that, at least in the space business, the

U.S. was not a reliable partner. The two driving forces of European collaboration again experienced the effects of American supremacy in space. Being subordinated to American interests pushed Europe further to develop its own space programs in order to gain autonomy.[27]

European efforts to cooperate in space did not come easily. On the contrary, experts and governments had to go through a painful process of learning how to organize institutional collaboration in such a way that the fragile equilibrium between national and transnational interests could be preserved. With ELDO, the learning curve was particularly steep, and the cost, which ultimately had to be covered by European taxpayers, was enormous. The question of which projects should be organized nationally and which transnationally remained contested. In West Germany, for example, the scientific community as well as industry never ceased to complain about the size of the national program, which in their view was much too small in relation to the expenditures for transnational programs. The argument was that only a strong national knowledge base allowed successful involvement in transnational projects on an equal footing. The primary reference point was France, which to West German science and industry represented a successful national program that enabled the country to exercise leadership in Europe-wide projects. This controversy exemplifies again that Space Europe was never free from vested national interests.

How did Space Europe respond to the *technical* disaster of the Europa launcher program? What did European actors learn from the *political* disasters of ELDO and Symphonie?

Space Europe Western Mode: ESA

November 10, 1987 had been a long day for Europe's ministers of research and for the director-general of the European Space Agency (ESA) and his support staff. The tables of the conference room in The Hague were still packed with thick folders of material, draft versions of programs and resolutions, empty coffee and tea cups, and beverage containers. Only two people were left in the room: Reimar Lüst, the ESA's director-general, and Heinz

Riesenhuber, West Germany's minister of research and technology and president of the ESA's council of ministers. After many hours of heated debate, both men were exhausted but perfectly happy. They felt as though they had experienced something unique, something that happens once in a lifetime.[28] An unexpected success had materialized; they had achieved much more than they could have hoped. Both of them, and in fact other participants as well, recall the November 1987 ESA council meeting as a magic moment, a seminal day in Europe's space history. For the European space community, The Hague is a site of memory, where some perspicacious actors took brave decisions to offer Space Europe a bright future.

Lüst and Riesenhuber had convinced the ministers of the ESA member states to agree to a package deal that allowed for the long-term planning of programs. In the previous January 1985 council meeting in Rome, Lüst had already succeeded in launching a new program under the forward-looking name "Horizon 2000." That step had literally opened a new horizon for Europe in space. The member states had agreed to an ambitious long-term budget plan for the years 1985 to 1995, increasing annual expenditures more than 50 percent. Much to Lüst's satisfaction, ESA's science program had been boosted by a net rise of 5 percent per year, the first increase of the science budget in 15 years. European experts could now plan small, medium-sized, and even large-scale projects with a time span of 20 years. Horizon 2000 laid the foundation for remarkable record missions of European space experts such as the Cassini–Huygens project that eventually, in 2005, landed the space probe *Huygens* on the Saturn moon Titan. These missions would understandably evoke jealousy on the part of their U.S. colleagues—no doubt the highest form of scientific recognition. Furthermore, with a mandatory charter compelling every member state to contribute a defined share, Horizon 2000 proved a "source of stability."[29] In contrast to the stable science part, the large-scale technical elements of the ESA's program were constantly in flux and had to be negotiated anew whenever member states wished to alter their degree of participation.

In Rome, Lüst had been supported by the smaller member states. In The Hague, Lüst and Riesenhuber were confronted with the much more difficult task of balancing the conflicting interests and priorities of the big three: France, West Germany, and Britain. The

Fig. 8.5 **A Training Camp for European Astronauts:** *To integrate ESA's astronaut training facilities in a joint, truly European center was one of numerous achievements that Reimar Lüst recorded during his mandate as ESA's Director General. In 1987, Lüst and the German minister of research and technology Heinz Riesenhuber, by then also president of ESA's council of ministers, convinced the member states to overcome national interest conflicts and to agree on "Horizon 2000," probably ESA's boldest program ever. The picture shows training facilities of the European Astronauts Centre which is located at the German Aerospace Center in Cologne.*

least disputed part of the package was the new launcher Ariane 5. French minister of research Hubert Curien could point not only to the undisputed technoscientific achievements of the Ariane launcher family since the first take-off from Kourou on Christmas Eve 1979. Moreover, in the economic sphere, the European launcher had overtaken its U.S. competitors in the number of commercial satellites placed into orbit.

Whereas the French had their stakes in Ariane, the Germans focused on *Columbu*s, a manned space station originally planned to be an independent, solely European undertaking, but which was eventually integrated into the International Space Station (ISS).

Manned spaceflight, as a new focal point in West German and European space expertise, dates back to the early 1970s. The technoscientific heart of Europe's activities in manned spaceflight was Spacelab. This pressurized laboratory module was rooted in the post-Apollo era of NASA, when European nations were invited to participate in the Space Shuttle program. Spacelab started as a West German project and was Europeanized in 1972 when ESRO took over responsibility for it. For West Germany, the reusable laboratory offered a convenient way out of the principal dilemma of choosing between European and transatlantic alternatives for space collaboration. Contributing to this scientifically and technically sophisticated project provided West Germany the opportunity to enlarge national expertise in technically advanced launcher construction, to gain experience in the management of complex space missions, and, finally, to take a leading position in micro-gravity research which, it was thought, would become a key technology in the future. In August 1973, NASA and ESRO signed a memorandum of understanding to build the science laboratory and to use it on Space Shuttle flights.

When European astronaut Ulf Merbold returned from the first Spacelab mission in December 1983, he and his five crew companions from Space Shuttle *Columbia* had performed as many as seventy-two experiments in a variety of scientific disciplines. The mission opened the future for the ESA in manned spaceflight, all the more so as industrial contractors such as MBB and Alitalia had started examining possible follow-up projects, including plans for an independent European space station under the label Eureca.[30]

Spacelab became top priority, and more and more resources were poured into this prestigious ESA project, in cooperation with the U.S. Consequently, it was the West German government that pushed the ESA to continue working with the Shuttle–Spacelab system by participating in NASA studies on a future space station, which led to *Columbus* as the European module for the U.S.-led International Space Station.

The Europeans also investigated space transportation elements to complement the *Columbus* module. Candidates included a winged reusable booster rocket for the Ariane 5 and a multipurpose "space tug" for servicing satellites and transporting cargo to *Columbus*. The French proposed *Hermes*, a reusable shuttle, imitating those of the Soviet Union and the U.S., which had been designed by the *Centre National d'Études Spatiales* (CNES) as early as the mid-1970s and was offered to the ESA for Europeanization in 1985. France gained support from West Germany, which in return received backing for *Columbus* by France.

But both *Columbus* and *Hermes* met fierce resistance from Britain, the third party of the big three of Europe in space. With Hotol (**Ho**rizontal **t**ake-**o**ff and **l**anding), an air-breathing spaceplane conceptualized by Rolls Royce and British Aerospace in 1986, Britain had already developed its own ideas for a European shuttle. Furthermore, Britain again demonstrated that it was one of the most reluctant and unreliable partners in European space collaboration. While at the Council meeting in Rome, Downing Street had been supportive of the ESA's future program; however, Margret Thatcher's cabinet afterwards adopted a very critical stance towards the costly manned spaceflight activities. When the ESA Council ministers convened over dinner for a first exchange of thoughts the evening before the decisive meeting in The Hague, British Research Minister Kenneth Clarke made no secret of his country's unwillingness to agree to the intended package deal. Famous at home for his controversial pro-European-integration stance, he furiously rejected every effort to expand the ESA budget as proposed. He even went so far as to call his European colleagues "mollycoddlers" in the event that they refused to follow his example.

ESA's director-general Reimar Lüst spent the night working out a plan to minimize the effects of British resistance. He reversed the alphabetical speaking order, which usually started with Belgium and ended with the United Kingdom. Being first to speak the next morning, the Cambridge-trained British minister used his full eloquence to try to convince his colleagues that the package deal was a bad idea for Europe. However, all the other representatives remained untouched by British opposition and supported the ESA directorate's long-term planning.

Lüst has been characterized as a man of action, a mover and shaker in the field of science and science policy alike.[31] In fact, the outcome of The Hague council meeting, which came as a surprise to all participants as well as to outside observers, displays evidence of his fine diplomatic acumen and managerial capabilities. His ability to mobilize the resources of the ESA member states for a leap in European space activities was based on his status as an undisputed technoscientific expert. Lüst combined personal virtue, familiarity, and charisma, a typical example of the key figures that have shaped "late modern" technoscience.[32] The political representatives of Space Europe had full trust in his expertise, for he had earlier shown that he was not only able to win state support for large-scale space missions but also to carry them out effectively.

Towards European Integration in Space

After the ESA Council meeting in The Hague, the British government withdrew from the programs of the package deal, whereas

Table 8.1 Participation of the member states in the large-scale technical projects of the ESA, 1987

Member state	Ariane 5 (in %)	Columbus (in %)	Hermes (in %)
Austria	0.4	–	0.5
Belgium	6	5	6.4
Denmark	0.5	10	0.5
France	45	13.8	45
Germany	22	38	30
Ireland	0.2	–	–
Italy	15	25	12 – 15
Netherlands	2.0 – 2.5	1.0 – 1.5	1.5 – 3
Norway	0.4	0.4	–
Spain	3	6	–
Sweden	2	–	–
Switzerland	2	–	1.5
United Kingdom	–	–	–

all the other member states subscribed to it. The differences in the members' participation (see Table 8.1) display the constellation of interests, hierarchy, and power balance in the 1980s and hence allow us to gain some general insights into Space Europe both as a technoscientific institutional entity and as an expert community. Additionally, these differences can serve a starting point for identifying changing patterns of Space Europe since the end of the Cold War, which will be briefly summarized here in half a dozen points.

(1) The 1987 package deal was a turning point in the history of Space Europe. On the one hand, the Council meeting's decisions stabilized ESA's institutional collaborative patterns and enabled the identification of a common purpose in Europe's overarching interest vis-à-vis the U.S. On the other hand, it alienated Britain from its continental partners. From a technoscientific perspective, the balance shows ambivalence as well. The Hague meeting paved the way for Ariane 5, which evolved into the ESA's workhorse and became the most successful commercial launcher ever. The space laboratory *Columbus* was launched into orbit by Space Shuttle *Atlantis* and attached to the International Space Station in February 2008, some twenty years after the meeting at The Hague. It is expected to operate for ten years, enabling European

scientists to perform a multitude of experiments ranging from fluid science to physiology. The reusable shuttle *Hermes*, on the other hand, was never realized. After a long series of changes to the concept and design, the overall project was stopped in 1992, giving the ESA another opportunity for "learning from our mistakes."[33]

(2) Two decades earlier, Space Europe had already had an opportunity to learn from earlier attempts, and it used this opportunity with remarkably consistency. In two package deals of December 1972 and July 1973, ELDO, ESRO, and CETS, the European Conference on Satellite Communications, were transformed into a single institution, knitting together the expert communities of space science, launcher technology, and communications satellite development. The ESA, which was officially founded in April 1975, also absorbed the European Space Conferences, a formerly independent political body of the member states, using it as the basis for its institutional decision-making structure, the ESA Council. Through this reorganization, the ESA gradually emancipated its technoscientific knowledge capacities from domination by the member states' fragmented political interests.

Parallel to these fundamental structural shifts, the ESA's culture of producing and applying technoscientific knowledge also underwent three major alterations in order to correct the failures of the ESRO and ELDO.[34] First, the overall program was divided into a two-tier system. The mandatory science program and administrative costs were covered by the general budget to which all member states contributed according to the size of their GNP. All other programs, in particular the development of launchers and satellites, were run as essentially an open shopping list from which the member states were free to decide whether to buy, and in what amount. Second, projects that originated from national programs could be Europeanized by offering member states the option to contribute to them. The most visible examples of the Europeanization of national projects are Ariane and Spacelab. At the same time, both projects reflect a new ESA strategy rather than dividing the work equally among all members, one country was assigned the technoscientific leadership and ultimate responsibility for the project. In this case France led launcher development and West

Germany manned space research. Third, the ESA preferred to award subcontracts to transnational consortia, thus fostering European integration on the industrial level and improving the competitiveness of European industries in the globally highly concentrated aerospace business. This strategic move led to the formation of Arianespace in 1980, the world's first commercial space transportation company, which in 2008 had twenty-four shareholders from ten European countries. With Arianespace, "the Europe of big technological projects with continental and global ambitions was continued."[35]

(3) Space Europe shows a discrete geography that, on the one hand, overlaps with the geography of the political integration of Europe and, on the other hand, differs from the polity of integrated Europe. A first obvious difference is marked by the inclusion of some non-European territories. Mapping the geography of Space Europe means including the Australian rocket range Woomera and the equatorial spaceport Kourou in French Guyana. Both sites developed into technoscapes decoupled from their sociocultural environments and were shaped by technoscientific experts under the sheltering umbrella of military forces. In Kourou the French Foreign Legion had to protect the spaceport, which was contested by indigenous people, as recurrent clashes between the Legion and the regional Creoles showed.[36]

A second difference is marked by the participating nations. Not all member countries of the European Union are members of the ESA, and not all ESA member states are members of the EU. When Romania signed its accession agreement to the ESA Convention in January 2011, the member states (nineteen at the time) included the non-EU members Switzerland and Austria; even Canada sits on the ESA Council and takes part in some projects under a cooperation agreement. Furthermore, the relationship between the ESA and the European Union has been contentious despite all official claims that the two organizations work in harmony (e.g. they "share a joint European Strategy for Space and have together developed the European Space Policy.")[37] After the Single European Act of 1987, the European Commission, as well as the European Parliament, increasingly pushed for a more active EU role in space. In 1989, Commission President Jacques Delors and ESA Director-General Reimar Lüst met in Brussels for an exchange of ideas and left by mutually ensuring solidarity and "respect for [one] another's competencies."³⁸ The creation of the Space Advisor Group (SAG) as a high-level discussion forum for representatives from the ESA, national governments, and the EU Commission only briefly stemmed the growing EU ambition to establish its own autonomous space agenda. In 2003, a White Paper of the EU Commission identified space as "a new frontier for an expanding union" and claimed "new responsibilities for driving, funding and coordinating activities within an extended space policy."³⁹ In the same year, the Galileo project to build a European satellite radio navigation system aimed to guarantee technical, and thus political, independence from the U.S. GPS system opened the door wide for the European Union, which took control of a large-scale space project for the first time. In 2004, the ESA and the EU adopted a framework agreement that created a joint Space Council. Consequently, in a recently published strategy paper, the European Commission has underlined that "space serves to cement the EU's position as a major player on the international stage and contribute to the Union's economic and political independence." The same paper criticizes the fact that space management in Europe "remains fragmented and international investments segregated," and calls on the ESA to gear its activities "solely towards EU programmes."⁴⁰ The Commission's attack on the ESA indicates once more that Space Europe is being driven, by different institutional bodies and agendas, based on different expert cultures. Although these cultures ought, in theory, to harmonize, in practice, they diverge, underlining the dichotomies and ambivalences of Knowledge Europe.

(4) Space Europe originally grew from a sense of concern about increasing U.S. and Soviet dominance in the scientific and public realm. Since the fall of the Soviet Union, the U.S. has continued to serve as a reference point. Space Europe has oscillated between collaboration and conflict with the U.S. Space Europe can thus also be framed as part of a transatlantic debate about how to manage technoscientific projects, infrastructures, and institutions that has endured throughout The Long Twentieth Century. Already, the early history of ESRO and ELDO can be narrated as an effort to bridge the technological and management gap between the European and the U.S. space programs.⁴¹ In fact, in the late 1960s, when a French

Fig. 8.6 A British Vision for Europe: *When the ESA envisioned a future European space vehicle, British Aerospace and Rolls Royce jointly proposed Hotol. The reusable winged launcher was to be fitted with an air-breathing engine. Development began in 1986, when the British government provided funding for the project. After Britain failed to Europeanize Hotol in 1987, the project came to an end. In 2009, however, the much smaller version Skylon gained support from ESA. The British company Reaction Engines Limited developed a prototype of Skylon, which would be able to land on a runway like a conventional aeroplane.*

journalist invented the concepts of the "American challenge" and the "technological gap" between the U.S. and Europe, concepts that became dominant in public discourse and European politics, he also referred specifically to the management gap in space technology. Politicians all over Western Europe took this seemingly widening gap as a given and, like the hard-nosed former West German Minister of Defense and of Finances Franz Josef Strauss, called for a concerted European response.⁴² While *Sputnik* and the landing of Apollo 11 on the moon were turning points in political and public discourse, for the space community, the post-Apollo program of NASA became a formative experience. European space experts came to know the U.S. as an unreliable partner in transatlantic collaborations, an experience their colleagues from the nuclear sector had already encountered in the postwar reconstruction period. They responded by strengthening their own integrative efforts to gain independence from American hegemony. NASA,

Fig. 8.7 A Post-Colonial Spaceport for Europe: *Kourou in French Guiana was selected in 1964 to become the spaceport of France and in 1975 was offered to ESA as well. Located close to the Atlantic coastline and surrounded by tropical forests, the launch site is a highly artificial technoscape, and for the indigenous people, it is a brute manifestation of European colonialism. The image shows the rollout of the first Ariane 4 rocket at Kourou in April 1988.*

once an object of admiration and imitation, turned into a source of hindrance and vexation, again fostering efforts to invigorate European collaboration.

Space Europe opened new forms of transnational cooperation. Years before the Berlin Wall was dismantled, experts from the ESA, as well as from individual member states, started to negotiate bilateral programs with the Soviet space complex. With the Cold War coming to an end, Soviet spaceflight went through a severe transition phase, suffered from drastic budget cuts, and desperately searched for new funding opportunities. Experts on both side of this burgeoning collaboration initially encountered opposition and interference from state authorities, yet they were strongly supported when collaboration developed as a main political goal, both in Eastern and Western Europe. The 1990s witnessed an accelerated integration of Western and Eastern technical and economic space complexes, which fundamentally changed the geography of Space Europe. In 1994, the Eastern Interkosmos program was disbanded, and Russia became a formal member of Eutelsat, the European telecommunications satellite organization. At the end of the century, an inventory of the Centre for the Analysis of European Security listed no less than eighty-seven collaborative projects and joint ventures in space and communication technologies.[43] The shifting geography of Space Europe after the Cold War even allowed for the integration of the spatial core of the Russian space complex, which was part of the country's industrial research and defense ministry system, into Space Europe's knowledge society.[44] "Astronauts" who were designated to fly on joint ESA–Soyuz missions were now trained at the Yuri Gagarin Cosmonaut Training Center in the "Star City" Zvyozdny Gorodok, located near Moscow, a military facility previously closed to the public. As in the formative period of Space Europe, the European–Russian space collaboration saw experts as initiators while state and business actors later integrated technoscientific cooperation into the European political economy. Cooperation between the ESA and its member states and Russia can thus be understood as a multi-level exchange of resources to enhance mutual benefits, which aimed both at augmenting the knowledge base and the global economic competitiveness of Space Europe. It can also be understood as a signal to the U.S. that post-Cold War Europe could combine the Continent's Western and Eastern resources in joint programs.

(5) Space Europe was a child of the Cold War. When Edoardo Amaldi and his fellow scientists proposed to pool European expertise in space sciences and rocket technology, they responded to the emerging superpower space race. But compared to the U.S. and the Soviet Space programs, Space Europe charted discrete pathways in two directions. First, the European efforts were neither directed to a single overarching mission such as reaching the moon, nor were they driven by interests in military and security

affairs in the first place. On the contrary, military activities were deliberately excluded by the ELDO, ESRO, and ESA programs, and this exclusion was carefully observed by the expert communities. In reality, the dual-use character of the aerospace sector led Europe to develop technologies potentially had military purposes as well.[45] European space experts, however, began to devote themselves to problems of civilian and international security only later, in the wake of the terrorist attacks in the U.S. of September 11, 2001 ("9/11") and the new fears that they brought to light. To meet this threat, the European Union called for new programs in satellite-based research and remote sensing, distressing many members of the ESA's expert communities. Thus, it was not ESA, but the policy-driven arm of Space Europe, the European Union, that drove this regime change.

Second, the scientific impulse which had launched Europe into space in the early 1960s continued to work as a core driving force. No doubt the European space programs also served political and economic ends, and the ESA developed into a major player to prove European competitiveness in high-tech sectors. Primarily, though, European space programs grew out of technoscientific expert knowledge, and their evaluation was based on the principles of peer-review. Consequently, it was the mandatory science program that knitted both the large and the small member states into a unified community, still based on national interests, but tracing an overarching agenda defined by ESA experts. The ESA's principal technical center, ESTEC in Noordwijk, is a prime example of European collaboration. The center's daily work was cooperation after all, notwithstanding structurally-distinct individual "interests, abilities, allegiances, identities, knowledge, and goals," thus substantiating the European ideology of "unity in diversity."[46] Experts cooperated in the day-to-day work at numerous ESA research centers widely dispersed across Europe. These stretched from Frascati in Italy to Kiruna in Sweden, from Noordwijk on the Dutch coastline to Oberpfaffenhofen near Munich, with the headquarters located center-stage in Paris. These were the physical sites, whereas the hidden integration of Europe via expert networks in space science and technology took place on long-term projects that united thousands of experts from all member states. The working culture of these European expert networks displays every feature of technoscientific practice. Following the established principles of dispute and controversy, projects often fragmented experts into opposing networks with discrete agendas. Thus, for example, at least since the early 1970s, when manned spaceflight with the resource-intensive Spacelab project gained momentum, a deep split separated the European space community into rival factions, who argued over priorities in the ESA's agenda. It was precisely the politically tinged character of manned spaceflight that caused such vigorous protest from space scientists, because it seemed to endanger the apolitical mission of the European space community. Even Reimar Lüst, who in his tenure as ESA director-general had pushed the space laboratory *Columbus* and the spaceplane *Hermes*, could not resist strongly criticizing Europe's participation in the International Space Station. No convincing concept for its utilization, he argued, "has as yet been developed."[47]

(6) European collaboration in space has given political scientists headaches. Their theories have hardly matched the development of the ESA. When David Mitrany, both "father" of functionalist theory and mastermind of European integration, summarized the effects of European collaboration in satellites and space travel in 1971, he found evidence that it had "in truth reached the 'no man's land' of sovereignty."[48] While Mitrany and his followers believed that technoscientific cooperation should be politically prioritized because it was a less contentious field than other political arenas, the first phases of European space collaboration were contested at all levels. Furthermore, neo-functionalists like the American political scientist Ernst B. Haas, who stressed the importance of economic self-interest for any successful integration project, would have had a hard time identifying the procedures that enabled ELDO and ESRO to clearly meet the national economic self-interests of its member states.[49] Neo-realist perspectives, on the other hand, which point to rational state behavior, the protection of national autonomy, and the centrality of intergovernmental negotiation have been unable to cope with the significance of non-state actors such as technoscientific experts and institutions like the ESA as "an independent variable" for space Europe.[50]

Fig. 8.8 A Space for Space Europe: *At the European Space Research and Technology Centre (ESTEC) at Noordwijk, the Netherlands, technoscientific experts from the various ESA member states jointly develop and test European space artifacts, thus materializing the European ideology of "unity in diversity." The picture shows vibrational tests of the Rosetta space probe at ESTEC. The robotic spacecraft Rosetta was launched on March 2, 2004, on an Ariane 5 rocket with the mission to study the comet 67P/Churyumov–Gerasimenko.*

Space Europe, to summarize, has to be understood not as a monolithic entity, but rather as a multifaceted technology combining multiple political entities and expert networks, each with different professional agendas, economic interests, technical orientations, and conceptions of transnational collaboration and European integration. Like Nuclear Europe, Space Europe shows a rich variety of transnational collaborations that range from the local to the global level. Its history is far from being a linear chronology of progress and bold achievements, as often depicted. Despite the impressive position of Ariane in the market for launching commercial satellites and a great number of effective space missions that have given European experts a lead both in space sciences and remote sensing, this history is full of conflicts and dichotomies. Yet it is also a history of transnational collaborations initiated and driven by expert communities.

L'EXPRESS

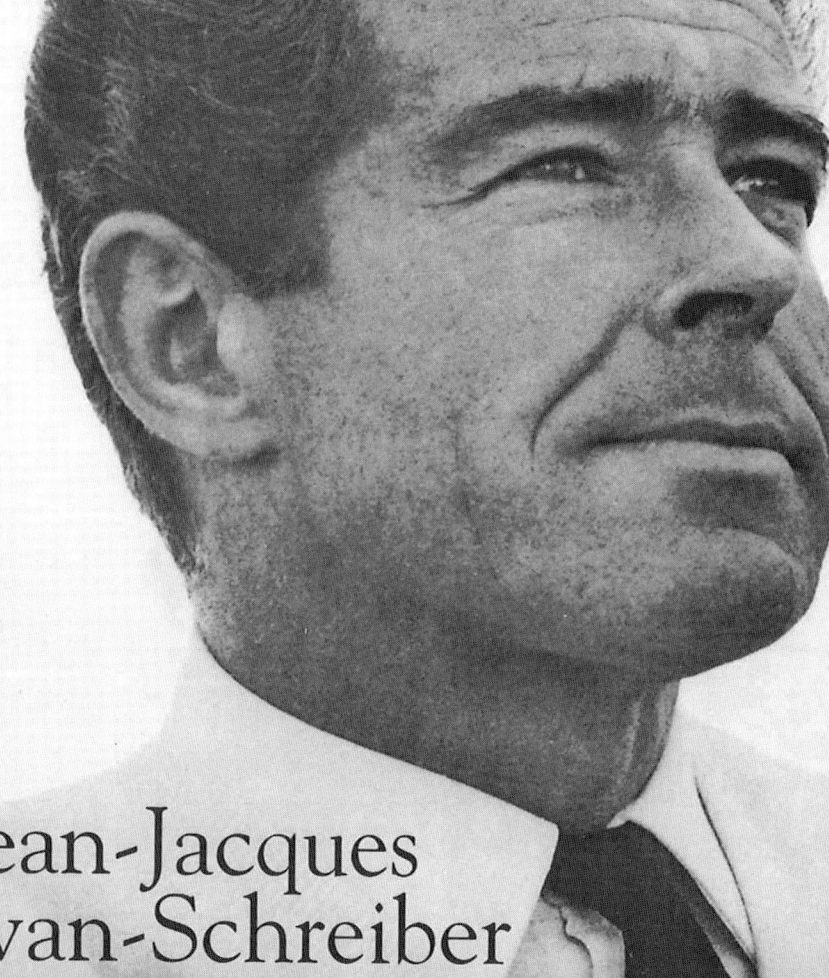

Jean-Jacques
Servan-Schreiber
1924-2006
Le pionnier

L'URGENCE ÉCOLOGI[QUE]
L'appel de Nicolas H[ulot]

JEAN-JACQUES SERVAN-SCHREIBER, FONDATEUR DE L'EXPRESS
« Il faut dire la vérité, telle que nous la voyons »

9
Experts' Europe from a Bird's-Eye View

Lisbon 2000: Building Europe on Knowledge

The early years of the nineteenth and twentieth centuries saw creative outbursts of future visions. In both 1800 and 1900, utopian and dystopian ideas flourished. Some people envisioned a bright future, while others anxiously expected crises and catastrophes.[1]

In March 2000, it was perhaps again the optimistic climate of a new century and a new millennium that spurred European leaders to formulate a vision for the European Union. Lisbon, Portugal's scenic capital on the banks of river Tejo, hosted a special meeting of the European Council that aimed at no less than a paradigm shift in the history of Europe as a social and political entity. The Portuguese Presidency of the European Union succeeded in securing a compromise that interwove numerous single political actions and diverging interests into one coherent program. The Lisbon Agenda, as it came to be known, was a ten-year plan for making Europe "the most dynamic and competitive knowledge-based economy in the world, capable of sustainable economic growth with more and better jobs and greater social cohesion, and respect for the environment." The Lisbon Agenda prioritized the strategic triangle of strengthening employment, economic reform, and social cohesion. The key target of this ambitious plan was the "shift to a digital, knowledge-based economy."[2]

In Lisbon, political leaders embarked on an idea that the European Commission had circulated some months earlier. The goal was to turn Europe into a fully integrated space for knowledge production, the "European Research Area" (ERA). A high standard of knowledge and expertise would anchor Europe's new identity. The previous year, European leaders in Bologna had already agreed to ensure comparability in the standards and quality of higher education credentials and thus to enhance cross-national students mobility within the European Union and beyond. The long list of actions to be launched by the European Union included structural steps such as strengthening the relation between the various scientific and technical organizations; it also included situational initiatives such as creating grants for researchers from outside Europe. The political agenda culminated in a quantifiable target: raising European-wide expenditures for research and development to 3 percent of the overall Gross National Product. With the Lisbon Agenda, the quest for a transnational European identity shifted to the realm of science and technology.

Unlike many of the initiatives discussed earlier in this book, the inspiration for this novel political vision of Europe's future originated not from the technosciences but rather from the social sciences. The European Commission had consulted various prominent experts, especially from science and technology studies and from innovation research. Manuel Castells, for example, the author of the widely discussed trilogy *The Information Age* (1996–98),[3] contributed his ideas on fostering the shift to a digital economy based on information and communication technologies. The evolutionary economists Christopher Freeman, Giovanni Dosi, Bengt-Åke Lundvall, Richard Nelson, Carlota Perez, and Luc Soete emphasized the importance of innovations resulting from technoscientific knowledge. The most influential expert, however, was Maria João Rodrigues, later named "mother of the Lisbon Strategy." Rodrigues was a connecting link between academia and politics, a highly qualified expert and an experienced political administrator at the same time. Professor of Economics at the University Institute of Lisbon and Minister for Qualifications and Employment under Portugal's Prime Minister António Guterres from 1995 to 1997, Rodrigues succeeded in

Fig. 9.1 Envisioning a Greater Europe of Higher Education: *When European leaders met in Bologna in 1999, they agreed on joint standards for academic education. At present, forty-seven nations are participating in the Bologna process to harmonize academic training based on the European Credit Transfer System (ECTS). The black area on the map depicts the European Higher Education Area with its vision of a greater Europe. The two gray spots in Europe mark Belarus and Macedonia, which have not yet signed the Bologna declaration.*

convincing European leaders that knowledge creation was crucial for Europe's future.[4] Later, in 2007, when Portugal again chaired the European Union, Rodrigues managed to forge the Lisbon Treaty that established a constitution for Europe. Coming into force in December 2009, Europe's new constitution replaced the foundational Treaties of Rome (1957) and Maastricht (1993).

According to the Lisbon Agenda, the challenges of the twenty-first century demanded far-reaching alterations to Europe's research and innovations policy. The Commissioners proposed several goals that, they hoped, would create a uniquely European response to future challenges. This included networking existing "centers of excellence" and creating new ones; improving instruments and resources to encourage investments in research and innovation creating ties among scientific communities, companies, and researchers in Western and Eastern Europe; and finally, promoting "common social and ethical values in scientific and technological matters."[5]

European leaders' ultimate political aim was to create a robust and socially cohesive knowledge society, thereby spurring the transition to an economy and a society based on technoscientific knowledge. Transforming Europe's national resources in research and technology into a single space of knowledge creation would provide impetus to the stagnating project of European integration. With the Lisbon Agenda, the European Commission positioned science and technology as vanguards of political and social integration. Moreover, the Commission strengthened its position as "central soundingboard of a neo-liberal vision of the future"—a future to be based on technoscientific expertise.[6]

Flying High: Is there a European Knowledge Society?

The frequently repeated proposal for an integrated European Research Area that would then lead to a powerful European knowledge society generated a new master narrative. For policymakers, master narratives provide a comprehensive explanation of historical change and serve both as starting point and justification for programs and future projections. They make sense of history, as the French philosopher Jean-François Lyotard has noted.[7] Since the Lisbon Council, the traditional master narrative of a Europe built on political and economic integration has been gradually replaced by a new narrative: that of a unified European knowledge society, an idea that emerged from the tradition of the European university. By referring to the European origins of the university, policymakers drew a direct historical line from the Middles Ages to the present and extended it into the future.[8]

Configuring the new master narrative of the European knowledge society around the university also involved searching for benchmarks outside of Europe. In February of 2005, when José Manuel Barroso, president of the European Commission, proposed establishing a European Technology Institute, finally inaugurated in 2008, he had the famous Massachusetts Institute of Technology (MIT) as his model. Barroso was not the first to advocate a European MIT. In the early 1960s, amid a dense political and public debate on Americanizing the European system of research and technology, scientists and policymakers discussed creating a Europe-based "Atlantic Institute" for science and technology. In the nineteenth century, the modern European university—focused on acquiring and teaching knowledge through laboratories and seminars—had deeply influenced the shaping of the American academic system in general and its research-oriented universities such as Harvard and MIT in particular. During the twentieth century—the "American" century—European elites looked across the Atlantic in search of orientation.[9] The U.S. continued to be a model at the beginning of the twenty-first century, as European leaders invented the tradition of a European knowledge society.

History has always been used to support political goals, from Herodotus to Barroso, from antiquity to the present. So EU leaders and their supporting experts referred to longstanding historical traditions in Europe's culture of knowing and emphasized the novelty of the idea of a European knowledge society at the same time.

Was the Lisbon Agenda truly a paradigm shift in the history of Europe? Was the knowledge society vision novel or deeply embedded in European history? We argue that though trumpeted as new, it had a longer history. Yet this history did not trace a smooth line, crowned by the EU initiative. On the contrary, it is full of complexities and ambivalences across The Long Twentieth Century. We have stressed the multiplicity of knowledge societies in Europe between the poles of national competition and transnational collaboration, between divergence and convergence. The plurality of knowledge societies is crucial here, a continuous flow of different efforts to create and use expertise for societal ends is characteristic of Europe since the mid-nineteenth century. What, then, was new in projecting a Europe founded on expertise, resulting from the Lisbon Agenda?

Much like Barroso, Michel André, a European Commission historian and advisor for research policy, has emphasized the uniqueness of the concept of an integrated European research area on the one hand, and its embeddedness in history on the other. For André, the concept presents one of the rare cases of a "really important" idea that "appeared and disappeared several times" in Europe's history.[10] In the search for its origins, André identifies the "working program in the field of research, science and education" that Commissioner Ralf Dahrendorf proposed in May 1973 as its first manifestation. Lord Dahrendorf's suggestion was "to create an effective single area for European science in which cooperation and competition complement each other in a sensible way."[11] The idea then disappeared and was rediscovered independently in the 1990s by Dahrendorf's distant successor, Antonio Ruberti. In 2000, yet another Commissioner, Philippe Busquin, used the Lisbon Agenda to turn the idea into an official EU policy project.

Fig. 9.2   European Transnationalism, French Style: *A graduate from the* École Polytechnique, *Jean-Jacques Servan-Schreiber was fascinated by science and technology. In 1953, he co-founded the weekly news magazine* L'Express, *which developed into the French equivalent of* TIME *magazine. This picture shows the cover of* L'Express *no. 2888, a special issue in honor of Servan-Schreiber after his death in November 2006. In his political essay* Le Défi Américain *from 1967, Servan-Schreiber saw Europe and the United States engaged in a silent economic war, in which Europe was falling completely behind in management, technology, and research. His international bestseller advocated transnational cooperation in Europe.*

André usefully historicized the European knowledge society's political master narrative. He correctly noted that the history of European collaboration in research and technology is "richer and more complex than is generally acknowledged."[12] However, he did have shortcomings. First, he only attributed the idea of the European research area to prominent political actors. Second, he traced European technoscientific collaboration back no further than the 1970s.

As our earlier chapters have detailed, Europe witnessed a number of initiatives similar to the Lisbon Agenda in the 1950s and 1960s. A particularly interesting case was the debate about a "technological gap" between the Unites States and Europe, launched by the French journalist Jean-Jacques Servan-Schreiber.[13] In 1967, he published his essay *Le défi américain* (*The American Challenge*), which quickly became an international bestseller. Today, we know that the postwar transatlantic gap in technology had already began to close by the time when Servan-Schreiber noticed it. His thesis, however, developed into a driving force that motivated European technoscientific and economic integration. In that same year, 1967, for example, a high-level expert group on *Politique de Recherche Scientifique et Technologique* (PREST) proposed that the European Council could close the gap by fostering European-wide cooperation in information technologies, telecommunications, transport, oceanography, material sciences, environmental sciences, and meteorology. When the Commission declined supporting a transnational system of governance in science and technology, the Council of Ministers took the initiative. In 1970, the Council created the Cooperation in Science and Technology (COST), a loose administrative body of national representatives with high ambitions but without its own budget. From the 1970s onwards, Europe experienced numerous political efforts to stabilize an intergovernmental science and technology policy beyond the subsidiarity principle which narrowly limited the Commission's space for maneuvering.[14] These efforts spread, from the creation of the European Science Foundation in 1974, an umbrella administration that today includes more than seventy organizations devoted to basic research in twenty-one European countries, to the Maastricht Treaty of 1993, which authorized the Commission to promote all research activities deemed necessary to reach its declared goals.

Long before Maastricht, however, the European Commission began continuously expanding its role in creating a Europe based on expertise. In 1984, the Commission launched its first multi-year "Framework Programme for Research and Technological Development," a landmark event. Framework Programmes served as a rapidly growing bureaucratic tool to foster research and economic growth in Europe. While the First Programme (1984–87) budget amounted to 3.3 billion euros, Seventh Programme (2007–13) spending rose to 50.5 billion euros. The Eighth Programme (2014–20), with the telling name "Horizon 2020," will fuel European innovation systems with 80 billion euros. Every project in the Framework Programmes has to include researchers from a minimum number of eligible European countries—interestingly enough including Turkey. Because the primary political aim has been to foster Europe's economic competitiveness, these teams have largely focused on technical applicability. Starting with the Fourth Framework Programme (1994–98), however, the social sciences and even the humanities were included.

The Sixth Framework Programme (2002–7) introduced a number of new tools to promote the creation of an integrated European knowledge society, including the European Technology Institute and the European Research Council (ERC). The latter offers the Commission, for the first time, a way to fund basic research that is not immediately applicable to economic interests. With "Starting Grants," Consolidator Grants," and "Advanced Grants," the Council aims to support Europe's scientific and engineering elite at both the postdoctoral and the senior level. In 2012, the Council created new "Synergy Grants" that aim to constitute multidisciplinary teams of investigators to perform "pioneering frontier research in any field of science, engineering or scholarship."[15]

It is no coincidence that one of most outspoken advocates of integrating Europe's intellectual resources, Helga Nowotny, is currently president of the European Research Council. This *grande dame* of science studies in Europe helped to create the Council, served as its first vice president, and was promoted to president in 2010. Nowotny is "a true European," in the sense that she fundamentally believes in Europe's future as a political and societal entity shaped by knowledge. She personally incarnates the vision of Europe as an integrated research area, a unified knowledge

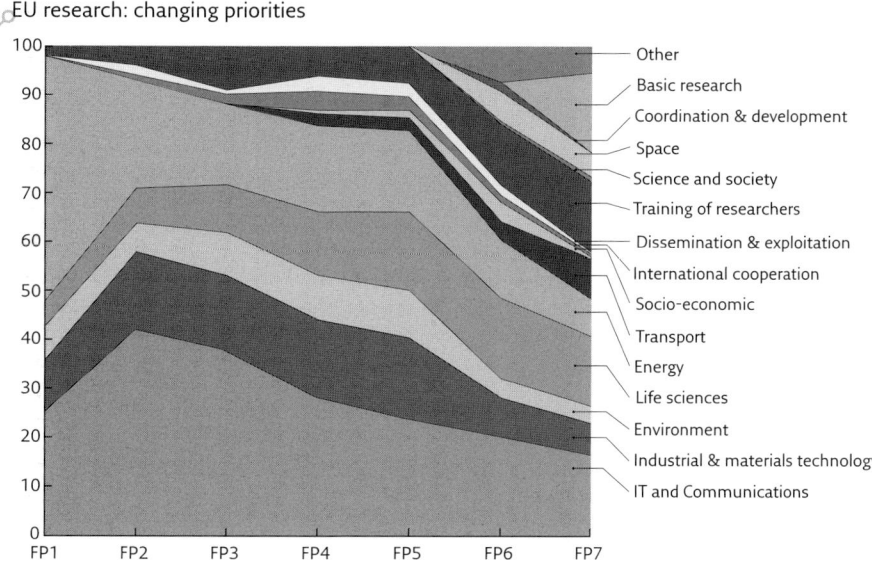

Fig. 9.3 The Changing World of Funding Priorities: *When the European Commission launched its First Framework Programme in 1984, the study of "Science and Society" had high priority. Gradually, as the diagram showing the Research Council budget depicts, the field lost visibility and will continue to shrink in importance in the Eighth Programme, "Horizon 2020," that will span the years 2014–2020. However, the social sciences, and for the first time the humanities as well, are represented in the funding category "Basic Research."*

society that is based on "socially robust" interrelations between science and society, experts, and the public. The European Research Council is "considered a success story in European research policy"; one, if not the only, case of European integration policy that has not met severe criticism from Euro-skeptics. In less than a decade after its inception, the Council succeeded in emerging as Europe's most prestigious funding body. Those lucky few who manage to win grants acquire both a substantial pot of research money and an invaluable helping of social capital and recognition. According to Nowotny, a major reason for the Council's success is that it funds research "in the 19th century, German conception of *Wissenschaft*," that includes the sciences, the social sciences, and the humanities.[16] Nowotny's reference to the nineteenth century again demonstrates the role of history in the politics of building Europe on knowledge.

Nowotny's colleague in science studies, Alfred Nordmann, also understood the Sixth Framework Programme as a paradigm shift in European politics towards science, technology, and innovation. For Nordmann, this Programme, which started in 2006, was nothing less than an experiment with new technologies "in the laboratory of society," where all Europeans contributed in the quest for identity.[17] As rapporteur of the High-Level Expert Group on "Converging Technologies for the European Knowledge Society," Nordmann knew what he was talking about. His group was the European response to a 2002 joint report by the U.S. National Science Foundation and the Department of Commerce urging the amalgamation of nanotechnology, biotechnology, information technology, and cognitive science.[18] When comparing the U.S. and European reports, both Nordmann and outside observers noticed a sharp contrast, even a "clash of cultures."[19] The U.S. analysis featured technological determinism, individualism, and engineering *of* body and mind. There, technological innovation was seen as a means of overcoming limits to achieve human salvation. The European report emphasized the co-construction of technology and society, societal welfare, and engineering *for* body and mind, while social innovations help to realize technical potential. According to Nordmann, the Lisbon Agenda and the experimental tools for fostering new technologies that followed from it have offered Europe a distinct way of innovating deeply embedded in European cultural traditions with the potential to foster a socially-permeable knowledge society. Has Europe, then, finally found its way towards becoming a unified society built on knowledge, and therefore creating a new identity for itself? The "Expert Group on Science and Technology," a prominent cluster of academics in science and technology studies, expressed serious reservations in their January 2007 report to the Commission. *Taking European Knowledge Society Seriously* argued that the European master narrative manifested in the Lisbon Agenda and the related policy practices neglect the public as an indispensable resource for a "robust and sustainable European knowledge society." The Lisbon Agenda, they believed, also "disconnect(s) our knowledge, experience and imaginations from history." In particular, the political notion of a global race to achieve innovations detaches the search for technoscientific progress from a "larger historical trajectory" that locates initiatives in a wider perspective.[20] The initial enthusiasm from Lisbon was quickly followed by a more pessimistic phase of European politics, as it became clear not only that the ambitious goals were unreachable, but also that the suggestive equation of Europe and knowledge drew a somewhat one-sided and simplistic picture.

In this sense, the Lisbon Agenda illuminates how the politics of expertise in Europe did not necessarily seek merely to increase expertise or to share it more easily; it went much further. The above quotation from Nordmann expresses widespread hopes that technoscientific expertise could change and improve society, sometimes radically. Moreover, the idea of building Europe on knowledge strengthened European self-assurance and identification. The U.S. was both a model and a point of comparison in this endeavor. Clearly, the European Commission also had such larger goals in mind when it prioritized knowledge-building as the heart of the Lisbon Agenda.

Flying Low: Are There European Expert Communities?

Starting in the 1950s, Europe undertook a continuous series of political efforts to strengthen cross-national scientific and technical cooperation, as we have shown. In addition to establishing single-issue organizations such as CERN or ESA, the EC developed the Framework Programmes to promote research and thus economic growth. Although individual projects within the Framework Programmes were limited in duration, they created an ever-growing network of experts across technical fields and disciplines. Since 1984, the Programmes have established hundreds of thousands of transnational links. The Fifth Framework Programme alone, which ran from 1998 to 2000, created 180,000 cooperation links. Each funded project included on average fourteen partners from as many as seven European countries.[21] Socio-economic growth assumed an even more central role as the foundation for the new Europe in the late 1990s when the idea of the "European Paradox" gained currency. Europe's overall performance in science in relation to investments in research was seen as excellent, while its technological and commercial performance had continuously declined since the 1980s. Edith Cresson, the European Commissioner for research, asked Paraskevas Caracostas and Ugur Muldur, two economists in her directorate, to analyze why. On the basis of their analysis, Caracostas and Muldur drafted a new innovation policy for the twenty-first century, which they entitled *Society, the Endless Frontier*.[22] The two experts deliberately paraphrased the title of Vannevar Bush's 1945 report to U.S. President Harry Truman, *Science, the Endless Frontier*, which became a kind of Cold War bible for generations of science policymakers. Much like Bush half a century earlier, Caracostas and Muldur advocated a continuous expansion of public research expenditures in order to create technical innovations and consequently to address social issues such as jobs, health, and environment. The European Commission's *Green Paper on Innovation* (1995) and *First Action Plan on Innovation in Europe* (1997) were meant to cure the "European Paradox." By forcefully investing in transnational technoscientific activities, these plans would pave the way for the unified European Research Area built on expertise.

No doubt, these collaborative visions and programs strengthened European integrative forces. They formed transnational teams and communities of scholars and, as such, pushed the Europeanization of science and technology. In 1998, the journal *Nature* concluded that "collaboration is the name of the game" for innovation activities and reported that EU programs stimulated more than 200,000 such links between 1990 and 1996 alone; more than 90 percent cross-national.[23] In fact, these scholarly networks, initiated from the top down, but presented as a bottom-up process based on peer-reviewed quality control, engaged experts in cross-national research with the deliberate goal of invigorating the European knowledge base. After having spent hundreds of billions of euros on these programs over the last few decades, European bureaucrats have been keen to evaluate their effectiveness, particularly their impact on Europe's economic performance. Stacks of research reports have been produced to measure both the quantity and the quality of this impact. The resulting studies are often shaped by ostensibly political agendas, full of mono-causal argumentations, lacking a sound methodology, and struggling with contradictory results. Basically, policymakers, bureaucrats, and the interested public can read into these studies whatever they wish to get out of them. However, most EU officials would agree with the assessment that the Framework Programmes have had "a clear positive impact" in the sense of "improving Europe's scientific, technological, and innovative performance...strengthening Europe's human capital and integrating European research infrastructures," and generating "wider economic, social, and environmental benefits."[24]

In addition to these official assessments, a different body of studies on the Europeanization of science and technology stimulated by the Framework Programmes and related political initiatives has emerged. These latter studies have followed the actors, to use a dictum of Bruno Latour, the famous French sociologist of science. Latour's colleagues from the social sciences have focused on the actual working of these international projects and particularly on how and where the respective research teams publish. Studies of this kind are usually based on scientometric methods to measure the inputs into and the outputs from science. Rather than flying high up in the skies of high-level European research politics, they fly low by focusing on the task of concretely measuring the results of European research activities. Moreover, they are not interested in financial inputs, but in scholarly outputs of Europe's system of expertise.

We now will also fly low by having a closer look at these studies' findings. This sheds light on one of this book's overarching questions, namely whether Europe has experienced the rise of cross-national knowledge societies and expert cultures, or more bluntly: are there European communities of technoscientific experts today? A straightforward question such as this deserves a nuanced answer, which we will offer in four parts:

First: Globalization and Europeanization of knowledge production are closely connected.

The worldwide order of knowledge in the postwar era reflected a center–periphery asymmetry in which the U.S. and the Soviet Union were hegemonic. Globalization and the end of the Cold War made the location of "centers" and "peripheries" less and less evident and facilitated a new knowledge, heterogeneous and less centralized. Communities of experts tend to be increasingly organized on a continental scale. The Europeanization of science and technology is thus an integral part of a global transformative process, which for the lack of a better term has been designated as "continentalization." When considering the collaborative publication of research results, the global community is divided into cross-national groups often constituted along continental lines. Expert networks in Europe count increasingly on their own intellectual resources and strive to strengthen their technological independence from networks on other continents. From 1985 to 1990 alone, for example, in engineering and technology collaboration with colleagues from non-European nations experienced a decline of about 11 percent, while transnational collaboration within the European Union increased 5.5 percent for all fields of science and technology.[25]

Comparing Eastern Europe with Scandinavia indicates that the geography of expertise in Europe exhibits partly vanishing, partly lasting sub-continental, that is, regional, structures. During the Cold War, international collaboration in Eastern European countries was strongly dominated by Comecon (Council for Mutual Economic Assistance) relations, except for Hungary and Poland, which had been open to external cooperation since the early 1980s. After the fall of the Iron Curtain, the strong collaborative bonds in Eastern Europe quickly faded away, replaced by rapidly expanding links with Western countries inside and outside Europe. Consequently, expert transnationalism in Eastern Europe quickly surpassed that of Western Europe. In Hungary, by 1993 every second published research paper was internationally co-authored, while in most Western countries the proportion was only one in four or five.[26] If Eastern Europe has discarded its distinct pattern of knowledge creation over the last two decades, Scandinavian Europe has remained a strongly integrated knowledge space. The Swedish, Danish, Norwegian, and Dutch innovation systems have retained a distinct identity. Based on long-term cultural and political bonds, the Nordic countries have sustained their relatively autonomous configuration in global expertise.[27] However, in absolute numbers their cross-national links with European Union member states are starting to outnumber research collaborations with their Scandinavian neighbors. A similar pattern emerges from the three core continental powers of knowledge creation: Germany, Italy, and France. Long before the year 2000 these nations were closely tied together. Despite the existing high level of integration, collaborations both among these nations and with neighboring Switzerland continued to grow. Significantly, in terms of globalization, all three nations also increased their collaborations with the U.S. during the same period.[28]

The European case also demonstrates that, in a globalized world of expertise, space and regional proximity still matter. An analysis of some 21 million co-authored research publications has revealed that the average collaboration distance per publication increased from 334 kilometers in 1980 to 1,553 kilometers in 2009.[29] The big picture of a globally-interconnected system of knowledge changes, however, if one focuses on regional networks. By examining all co-publications among 313 regions in 33 European countries between 2000 to 2007, a large-scale study found "that the bias to collaborate with physically proximate partners did not decrease, while the bias towards collaboration within territorial borders did decrease over time."[30] In other words, while European national boundaries have become more and more permeable, border-crossing regional knowledge clusters have emerged which often are based on long-term collaboration. Such a spatially-confined knowledge society can also be characterized as a regional creative milieu or regional innovation culture that brings together experts from science, technology, and industry in order to trigger innovation.[31]

Second: The dynamics of the Europeanization of knowledge production vary from discipline to discipline, from technology to technology.

For a long time the primary goal of the European Union's research policy was to enable economic growth by strengthening the scientific and technological base of European industry. Technical innovations resulting from transnational collaboration were to spur the productivity and competitiveness of European businesses. Therefore, many programs focused on high-tech industries and so-called new or key technologies with the potential to raise Europe's overall portfolio. The information and communications sector seemed crucial to Europe's global competitiveness. Industry-specific programs such as the European Strategic Program on Research in Information Technology (ESPRIT), which ran in five consecutive segments from 1983 to 1998, sought to keep microelectronics, computer technology, and software production in Europe to maintain employment, healthy trade balances, and innovation capacities. In the late 1960s and early 1970s, efforts to respond to the "American challenge" by pooling industrial resources in an integrated European trust for data processing and microelectronics failed. Two decades later ESPRIT and related national programs provided a technology boost. They improved the knowledge base for information and communication businesses, and definitely led to an impressive rise of international collaboration.[32] In this sector, the number of science and technology alliances skyrocketed during the 1980s and 1990s. Two-thirds combined two or more partners from Europe. In Germany, 90 percent of all technological collaborations involved international enterprises. In 1991, the "national" champion Siemens alone was engaged in 1,400 cooperative projects, with only two domestic partners in the company's vast transnational network.[33]

At the turn of the millennium, Europe's information and communication sector was thus densely connected, both within Europe and globally. Experts in academia and industry collaborated in tens of thousands of research projects and technical alliances. They profited from the European Union's funding programs, which undoubtedly increased Europe's knowledge base. However, did these programs also enable economic growth, their primary goal? They did not, critical evaluations unanimously agree.[34] Government programs pushed the European electronics and computer industry into highly contested market segments, where firms struggled to compete in the long run with the U.S., Japanese, and other Asian enterprises, not least because the programs favored producer over user-oriented industries. The EU's investment in technology did not cure but rather exacerbated the Continent's economic problems. Even the European Commission had to admit in 1995 that despite heavily investing in strategic support programs for more than a decade, "the technological and economic performance in high-technology sectors such as electronics and information technologies has deteriorated."[35]

Third: The level of European connectedness in knowledge production differs from nation to nation.

Technoscientific experts in small countries are more dependent on international collaborative partners than their colleagues in larger countries, who are less motivated to seek international collaboration. In their study of emergent collaborative patterns in Europe between 1985 and 1995, an international team emphasized that "scientists in small countries have practically no other choice than to find collaborating partners from outside their borders."[36] In Luxembourg, for example, two-thirds of all papers in the natural sciences, medicine, and engineering technologies were internationally co-authored, vs. about one-third in Germany, France, and Italy.

In 2004 eight Eastern Europe an countries joined the EU. They had been candidates for several years and thus had been able to profit from EU funding. The European Commission was keen to know whether access to EU funding had a substantial impact on these nations' scientific and technological profiles and therefore commissioned a scientometric study. Its authors selected six countries of similar size to compare changes in their profiles during the previous two decades: the old member states Finland, Ireland, and Portugal and the new members Hungary, Slovenia, and Estonia. The results were mixed. By closely examining national systems of higher education and innovation, the authors observed that the three new members all followed the Western model of strong collaboration between academia, state-sponsored research institutions, and industry. Despite the "European homogenisation and convergence" on the institutional level, the Eastern European countries "maintained their individual peculiarities and preferences" regarding collaboration with other European and non-European nations.[37] No doubt institutional systems can change much faster and more easily than culturally grounded patterns of knowledge exchange and transnational collaboration. How big the national differences still are in Europe after decades of heavily promoting integration is expressed in Figure 9.4.

Language is a crucial factor in explaining differences in the level of European connectedness. It has been argued, quite convincingly, that Europe in early modern times was a well-integrated society of experts.[38] Knowledge could circulate easily in the European *république de lettres*, because scholars had a shared *lingua franca*, first Latin and later French, in which they could communicate. With the rise of the nineteenth-century nation states, the *lingua franca* weakened and gave way to a multitude of vernaculars. Language barriers began to have an ever-growing importance for expert communication and collaboration patterns in Europe. Only in recent decades has a new *lingua franca* developed as a result of U.S. dominance in world politics: English. However, English has not yet become the universal language of communication for European experts. To the contrary, in many fields of scholarship, knowledge exchange still happens to a large extent on the basis of national languages, in particular in countries such as France or Germany that are big enough to sustain relatively autonomous markets of both scientific publications and technical innovations.

The Dutch innovation researcher Loet Leydesdorff thus touched on a crucial point when he posed the critical question: whether Europe has developed into "a single publication system." Leydesdorff's answer, which he based on the findings of recent scientometric studies, was twofold. Clearly countries like Spain and Italy, which, in EU phrasing, belong to the "less favored regions in Europe," have considerably increased their international connectedness since the 1990s. Moreover, intra-European collaborations have grown quickly and even outdo those between European and non-European countries. Still, compared to the financial sector, a widely integrated European monetary system after the transition to the Euro in 2000, European knowledge "does not yet exhibit systemness."[39]

Comparing the monetary system with the knowledge system highlights a fundamental difference between the two. While the European Union could politically command monetary unification, it can only partially influence the integration of expertise by providing financial support through the Framework Programmes and other activities. Cooperation most often emerges from experts' functional needs rather than from political incentives. Experts often created institutional networks in order to exchange knowledge. We will now turn to the patterns of Europeanization that have emerged from these networks.

Fourth: European knowledge societies frequently emerge from within expert communities rather than through official government resolutions.

European experts' transnational aspirations predate modernity. Early modern Europeans shared knowledge through a border-crossing network among a community of experts, however small and elitist this community might have been. Yet, it was not before the second half of the nineteenth century that experts came together to found organizations with members from several countries. Between 1880 and the First World War a wave of new institutions sought to stimulate the international exchange of knowledge in both established and newly created fields of technoscientific inquiry. The emerging cosmopolitan polity of knowledge essentially originated in Europe. European experts began to advocate formalized exchange in cross-national organizations for practically

Fig. 9.4 Language Matters: *The graph displays the percentage share of internationally co-authored papers in various European countries, the U.S., and Japan, 1985–95. Considering Japan's leading position in many fields of technology and innovation, its low value requires further explication. An explanation can be found in the country's geography and, even more important, in language barriers that make it difficult for Japanese experts in science and technology to engage in international collaborations. They cooperate intensively with their colleagues in South Korea, with whom they can communicate quite comfortably, but much less so on a global scale.*

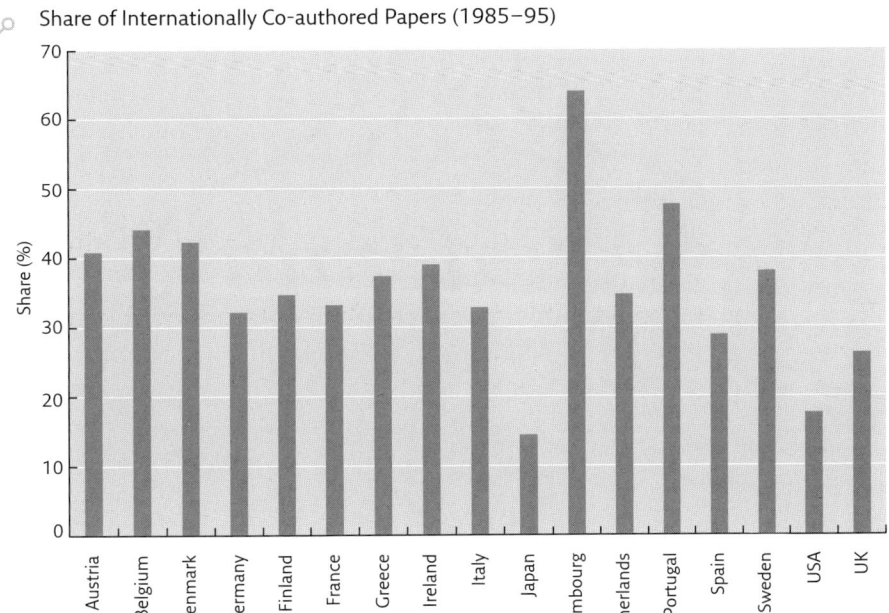

every discipline. In technoscience, the "mechanics of internationalism" originated from Europe.[40]

A century later, Europe experienced another wave of novel transnational organizations devoted to lobbying for, performing, coordinating, and communicating research. This time, the newly emerging institutions aimed specifically to create European networks, not global ones. In 1959, only two years after the Treaty of Rome, universities founded the Conference of European Rectors and Vice-Chancellors, later renamed the Association of European Universities. In 1973, some member organizations decided to set up a separate body, the Liaison Committee of European Union Rectors' Conferences, to jointly strengthen academia within an ever-growing European Union. The Academia Europaea and the All European Academies, the European Association of Contract Research Organizations, and the Confederation of European Rectors' Conferences mirror the respective organizations on the level of Europe's national systems of science and technology, in these cases academies, contract research organizations, and universities. In a way, these institutions brought together their national members' interests at the European level.[41]

Fig. 9.5 An Academy of Europe: *Founded in 1988, the* Academia Europaea *revitalizes the traditional European concept of the academy to bring together Europe's leading experts from the "Sciences, Humanities, and Letters" to encourage interdisciplinary research and to identify topics of trans-European importance. The Academy is one of many institutions within Europe's community of experts that emerged during the 1980s in response to the European Commission's rapidly expanding governance of science and technology. The tree appears both in its logo and the name of its newsletter, depicting the idea of unifying all branches of European expertise under one leafy umbrella.*

With time, the network of European institutions of science and technology, many of them referred to by hard-to-memorize acronyms, became ever more dense, with industry contributing to this exorbitant growth. In addition to the Union of Industries of the European Community and the European Round Table, which aimed to represent the interests of European industries in general, numerous organizations dedicated to specific branches were founded. These industrial organizations included the European Information Technology Industry Round Table, Collaborative Research and Development in Sustainable Technologies, and EUROPABIO, representing Europe's biotech industries. Furthermore, industrial research developed its own network of European organizations such as the European Association of Applied Science Academies, the European Industrial Research Management Association, the Federation of European Industrial Cooperative Research Organizations, to mention but a few.

In a way, the complex network of bodies and institutions devoted to performing, promoting, and coordinating research on the European level often responded to the expanding role of the European Union in building Europe on expertise. More often, however, transnational organizations resulted from experts' bottom-up initiatives to further their fields of knowledge. Experts' transnational collaboration and the border-crossing flow of knowledge resulting from cooperation has been one of the most powerful forces driving Europe's hidden integration during and after the Cold War.

Conclusion

The setting was Europe, pre-1800, where images and evidence of state power abounded. In France, one icon of state wealth and influence was Louis XIV's opulent court, with palaces from the Louvre to Versailles. In the British Empire, naval fleets bristled with the state's military might. And in Prussia, successive monarchs' oversized armies signaled the kingdom's impressive strength.

Predictably, the reality of state power was far more complex. Namely, stores of specialized knowledge and legions of qualified experts were required to create and maintain everything from states' economies to their construction projects, from their academies to their armies. But despite their significance, experts became visible—and their relevance undisputed—only in the nineteenth century. Earlier, specialized schools had provided training for work in various areas of mining, for example—from the physical task of wielding a pickaxe to the intellectual labor of devising systems for coal extraction. Institutional training for the navy, and for artillery handling, was another area where highly specialized knowledge left its mark. However, only in the nineteenth century were these schools augmented by numerous others. These novel institutions were broadly conceived to train new elites armed with the breadth of knowledge to secure success for their states. The ability to command these experts was now in itself a sign of power. At issue was not merely who possessed the lushest palaces, but who enlisted the smartest experts—who controlled the knowledge and expertise vital to nation-building.

It was during the mid-nineteenth century—when industrialization gained traction, and technical knowledge circulated much more intensely and widely—that experts moved to center stage. The second half of the century spotlighted experts, who previously had operated largely behind the scenes. When, in 1851, those in science, technology, and industry gathered in London for the Great Exhibition of the Works of Industry of all Nations, these experts made their existence known to the established elites and to the public, which visited the spectacular show by the millions. The London Exhibition indeed set the stage for a Europe of unprecedented integration: a Europe built on expertise.

Historically, some thinkers had believed that scientific and technical knowledge would constitute Europe's unifying forces. French utopian socialist Henri de Saint-Simon is perhaps the most notable. In his 1814 treatise, *De la réorganisation de la société européenne* (The Reorganization of European Society), Saint Simon in 1814 foresaw a unified Europe based on the galvanizing influences of science and technology. Enlightened philosophers, along with ingenious scientists and creative engineers, would liberate Europe from the bitter conflicts of the Napoleonic Wars and create a consolidated Europe built on scientific and technical expertise.[1] Saint-Simon assessed the potential of technology and science with uncanny accuracy. London's Crystal Palace of 1851, then, already embodied technical expertise as well as a forum for transnational communication. And, indeed today, Europe is more integrated than ever before. Considering Saint-Simon's vision in a nuanced way, however, we find a pronounced idealism: specifically, Saint-Simon overlooked the destructive power of science and technology. For Saint Simon's vision was only fulfilled after Europe had overcome two more periods of cruel conflict—the First and the Second World Wars. Technology and science have proven their power to both unite and fragment Europe.

Building Europe on Expertise has offered a cutaway view of European history in which experts engaged in cross-national collaboration, via world exhibitions, joint scientific endeavors, congresses—and countless other interactions. These experiences of border-crossing cooperation created a shared European experience, a true European history. Just as often, cooperation derived from experts' lofty ideals—ideals of scientific cosmopolitanism and technical universalism. As often, collaboration was motivated by banal, individual interests, such as career-building. Another enabling factor was collective interests, such as the wish to achieve something by pooling resources on a transnational level.

More and more, experts were able to transcend their individual nation states via scientific projects. For a century after the London Exhibition, however, most experts did not foresee the process of Europeanization, the formation of a European identity that followed from cross-border knowledge exchange. The awareness of Europeanization became widespread only in the second half of the twentieth century. Moreover, Europeanization was often tied inextricably with internationalism, part and parcel of the emerging cosmos of scientific and technical globalism. Rarely did experts pursue a European agenda, nor did they apparently give much thought to the concept of a united Europe. Only after the 1940s did Europe develop a conscious, deliberate, and planned program of technoscientific collaboration, a multifaceted program driven by vested interests in political integration. Even then, when cooperation coincided with the clear-cut political mission to spur integration, experts in Europe very rarely coalesced into a single European knowledge society.

Scientific and technical collaboration was not an easy process, nor was it linear, involving ever-growing transnational networks and institutions. On the contrary, the history of building Europe on expertise was fraught with the tensions of the Age of Extremes, which began with the First World War. In fact, Europe's Great War ended a long period of vivid globalization and a rather productive spirit of national competition. Cross-national knowledge circulation and economic exchange shrank drastically. And the scientific internationalism and free trade that had bourgeoned in the decades after the Great Exhibition now collapsed. Experts "took sides" in the war; they contributed their technical expertise to the military force of belligerent nations. In addition to increasing the war machinery by helping to build airplanes, mines, and fortifications, experts engaged in nationalistic propaganda, vociferously trying to justify military campaigns. In 1914, for example, scientists were prominent among those who signed The Manifesto of the Ninety-Three, the proclamation meant to justify Germany's military invasion of Belgium. In essence, experts on both sides of the war became special battalions in the "war of minds."[2]

The radicalization of expertise led to political distortions and massive destruction. The Great War also stimulated research and technical innovations in the military arena, offering experts new opportunities—dark as those opportunities were. Wartime turned technoscientific experts, even more so than before, into key actors who were widely accepted by society and deeply linked to their nation states. The founding of the Technical University of Warsaw in 1915—in the very midst of the Great War—exemplifies this. France, as the European continent's first and foremost nation state, had led the way in demonstrating the need to educate and integrate the new expert elites. Following the French example and the *École Politechnique*, other European nations transformed their experts into agents of change who operated far beyond the laboratory, the workshop, and the factory.

Nation states and experts remained ambivalent about one another. Experts could indeed profit from the relationship, which guaranteed them scope; some degree of freedom; and, in the case of the best-known experts, public support. Many experts received accolades and gained prominence as icons of the new nation states. Ever greater in number, these experts did not displace the generals and admirals, the lawyers and administrators who had dominated the European scene to date. Rather, they forged new fields for technology and science, responding to the new demands of an industrializing Europe. Increasingly, experts demanded a voice in political affairs. By turns, they imitated, challenged, and submitted to the established elites, making their way into European state administrations. Nation states depended on experts to run their ever-more-complex technical systems. Governments also needed "public experts" to demonstrate their capacity to modernize; to compete with other nations; and to thus secure and sustain legitimacy both domestically and across borders.

The nation state maintained—if not increased—its standing as the most likely context for expertise well into the twentieth century. Europe's nation states also competed with one another for the services of experts, thus spurring open knowledge exchange. Universalism and nationalism, we showed, were not mutually exclusive domains. There were no battle lines drawn between a camp of closed-minded, nationalistic experts and a contingent of broad-minded internationalists who thought and acted in the spirit of universalism. Rather, we identified overlapping territories and unexpected consequences of knowledge exchange among experts. We analyzed and redefined "nationalist" and "universalist" by citing the tensions between both categories—tensions that were present long before 1914 and that re-emerged more markedly after 1918. In the context of far older traditions, ideas, and cultures, the First World War prompted a number of European experts to view universalism as a more radical ideology than previously. Technological progress was the starting point for very different kinds of thinkers. Experts rallied around the questions of how to order the vast new stores of knowledge—and how to apply this knowledge so as to heal the wounded world. For some experts, these goals invoked a more technical route to process, which was believed to bring about social progress when rationally applied. For other experts, exemplified by Le Corbusier and even more so by Karel Teige, only active political change would yield a new society.

Some of such experts' independent projects were so ambitious that, to our contemporary sensibility, they appear doomed from the start. These represent far deeper attempts by experts to shape society, motivated by what is most aptly described as an individualistic—if not eccentric—strain of idealism. Consider, for example, the Mundaneum project, the attempt, in 1910, to gather and classify all of the world's scientific knowledge. Not only Mundaneum's creator Paul Otlet, but also the inventor of Isotype, Otto Neurath, personified the lessons drawn by many from the First World War. Neurath believed in the need for social economy. He, like many other experts, aimed to improve society by actively speeding up the trickle-down effects of technical innovations. This meant organizing society according to insights gleaned from the factory and the laboratory. In Neurath's universe, a Europe built on expertise was not to be an elite project, but a project in which everyone could participate. Through this inclusiveness even the countryside at the Continent's periphery would be transformed. This latter idea became the main goal of the CIAM (*Congrès internationaux d'architecture moderne*), the organization of modernist architects focused on technical progress. CIAM members endeavored to implement what Otlet and Neurath had envisioned: the use of expert knowledge to improve society. They redefined themselves as technoscientific experts; and they strove to occupy the key positions necessary to overseeing the transformation of whole states and societies into rationally functioning entities. In fact, Neurath influenced CIAM profoundly through his personal participation and by fostering the development of a universal graphic vocabulary. Eventually, this visual language became the framework for systematizing the politics of comparison that not only CIAM regarded as the formula for progress. Technology was presented as the healing force that would cure the ills of the modern era and the past alike. Experts transformed themselves from playing a narrowly interpreted role to being, like the architect Le Corbusier, people who held some of society's most visible and powerful positions—people who changed society.

With the Nazi seizure of power in 1933, and the turn towards authoritarian regimes evident elsewhere, CIAM members cruised the Mediterranean together in a dilapidated ship. Images like these—of experts at sea, far from the conflicts plaguing East Central Europe—could mistakenly portray such experts as elitists, and marginal ones, at that. In truth, these architects-turned-technoscientific-experts not only embraced the power of planning, but by the time they set off on the trip, their reasoning had achieved concrete results not least in the Eastern part of the Continent. Modernist ideas prominent in CIAM's congresses helped to shape Europe's first functional city, created during the 1920s and 1930s, and located in the vicinity of the Baťa shoe company's factories in Zlín, Czechoslovakia. This is where technoscientists' ideas converged with the thinking of the Baťa brothers to transform their factories, the town, and the surrounding landscape—and not least the workforce itself. The Baťas sold their vision of a new world of technology. This universe embraced everything from airplanes to built environments (whole regions constructed according to the new—purportedly rational—insights of technology and science) to ways of educating new technical experts who could realize this vision. Thus in some ways they even outplayed their great inspiration Henry Ford, of course testifying to the intense transatlantic exchange of ideas. To some, the Baťas were indeed visionaries. To critics, however, Baťa became a modernized incarnation of evil—"the Mussolini of shoes." According to this line of thinking, the Baťa factories dehumanized workers, manipulating them into ant-like creatures and forcing them to toil for the benefit of totalitarian regimes. Human became a mere commodity—if not the direct victims of elaborate schemes, genocidal in nature.

So, was there a direct path from the darker sides of rationalized technology evident in the Baťa factories to the backyards of totalitarian regimes? Certainly not. But, by the same token, the coincidence of experts and extreme ideologies gaining ground after the 1920s, and by 1940 dominating practically the entire continent, cannot be minimized or dismissed as a mere incident of opportunistic collaboration. Experts were essential to establishing National Socialism, Fascism, and Communism—all with disastrous consequences. The allure of these regimes was clear: they offered technoscientific experts apparently long-term stability; the power of direct decision-making; and, when ideological agendas harmonized, the willingness to implement radical measures. These were the power mechanisms that enabled masterminds of annihilation like Konrad Meyer as well as eager collaborators like Guglielmo Marconi to make technology available to the new strong men.

Experts' significant role in the World War era also draws attention to a particularly European legacy of societal dominance by the old elites. In part, it was the mulish endurance of the traditional social structures that provoked radical concepts fuelling the "great leaps forward," in both the Nazi regime and in Soviet Russia—with Trofim Lysenko being a particularly significant case in point. The new parties in power could draw on armies of ambitious experts eager to replace those sidelined, exiled, imprisoned, or killed by authoritarian regimes. Driving out experts on racial grounds (as in Germany) or for political reasons (as in the Soviet Union) triggered processes that ended with dramatic consequences for hundreds of thousands of experts. Both the Soviet Union and—far more dramatically—Nazi Germany pursued ideological policies at the cost of incurring massive brain drains that crippled pursuing national revival.

What had been a barely discernible trend in the nineteenth century coalesced into a fiercely extreme phenomenon during The Long Twentieth Century: commanding critical knowledge, experts both gained influence and became objects of coercion. This held for the Soviet Union, Nazi Germany, and, later, for large swaths of occupied Europe. Experts had to make decisions; neutral expertise—always an illusion—was now an obvious chimera. Many experts did refuse to engage with the totalitarian regimes or were deprived of any leeway for decision making due to forced emigration. Moreover, Europe experienced mass killings of the intelligentsia after the Nazis and Soviets entered the Eastern "bloodlands," the region between Warsaw and Moscow where the large majority of the killing took place during the Second World War.

After the Second World War, many of the most acclaimed German experts were forced into exile and found themselves serving the Allied powers through programs of intellectual reparations. The restrictions these experts experienced were far milder than those of colleagues in the occupied East. At least in Western exile, moreover, conditions were significantly better for these experts than for their countrymen at home.

Now, also critical voices emerged. Critics challenged experts' rise to social relevance and their powers, which appeared to be unchecked by either the state or the public. In 1949, Albert Einstein warned: "We should be on our guard not to overestimate science and scientific methods when it is a question of human problems; and we should not assume that experts are the only ones who have a right to express themselves on questions affecting the organization of society."[3] It is against this statement's intellectual background that we must also understand attempts to institutionalize experts' cooperation with each other—and indirectly with the state—in postwar Europe. As we demonstrated, these efforts reflected many goals, not least of which was to oversee and rein in experts, encouraging collaboration while preventing those with technical knowledge from pursuing the destructive agendas of the World War period. Here, again, we see at work societal agendas more far-reaching than "merely" building rocket launchers, particle accelerators, and nuclear reactors—European flagship artifacts. For better and for worse, however, most of the utopian visions that blossomed after the First World War would not come to fruition in post-1945 Europe.

As stated, Europe's experts did not constitute a single knowledge society—even in the Cold War era with its push for political integration. Instead, collaboration resulted in multiple manifestations of Europe. For example, a Nordic Europe of Scandinavian nations often co-existed with a Franco-German Europe. Forcefully defined was a Russian-led East Europe, of course; a British Europe or a Benelux-Europe. These various versions of Europe existed despite the need for technical uniformity and pooling of resources in such politicized technologies as atomic energy and space travel. Multiple patterns of regional collaboration corresponded with multiple knowledge societies and innovation cultures. Even in today's Europe, regional and national frameworks still guide innovation communities, despite all political efforts to rally the Continent into becoming an integrated place of research.[4]

Europe's many knowledge societies exist between the poles of national competition and transnational collaboration. We outlined four historical processes that have shaped them. First, globalization and Europeanization of knowledge production are closely connected. Worldwide, expert communities are increasingly organized on a continental scale. The Europeanization of science and technology is part and parcel of the wider effort to "continentalize" knowledge creation. Second, the dynamics of this process vary from technology to technology. For example, while the information and communication sector (including the computer industry) is widely integrated on both a European and global scale, the machine-tool manufacturing industry remains bound to national and regional innovation networks. Third, the level of European connectedness in knowledge production differs from nation to nation. Generally speaking, experts in small countries such as Luxembourg and Slovakia are more dependent on international collaboration partners than their colleagues in larger countries such as France or Germany. Fourth, regardless of political rhetoric, cooperation often emerges from within expert communities rather than via official government resolutions and programs. Experts' bottom-up initiatives to advance their fields of science and technology have been among the most powerful drivers of Europe's hidden integration.[5]

Looking ahead, what can we conclude about technoscientific experts' role in shaping Europe's future? How will experts contribute to Europe as a transnationally integrated space of innovation, organization, and networking knowledge? Will expertise prove to be the sturdy, sustainable foundation that will enable Europe to overcome both current and future crises? As this book goes to press, these questions are being hotly contested.

Given the current political and economic crisis on the Continent, one may be tempted to voice pessimistic predictions about Europe's future built on expertise. After all, the Lisbon Agenda has become yet another example of a now-infamous syndrome: the European Union's propensity for making ambitious plans that later clash with harsh realities and cannot be implemented. Recall the plan: in 2000, European leaders gathered in Lisbon set the goal for the

European Union to become the world's most competitive knowledge-based economy, bringing more and better jobs and greater social cohesion—all by 2010. The Agenda served as an attractive political program for legitimizing the European Commission's contested role in Europe's governance. Today, we see that most of the Agenda's aims have not been reached. Given Europe's deep crisis, the Lisbon plan has come to signify the naïveté—if not the incompetence—of the "Eurocrats" who first drafted it. For some, the Agenda's failure echoes that of the planners during The Long Twentieth Century in Moscow, East Berlin, and Warsaw—people equally out-of-touch with the economic realities and the limits of state action.

Even without the economic crisis that struck Europe in 2009, Europe would not have met the ambitious Lisbon goals, which included increasing expenditures on research and development to 3 percent of the European Union countries' combined Gross National Product. In recent years, the European economy has contracted rather than grown. Some countries have resorted to cutting their science, technology, and education budgets, as these are often perceived as easily reduced. Within the European Commission, that harbinger of integration, we observe a certain waning of belief in the role of science and technology as the catalyst of integration. Also being challenged is the long-held belief that, in the future, only a knowledge-based service economy would allow societies to transcend the old divisions of class, race, and gender. New ideas, including the theory that only re-industrialization can revitalize economies, now rival traditionally held beliefs about restoring economic balance. Today, the promise of the knowledge-based economy simply shines less brightly than ten years ago.

Moreover, momentum seems to be shifting to other parts of the world. The Gulf States and the rapidly growing Asian nations, for example, now make enormous investments—from prodigious funds to considerable human resources—in their innovation infrastructures. In 2013, Brazil announced its generously funded "Science without Borders" program, which will send some 100,000 engineering students from Brazil to other countries. This should allow Brazil to gather knowledge within a short time frame, leapfrogging over other countries in the race to become the world's top knowledge society. In a sense, this approach replicates on a large scale the cross-border circulation of engineers in nineteenth-century

Conclusion | 309

Fig. 10.1 When the Duke of Edinburgh (Prince Philip), attended the inauguration of the new premises of the University of Cambridge's engineering department in 1952, he was attesting to engineers' and technicians' remarkable rise in the preceding 100 years – in Britain and beyond. More telling than the visit itself was how it was commemorated: in oil on canvas, just like the cavalry attacks or monarchical family scenes which for centuries had marked what were regarded as truly important events, signifying the spirit of the time.

Europe, described earlier. Also familiar is the politics of comparison currently in play between Europe, the Americas, and Asia. This rivalry regarding science and technology is stunningly similar to the competition that existed between the European nation states during The Long Twentieth Century.

Europe as a world power of knowledge: from 1850 to the present, despite all conflicts and challenges, this idea has remained largely intact—and has been essential to the Continent's identity. With what we have just said, this might seem preposterous. Yet, to this day, European universities surpass Chinese and Japanese institutions, for example, in attracting more students. Europe continues to offer a variety of high-quality academic education programs, excellent research facilities, and efficient innovation systems—reflecting a remarkable capacity for change.

However, some commentators view as wrongheaded the entire enterprise of a European Union-inspired politics of science, technology, and innovation. Two critiques of the European Union outline the perceived threat to the European knowledge culture's positive traditions. The first critique: by proposing an integrated knowledge society, lifelong learning, and other commercial concepts in education, the European Union would pursue an economically motivated course that contradicts its own tradition of free and impartial inquiry to attain knowledge. By subsuming knowledge creation under the rubric of economic growth, the European Union would create nothing but "a kind of post-industrial reserve army." The second critique: the unifying imperative of European Union-led politics—the decision to standardize university curricula, for example—would destroy a desirable heterogeneity of the Continent's research and higher education systems.[6]

In the opinion of these critics, the European Union is in danger of reducing a rich history of technoscientific cooperation to a knocked-down version of the knowledge society. Recent "innovations" built into the European Union's research policy substantiate this concern. For example, the EU's "Impact" program demands proof of a clear benefit from research on social and economic improvement. Paradoxically, this new requirement also reflects a form of continuity; successful or not, hopes and dreams of changing society via expertise have persisted in Europe since the nineteenth century. Even today, we see some of these hopes and dreams taking the same form as Otlet's and Le Corbusier's Mundaneum. Consider as examples the aspirational "Science Cities" that Britain's former Prime Minister Gordon Brown designated in 2005. Think of the City of Culture, including a monumental library, planned—but only partially built—in Santiago di Compostella, Spain. And consider smaller projects like Luxembourg's *Cité des Sciences*. Such endeavors show that efforts to unify divergent European cultures of knowledge and higher education certainly did not begin with Europe's formal political integration. As we have illustrated, older traditions exist, traditions we referred to as the politics of comparison. Often, against a background of nationalistic dynamics, this politics of comparison led experts to emulate what were perceived as best practices. Such politics of comparison are still at play—albeit less vigorously than one hundred years ago—in today's Europe. The European Union has adopted expertise and knowledge as venues for bringing about integration, competitiveness, and change. This is strikingly reminiscent of the way in which European nation states functioned during the nineteenth century; it is part of the ongoing quest to build societies on expertise. Europe's future would depend on the ability to educate scientific and technical elites who are qualified to meet societies' demand for technical progress.

So, is this story of expertise a uniquely European narrative, after all? The answer is yes, in that the events described in this volume took place in Europe; reference European individuals; and describe European institutions. We also outlined *patterns* that are specific to Europe—from the intense competition for determining the Continent's technical training curricula, to the fusion of experts' national and transnational traditions, to the formation of supranational technoscientific institutions and programs such as CERN—collaborations on an unprecedented global scale. And there is another distinguishing feature that we defined for Europe's Long Twentieth Century: the strong role of knowledge in Europeans' self-understanding. This entails the belief, shattered and re-invented many times, that experts can achieve goals far beyond the scope of their scientific and technical disciplines. Experts have, in fact, functioned as a driving force of Europe's integration. Not least in this sense, Europe was built, is built, and will continue to be built on expertise.

Endnotes

Introduction

1. Yanni, "The Crystal Palace."
2. Edwards, "The Accumulation of Knowledge," and Kaiser, "Cultural Transfer of Free Trade."
3. Iriye, *Global Community*; Paulmann, "Reformer, Experten und Diplomaten," 175; Fuchs, "Wissenschaft, Kongressbewegung und Weltausstellungen."
4. Geyer, "One Language for the World."
5. Middell, *Imagining Europeans*; Bluche et al., *Der Europäer—ein Konstrukt*.
6. Friedel, *A Culture of Improvement*; see also: Epstein and Prak, *Guilds, Innovation, and the European Economy*; McNeely and Wolverton, *Reinventing Knowledge*; and Hilaire-Perez and Verna, "Dissemination of Technical Knowledge."
7. Popplow, "Europa wider Willen?"
8. See: Maier, "Consigning," 814; Raphael, "Verwissenschaftlichung," 168–70.
9. Mokyr, *The Gifts of Athena*; by focusing on Britain as the forerunner of industrialization, Mokyr stresses the age of Enlightenment as the burgeoning period of knowledge economies.
10. Mazower, *Dark Continent*.
11. Aly and Heim, *Architects of Annihilation*.
12. See: Kohlrausch et al., *Expert Cultures in Central Eastern Europe*.
13. Misa and Schot, "Inventing Europe," 8.

1 Educating Experts

1. Kamiński, "Inauguracja Wydziału Architektury," 27–28.
2. Ibid., 27.
3. On Comte's positivism as a Europe-wide *weltanschauung*, see: Lepenies, *Auguste Comte*.
4. Rolf, "Der Zar an der Weichsel."
5. Rolf, "Imperiale Herrschaft im Städtischen Raum," 140.
6. Żarnowski, "Learned Professions in Poland," 408.
7. Szyller was bestowed the order of Sw. Anna, third class for his design of the building. Omilanowska, *Architekt Stefan Szyller*, 46.
8. Muzeum Narodowe w Warszawie, *Architektura Plakatu*, 14.
9. Politechnika: Zakład Architektury Polskiej, *Warszawska Szkoła Architektury*, 34.
10. On the further development, particularly during the war period, see: Noakowski, "Powstanie Wzdyiału Architektury." On Noakowski see: Pągowski, "Architekt Architeków." See also Noakowski's memoirs: Noakowski, *Pisma*.
11. In a similar motivation the Germans had added a Vlaamsche Hoogeschool to the University of Ghent in Belgium in a largely unsuccessful attempt to secure the loyalties of the Flemish nationalist movement by offering education in Dutch.
12. Hanisch and Lange, *Vitenskap for industrien*.
13. Chionne, "Blok e Praesens," 170.
14. The term "foot soldiers of industrialization" we owe to Ruth Oldenziel.

15 Buchanan, *The Engineers*; Reinhard, *Geschichte der Staatsgewalt*.
16 Schweitzer, "Der Ingenieur," 68–70.
17 Grelon, "French Engineers," 107.
18 Disco, *Made in Delft*, 43–52.
19 Saint, *Architect and Engineer*, 458. In Greece the military remained the only institution for the training of engineers till almost the end of the nineteenth century. Antoniou, Assimakopoulos, and Chatzis, "The National Identity of Inter-War Greek Engineers," 243.
20 Grelon, "French Engineers," 108; on Vauban as an engineer see: Martin and Virol, *Vauban*.
21 Picon, *L'invention de l'ingénieur moderne*.
22 Belhoste and Chatzis, "From Technical Corps to Technocratic Power," 212.
23 Grelon, "French Engineers," 108.
24 Alder, *Engineering the Revolution*.
25 Grelon, "French Engineers," 109.
26 Ibid., 110.
27 Even today, the school's *élèves* receive the status of reserve officers, while the institution is still run by a general.
28 Belhoste, quoted in: Saint, *Architect and Engineer*, 438.
29 Smith, "The Longest Run," 659.
30 Picon, "French Engineers and Social Thought."
31 Ringrose, "Work and Social Presence," 295 and 298.
32 Ibid., 299.
33 Weber, *Peasants into Frenchmen*.
34 Schweitzer, "Der Ingenieur," 69. For an overview, see: Brown, Downey, and Diogo, "The Normativities of Engineers."
35 Peltcjean, "Scientific Development."
36 Myllyntaus, "Foreign Models and National Styles."
37 Surman, "The Local and the Global."
38 Diogo and van Laak, *Europeans Globalizing*.
39 Schweitzer, "Der Ingenieur," 84.
40 MacLeod, *Heroes of Invention*.
41 Buchanan, *The Engineers*.
42 Hirose, "Two Classes of British Engineers"; König, *Künstler und Strichezieher*, 173.
43 Saint, *Architect and Engineer*, 441. See also: Kahlow, "French Influence."
44 Saraiva, "A City for a Technical School," 321.
45 Gugerli, Kupper, and Speich, *Die Zukunftsmaschine*, 105.
46 Quoted in Saint, *Architect and Engineer*, 447.
47 Antoniou, Assimakopoulos, and Chatzis, "The National Identity of Inter-War Greek Engineers," 245.
48 Grelon, "Die deutschen Ingenieure aus französischer Sicht." On the phenomenon of "learning from the enemy" see: Paulmann, "Feindschaft und Verflechtung."
49 For a general problem of the gap between models of technical education and their application see: Karvar, "Model Reception."
50 Dittrich, "Experts Going Transnational"; Middell, "Kompatibilität oder Diversität europäischer Wissenschaftssysteme"; Diogo and van Laak, *Europeans Globalizing*.

51 Deicer, "Influence and Intercultural Exchange."
52 Pelc and Grötz, *Konstrukteur der modernen Stadt*, 318–24.
53 Radkau, *Technik in Deutschland*.
54 "Ludwig Mies van der Rohe†," 127.
55 Tomlow, "Introduction," 9.
56 Weiss, *The Making of Technological Man*, 123–39.
57 Ringrose, "Work and Social Presence," 295.
58 Gispen, "Engineers in Wilhelmian Germany," 105.
59 On the Wickenden commission: ibid., 105.
60 Fox and Guagnini, "Introduction," 7.
61 Gispen, "Engineers in Wilhelmian Germany," 104–5.
62 König, *Künstler und Strichezieher*, 227.
63 On the British perception of German technical education and vice versa see: Schulz, "The Promotion of Industry."
64 Siegrist, "The Professions in Nineteenth-Century Europe," 72.
65 Schweitzer, "Der Ingenieur," 70; Canel, "Maintaining the Walls," 132.
66 Piłatowicz, "Technicy Lwowa i Krakowa."
67 Disco, *Made in Delft*, 199–265.
68 Siegrist, "The Professions in Nineteenth-Century Europe," 68–88; König, *Künstler und Strichezieher*, 226.
69 Gispen, "Engineers in Wilhelmian Germany," 105–16.
70 Kahlow, "French Influence"; Radkau, *Technik in Deutschland*, 117.
71 Gispen, "Engineers in Wilhelmian Germany," 107; for France see: Weiss, *The Making of Technological Man*, 57–87.
72 König, "Technical Education and Industrial Performance."
73 Gispen, "Engineers in Wilhelmian Germany," 112.
74 Bolenz, "Baubeamte in Preussen."
75 Goldberg, *Honor, Politics and the Law in Imperial Germany*, 9 and 62. For France see: Nye, "Honor Codes and Medical Ethics in Modern France."
76 König, *Künstler und Strichezieher*, 223.
77 Gispen, "Engineers in Wilhelmian Germany," 105; Gispen, *New Profession, Old Order*, 187–219.
78 Späth, "Der Ingenieur als Bürger."
79 Żarnowski, "Learned Professions in Poland," 424.
80 Ibid., 119.
81 Jarausch, *The Unfree Professions*.
82 Chadeau, "Le contexte de la loi."
83 Schot, Rip, and Lintsen, *Technology and the Making of the Netherland*, 273.
84 Kranakis, *Constructing a Bridge*, 95.
85 Schweitzer, "Der Ingenieur," 80.
86 Saint, *Architect and Engineer*; Rodrigues, "The Debate between Engineers and Architects"; Pollak, *The Education of the Architect*; Pfammatter, *Die Erfindung des modernen Architekten*; Bolenz, *Vom Baubeamten zum freiberuflichen Architekten*; Heymann, *"Kunst" und Wissenschaft in der Technik des 20. Jahrhunderts*; Clark, "A Struggle for Existence."
87 Guillén, *The Taylorized Beauty of the Mechanical*, 123.
88 Kranakis, *Constructing a Bridge*, 95; König, *Künstler und Strichezieher*.
89 Canel, "Maintaining the Walls."
90 Berner, "Educating Men," 80.

91 Pursell, "'Am I a Lady or an Engineer?'".
92 Oldenziel, Canel, and Zachmann, "Introduction."
93 On students from Eastern Central Europe at Weimar's THs, see: Maasberg and Prinz, *Die Neuen kommen!*
94 Sigrist, "Les étudiantes étrangères"; Canel, "Maintaining the Walls," 135.
95 Gouzevitch and Gouzevitch, "A Woman's Challenge"; Schweitzer, "Der Ingenieur," 77.
96 Canel, "Maintaining the Walls," 132.
97 Berner, "Educating Men," 82.
98 Ibid., 83–85.
99 Musso, *Saint-Simon*.
100 For the "ideology of engineering" in the U.S. see: Layton, *The Revolt of the Engineers*, 53–78; Kline, "From Progressivism to Engineering Studies."
101 The continuing influence of the *École Polytechique* may be seen as a result of its ability to produce exactly this new kind of intermediary personae. Belhoste and Chatzis, "From Technical Corps to Technocratic Power."
102 Sinclair, "The Power of Ceremony." For this point see: Kaiser and Schot, *Writing the Rules*.

2   Technical Experts as New National Elites

1 See: van Meer, "The Nation is Technological."
2 However, the Emperor was regularly reduced to his function as King of the Bohemian lands; Agnew, "The Flyspecks on Palivec's Portrait."
3 Fittingly, the exhibition grounds housed an exhibition on Czech folklore from 1895 onwards; Janatková, *Modernisierung und Metropole*, 24.
4 Ibid., 44 and 25.
5 Lindqvist, "An Olympic Stadium of Technology."
6 Janatková, *Modernisierung und Metropole*, 26 and 29.
7 Ibid.
8 Schumpeter, *Imperialism and the Social Classes*, 89. This is somewhat too apodictic, as is the main argument of Mayer, *The Persistence of the Old Regime*.
9 F.A., "Kaisertum und Zeitalter."
10 König, *Wilhelm II. und die Moderne*, 113 and 115.
11 Gispen, "Engineers in Wilhelmian Germany," 119.
12 König, *Wilhelm II. und die Moderne*, 116.
13 Ibid., 123.
14 Spenkuch, *Das Preussische Herrenhaus*.
15 Quoted in König, *Wilhelm II. und die Moderne*, 118.
16 For the self-conception of engineers and the role they ascribed to their cultural mission, see: Dienel, *Der Optimismus der Ingenieure*.
17 Hortleder, *Das Gesellschaftsbild des Ingenieurs*, 84.
18 Raymaekers, "Between Capital and Labour," 7–8.
19 MacLeod, *Heroes of Invention*, 246.
20 Max Planck Gesellschaft, "Die Gründung der KWG."
21 Vom Brocke and Laitko, *Die Kaiser-Wilhelm-/Max-Planck-Gesellschaft*.
22 Nottmeier, *Adolf von Harnack*.
23 Charles, *Between Genius and Genocide*; Szöllösi-Janze, *Fritz Haber*.

24 Quoted in Späth, "Der Ingenieur als Bürger," 84; Guillén, *The Taylorized Beauty*, 9.
25 The following is based on Saraiva, "Inventing the Technological Nation."
26 Quoted in ibid., 267.
27 Saraiva, "A City for a Technical School," 317.
28 Schueler, *Materialising Identity*.
29 Blackbourn, *The Conquest of Nature*.
30 Grelon, "French Engineers"; Smith, "The Longest Run," 659.
31 Quoted in Smith, "The Longest Run," 692.
32 Ringrose, "Work and Social Presence," 293.
33 Rüger, *The Great Naval Game*.
34 MacLeod, *Heroes of Invention*, 91–124.
35 Van Laak, *Imperiale Infrastruktur*, 243–45.
36 Heywood, *Engineer of Revolutionary Russia*.
37 Lindenberger, "Strassenpolitik," 365–68.
38 On the relation of science and the nation in Europe see: Jessen and Vogel, *Wissenschaft und Nation in der europäischen Geschichte*, 7–37.
39 On Marconi, see chapter 5.
40 Siddiqi, *The Rockets' Red Glare*; Geppert, "Space *Personae*."
41 On the growing relevance of bacteriology—and Koch's role in it—see: Berger, *Bakterien in Krieg und Frieden*.
42 Latour, *The Pasteurization of France*.
43 Boudia, *Marie Curie et son laboratoire*.
44 MacLoad, *Heroes of Invention*, 231–32.
45 Goschler, *Rudolf Virchow*.
46 See: Crawford, *Nationalism and Internationalism in Science*. The general mechanism is described by Jessen and Vogel, *Wissenschaft und Nation in der europäischen Geschichte*, 7–37.
47 Van Laak, *Weisse Elefanten*.
48 Original: "Die Ingenieure sollen leben! In ihnen kreist der wahre Geist der allerneusten Zeit! […] Von Land zu Land, von Meer zu Meer—Der Ingenieur."
49 Bruisch, "Agricultural Experts."
50 Knoedler and Mayhew, "Thorstein Veblen and the Engineers"; Fischer, *Technocracy and the Politics of Expertise*.
51 Lintsen, *Ingenieur van Beroep*, 161–67, 193–96; Schot, Rip, and Lintsen, *Technology and the Making of the Netherlands*, 366–78, 374.
52 Mai, "Politische Krise und Rationalisierungsdiskurs."
53 Rathenau, *Die neue Wirtschaft* (with translations into English, Dutch, and Italian). See also: Volkov, *Walther Rathenau*, 160–2.
54 Generally on this phenomenon see: Otter, *The Victorian Eye*; Scott, *Seeing like a State*.
55 Quoted in Györgi, "Engineer Utopia," 87.
56 Quoted in ibid., 93.
57 Piłatowicz, "Technicy Lwowa i Krakowa"; Piłatowicz, *Kadra Inżynierska w II Rzeczypospolitej*, 123–28.
58 Quoted in van Meer, "The Nation is Technological," 102.
59 Piłatowicz, *Kadra Inżynierska w II Rzeczypospolitej*, 123–28. See also: Efmertová, "Les professeurs électrotechnicies Tchèques."

60 Rohdewald, "Mimicry in a Multiple Postcolonial Setting," 63–84; Drozdowski, *Eugeniusz Kwiatkowski*. Obviously, the most extreme example was the Soviet Union, with the grand expectations placed by Lenin onto the electrification of the country; see: Coopersmith, *The Electrification of Russia*.
61 Żarnowski, "Learned Professions in Poland 1918–1939," 413.
62 Loose, "How to Run a State."
63 Minorski, *Polska nowatorska Myśl architektoniczna*.
64 Żarnowski, "Learned Professions in Poland," 418–19.
65 Antoniou, Assimakopoulos, and Chatzis, "The National Identity," 244. For Italy, see: Bocquet, "Engineers and the Nation in Italy."
66 Rudberg, *The Stockholm Exhibition 1930*.
67 Marklund and Stadius, "Merging Modernity with Nationalism."
68 Janatková, *Modernisierung und Metropole*, 60. The Polish national exhibition in Poznań was similar in many ways, with equally striking modernist objects: Dybczyńska-Bulyszko, "Pawilon Polski," 143–62.
69 Janatková, *Modernisierung und Metropole*, 64. See, for a similar case, the General National Exhibition (*Powszechna Wystawa Krajowa*) in Polish Poznań in 1929: Störtkuhl, "Ausstellungsarchitektur als Mittel nationaler Selbstdarstellung."
70 Seipp, *The Ordeal of Peace*.
71 Steffen, "Wissenschaftler in Bewegung."
72 Fuchs, "Wissenschaftsinternationalismus in Kriegs- und Krisenzeiten."

3   Architectures of Knowledge

1 See: Švácha, "Before and after the Mundaneum," 119. The most complete reproduction of the three respective texts (Teige's text "Mundaneum", Le Corbusier's answer, and a reaction by Teige) may be found in Fabian and Winko, *Architektur zwischen Kunst und Wissenschaft*, in a German translation. The first two texts were published in an English translation in *Oppositions* 4 (1974), 83–108. The following quotations are taken from these editions.
2 Teige, "Mundaneum," 83.
3 Ibid., 86.
4 Fabian and Winko, *Architektur zwischen Kunst und Wissenschaft*, 231.
5 Le Corbusier entitled his notes "In Defense of Architecture." Le Corbusier, "In Defense of Architecture."
6 Ibid., 102 and 98.
7 Ibid., 101–2.
8 Fabian and Winko, *Architektur zwischen Kunst und Wissenschaft*, 233.
9 Ibid., 232.
10 Teige, "Antwort auf Le Corbusier," 283. Teige referred to the compromises embedded in the Treaties of Swiss Locarno in 1925.
11 Coppens, Derez, and Roegiers, *Leuven University Library*, 135–298; Schivelbusch, *Die Bibliothek von Löwen*; Kramer, *Dynamic of Destruction*, 6–15.
12 Clavin and Wessels, "Transnationalism and the League of Nations"; Schot and Lagendijk, "Technocratic Internationalism in the Interwar Years"; Fleury, "The League of Nations"; Steiner, *The Lights that Failed*, 349–84.
13 Pallas, *Histoire et architecture*.
14 Herren, *Hintertüren zur Macht*, 100.

15 Le Corbusier had been introduced to Otlet through Blaise Cendras in 1922. Van Acker, "Internationalist Utopias of Visual Education," 56.
16 Fuchs, "The International Catalogue of Scientific Literature," 165–93.
17 Ibid., 33.
18 "Intellectual Cooperation and International Bureaux Section, 1919–1946."
19 Long, "Who Killed the International Studies Conference?," 604–7.
20 Canales, "Einstein, Bergson, and the Experiment that Failed."
21 Quoted in Weitz, "Weimar Germany and Its Histories," 581.
22 Wijdeveld, Mendelsohn, and Ozenfant, *Académie Européene "Méditerranée."*
23 Ghils, "Fonder le monde, fonder le savoir du monde, ou la double utopie de Paul Otlet," 47.
24 Ibid., 36.
25 Van Acker, "Internationalist Utopias of Visual Education," 34.
26 Iriye, *Global Community*.
27 Van Acker, "Internationalist Utopias of Visual Education," 33.
28 Frohmann, "The Role of Facts."
29 Rayward, *The Universe of Information*, 47.
30 Abel and Newlin, *Scholarly Publishing*.
31 Feuerhahn and Rabault-Feuerhahn, *La Fabrique internationale de la science*.
32 Uyttenhove and van Pethegem, "Ferdinand van der Haeghen's Shadow on Otlet." Librarians, as experts of knowledge in the Otlet universe, had their own expert committee within the CICI.
33 Buckland, "On the Cultural and Intellectual Context," 54.
34 Union of International Associations, "The Union of International Associations," 116.
35 Nerdinger, "Der Traum von der Universalbibliothek."
36 Van Acker, "Internationalist Utopias of Visual Education," 36.
37 Otlet quoted in van den Heuvel, "Building Society," 145.
38 Otlet, *Traité de documentation: Le livre sur le livre, théorie et pratique*, quoted in ibid., 129.
39 Andersen and Hébrard, *Creation of a World Centre of Communication*. For a detailed though critical temporary assessment see: Edwards, "A World Centre of Communication." On the connection between this project and the Mundaneum see: Courtiau, "Le création d'une ville internationale."
40 Sonne, *Representing the State*, 241.
41 Early steps in this direction were the "Forum Orbis—Insula Pacis" of Josef Hoffmann in 1895 and Alfred Frenzl's "Peace Conference Palace on the Island of Lacroma" of 1899. See: Sonne, *Representing the State*, 242.
42 Garnier, *Une cité industrielle*.
43 La Fontaine and Otlet quoted in Sonne, *Representing the State*, 257 and 270.
44 Ibid., 262–63.
45 Ibid., 268.
46 Ibid., 265, 277, and 274.
47 Füeg, "Ordo ab chaos," 35.
48 Van Acker, "Internationalist Utopias of Visual Education," 73.
49 Ibid., 47; van den Heuvel, "Building Society," 134.
50 Welter, *Biopolis*, 124–27.
51 Van Acker, "Internationalist Utopias of Visual Education," 44 and 51.

52 Ibid., 35.
53 Meriggi, *Una città possibile*; Nerdinger, "Traum von der Universalbibliothek," 250–52.
54 On the relation of science and the city see: Hessler, "Science in the City," 211–15; Levin, *Urban Modernity*.
55 Geier, *Der Wiener Kreis*, 12.
56 Vossoughian, "The Modern Museum," 83.
57 ISOTYPE is an acronym for International System of Typographic Picture Education, but is also Greek for "the same sign." See also: the reprint of the original explanation of Isotype textes in Hartmann, *Bildersprache*, 116–19. On Isotype as an improved version of already existing pictorial languages: ibid., 65.
58 Neurath, "Bildliche Darstellung sozialer Tatbestände," reprinted in ibid., 6.
59 Ibid., 77.
60 Vossoughian, "The Modern Museum," 83.
61 Van Acker, "Internationalist Utopias of Visual Education," 74; Hartmann, *Bildersprache*, 26.
62 Hartmann, *Bildersprache*, 16.
63 Ibid., 41.
64 "Gerd Arntz."
65 Hartmann, *Bildersprache*, 23–28, 44.
66 Ibid., 91.
67 Von Debschitz and von Debschitz, *Fritz Kahn*.
68 Van Acker, "Internationalist Utopias of Visual Education," 66. On Otlet's inspiration, see: Vossoughian, "The Modern Museum," 84–86.
69 Van Acker, "Internationalist Utopias of Visual Education," 54 and 68.
70 The Polish Anne Oderfeld was Otlet's contact person at this institution; ibid., 59.
71 Hartmann, *Bildersprache*, 95.
72 Van Acker, "Internationalist Utopias of Visual Education," 60.
73 The ICIC worked closely with the International Educational Cinematographic Institute (*Istituto Internazionale del Cinema Educatore*), which was created in Rome in 1928 by the Italian government.
74 Van Acker, "Internationalist Utopias of Visual Education," 64.
75 Vossoughian, "The Modern Museum," 93.
76 Krajewski, *Restlosigkeit*, 67.
77 Couturat, *Étude sur la dérivation dans la langue internationales*.
78 Krajewski, *Restlosigkeit*, 78 and 82.
79 Ibid., 89.
80 Ibid., 94, 96, and 108.
81 Van den Heuvel, Rayward, and Uyttenhove, "L'architecture du savoir," 19.
82 Domschke and Lewandrowski, *Wilhelm Ostwald*, 98.
83 Krajewski, *Restlosigkeit*, 67 and 111.
84 Hapke, "Roots of Mediating Information," 318.
85 Rieger, "Arbeit an sich."
86 Hapke, "Roots of Mediating Information," 319.
87 Ibid., 319.
88 Todorov, *The Limits of Art*, 9.
89 Hartmann, *Bildersprache*, 103.

90 The term was coined by Staudenmaier, *Technology's Storytellers*, looking at the Journal *Technology and Culture* and how this shaped the history of technology.
91 Krajewski, "Paper Parasite," 296–306.
92 Ostwald, *Zur Geschichte der Wissenschaft*.
93 Heesen, *Der Zeitungsausschnitt*, 291; Wazeck, *Einsteins Gegner*, 309–29.
94 Ducheyne, "Paul Otlet's Theory of Knowledge," 110–16. Ostwald published a biography of Auguste Comte in 1914.
95 Uyttenhove and van Pethegem, "Ferdinand van der Haeghen's Shadow on Otlet."
96 Giedion, *Mechanization Takes Command*; Molella, "Science moderne," 374–89.
97 On the relation of the avant-garde to technology, see: Trommler, "The Avant-Garde and Technology," 397–416.
98 Dluhosch and Švácha, *Karel Teige*, 118.
99 Teige, *Vývoj sovětské Architektury*.

4 Expertise with a Cause

1 Somer, *The Functional City*, 36, 107, 110.
2 Gold, "Creating the Charter of Athens."
3 Mumford, *The CIAM Discourse on Urbanism*, 83.
4 Fred Forbat, Autobiography (manuscript), 184–85, Fred Forbat papers. Bauhaus archive. Berlin.
5 Steinmann, *CIAM: Dokumente*, 130.
6 For older traditions in this field, see: Saunier, "Sketches from the Urban Internationale."
7 See: http://www.moholy-nagy.org/category-s/29.htm. Accessed July 16, 2013.
8 Giedion, quoted in Steinmann, *CIAM: Dokumente*, 9.
9 Letter from Giedion to Szymon Syrkus, January 24, 1933, Giedion papers (42—K). Institut für Geschichte und Theorie der Architektur (Gta)—ETH Zurich.
10 On the self-proclaimed distinctiveness of CIAM, see: Misa, "Appropriating the International Style," 73, 74, 93.
11 Steinmann, "Political Standpoints in CIAM."
12 Le Corbusier, *La Charte d'Athènes*.
13 Mumford, *The CIAM Discourse on Urbanism*, 87.
14 Letter from Giedion to Le Corbusier, September 4, 1933, reprinted in Hilpert, *Le Corbusiers 'Charta von Athen'*, 179–82.
15 Vossoughian, *Otto Neurath*, 126–28; Gresleri, "Convergences et Divergences."
16 Letter of Le Corbusier to Giedion, August 29, 1933, reprinted in Hilpert, *Le Corbusiers 'Charta von Athen'*, 171 (author's translation).
17 Letter from Walter Gropius to Giedion, January 5, 1937, Giedion papers (42—K). Gta—ETH Zurich.
18 Weber and Le Corbusier, *Le Corbusier: A Life*, 259.
19 Bodenschatz and Post, *Städtebau im Schatten Stalins*, 111–15; Mumford, *The CIAM Discourse on Urbanism*, 71–73. Forbat, Autobiography, 150–52.
20 Švácha, "Before and after the Mundaneum," 129.
21 Straalen, "Empirische Stadtanalysen," 60–67.

22 Vossoughian, "Mapping the Modern City."
23 Van Eesteren and van Rossem, *Het idee van de functionele Stad*.
24 Mumford, *The CIAM Discourse on Urbanism*, 73.
25 Quoted in ibid., 79.
26 Chmielewski and Syrkus, *Warszawa Funkcjonalna*, 40.
27 Anděl, *The New Vision for the New Architecture*, 82.
28 Horňáková, "Baťa Satellite Towns around the World."
29 Kudláček, "Zlín Views," 29 and 31.
30 Švácha, "Tomáš Baťa and the Destruction of Old Zlín," 83.
31 Šlapeta, "Kulturelles und soziales Leben in Zlín," 73.
32 Ibid., 76.
33 Anděl, *The New Vision for the New Architecture*, 84.
34 Platzer, "Zlín—Ein architektonischer Sonderfall," 99.
35 Horňáková, "Der Aufbau Zlíns in der Zeit zwischen den Kriegen," 43; Šlapeta, "The Baťa Legacy," 57.
36 Anděl, *The New Vision for the New Architecture*, 82.
37 The term is used by Robert H. Kargon and Arthur P. Molella to describe the "techno-cities" of the 1920s. See: Kargon and Molella, *Invented Edens*, 25–46.
38 Szcepanik, "The Aesthetics of Rationalization."
39 Šlapeta, "The Baťa Legacy," 62.
40 Völckers, "Foreword," 7.
41 Moravčíková, "Social and Architectural Phenomenon of the Bataism in Slovakia."
42 Steinführer, "Uncharted Zlín," 109.
43 Nerdinger, "Zlín: Sozial gelackte Modernität," 17.
44 Quoted in Steinführer, "Uncharted Zlín," 114.
45 Devinat, "Working Conditions in a Rationalised Undertaking," 45–69, and 163–86.
46 Cohen, "'Unser Kunde ist unser Herr'," 113 and 121.
47 Platzer, "Zlín—Ein architektonischer Sonderfall," 98.
48 Cohen, "'Unser Kunde ist unser Herr'," 123 and 113; Bacon, *Le Corbusier in America*.
49 Cohen, "'Unser Kunde ist unser Herr'," 125.
50 Quoted in Szcepanik, "The Aesthetics of Rationalization," 206.
51 Rudolph Philipp quoted in Nerdinger, "Zlín: Sozial gelackte Modernität," 18.
52 Maier, "Between Taylorism and Technocracy."
53 Bonin, Lung, and Tolliday, *Ford*; Holden, "Fording the Atlantic"; Wilkins, *American Business Abroad*.
54 Fava, "People's Cars and People's Technologies," 107.
55 Quoted in Nolan, *Visions of Modernity*, 30.
56 De Grazia, *Irresistible Empire*, 76.
57 Frost, "Fordism and the American Dream"; Schweitzer, *Des engrenages à la chaîne*, 12 and 16; Shioni, *Fordism Transformed*, 67–173.
58 Nolan, *Visions of Modernity*, 35; de Grazia, *Irresistible Empire*, 104–5, 136–37.
59 Rohdewald, "Mimicry in a Multiple Postcolonial Setting."
60 Anděl, *The New Vision for the New Architecture*, 82.
61 Guillén, *The Taylorized Beauty of the Mechanical*; Bertrams, "Planning and the 'Techno-corporatist Bargain'."

62 For the U.S., see: Crawford, *Building the Workingman's Paradise*, 46–60.
63 Peaucelle, *Henri Fayol, inventeur des outils de gestion*.
64 Collins, *Werner Hegemann and the Search for Universal Urbanism*.
65 Rodgers, *Atlantic Crossings*, 373–79.
66 Gartman, "Why Modern Architecture," 86. See also: Gartman, *From Autos to Architecture*.
67 Pearlman, *Inventing American Modernism*, 157, 66–70; Alofsin, *The Struggle for Modernism*, 138–40.
68 Kentgens-Craig, *The Bauhaus and America*, 196.
69 "Bauhaus Man," *Time Magazine*, February 8, 1937.
70 Rodgers, *Atlantic Crossings*, 373 and 381.
71 Gropius, quoted in Kentgens-Craig, *The Bauhaus and America*, 118.
72 On Mumford, see: Miller, *Lewis Mumford*; Hughes and Hughes, *Lewis Mumford*.
73 Rodgers, *Atlantic Crossings*, 389.
74 See, for example, Léger, *Machine-Age Exposition*; Hitchcock and Johnson, *The International Style*; Gropius, *Internationale Architektur*.
75 Rodgers, *Atlantic Crossings*, 409.
76 Maier, "Between Taylorism and Technocracy," 27–29.
77 Bertrams, "Planning and the 'Techno-corporatist Bargain'."
78 Maier, "Between Taylorism and Technocracy," 30.
79 Bogdanov, *Essays in Tektology*; Wegner, *Imaginary Communities*, 104–8.
80 Christy, *The Price of Power*.
81 Herf, *Reactionary Modernism*; Rohkrämer, "Antimodernism."
82 Todorov, *The Limits of Art*, 42.
83 Teige and Dluhosch, *The Minimum Dwelling*; Barr, *Neues Wohnen*.
84 Etzemüller, "Social Engineering als Verhaltenslehre des kühlen Kopfes," 36.
85 Rabinbach, *The Human Motor*; Weindling, *International Health Organisations*.
86 Turda and Weindling, *"Blood and Homeland."*
87 Bucur, *Eugenics and Modernization*; Weindling, "Public Health and Political Stabilisation."
88 Weindling, *Epidemics and Genocide*.
89 Kühl, *Die Internationale der Rassisten*, 32–39. On eugenics before and after 1945 generally, see: the special issue of *Journal of Modern European History* 10 (2012).
90 Turda, *Modernism and Eugenics*, 72–79.
91 Nolan, *Visions of Modernity*, 40.
92 Guse, "Volksgemeinschaft Engineers"; Raphael, "Sozialexperten in Deutschland."

5 Faustian Bargains in Totalitarian Europe

1 Bernhardt, "Szkoła Wawelberga i Politechnika Warszawska."
2 Schenk, *Der Lemberger Professorenmord*.
3 Josephson et al., "Science and the Periphery," 212.
4 Schmaltz, "Luftfahrtforschung unter deutscher Besatzung," 407.
5 Best, as quoted in Lerchenmüller, "Das Ende der Reichsuniversität Strassburg," 116.
6 Benz et al., "Der Ort des Terrors," 31 and 33.

7   Kohlrausch, "Szymon Syrkus"; Roulet, "Le Corbusier in Vichy."
8   Josephson, *Totalitarian Science and Technology.*
9   Raphael, "Radikales Ordnungsdenken," 36.
10 Ash, "Wissenschaft und Politik"; see also: Trischler, "The Syndrome of Falling Behind."
11 Aly and Heim, *Architects of Annihilation.*
12 For the following, see: Heinemann, "Wissenschaft und Homogenisierungsplanungen"; Rössler, "Konrad Meyer"; Kegler and Stiller, "Konrad Meyer"; and Pyta, "Menschenökonomie."
13 Wildt, *Generation des Unbedingten.*
14 Heinemann, "Wissenschaft und Homogenisierungsplanungen," 55–64.
15 Beyerchen, "Rational Means and Irrational Ends."
16 Nelson, "From Manitoba to the Memel," 444; for the following, see also: Nelson, *Germans, Poland, and Colonial Expansion,* and Liulevicius, *The German Myth of the East.*
17 Sering, *Die innere Kolonisation im östlichen Deutschland.*
18 See, for example, Zimmerer, "The Birth of the Ostland"; Zimmerer, *Von Windhuk nach Auschwitz?,* and Langbehn and Salama, *German Colonialism.*
19 Stoehr, "Von Max Sering zu Konrad Meyer."
20 Meyer, "Ländliche Fördergebiete und ihre Sanierung."
21 Meyer, "Agrarprobleme des neuen Europa."
22 Meyer, *Die Landeserschliessung in den EWG Ländern.*
23 Marconi and Marconi, *Marconi My Beloved*; see also the intriguing study on the politics of the "communications revolution": Sussman, *Communication, Technology, and Politics,* 90.
24 Jolly, *Marconi*; Faliasecca and Valotti, *Guglielmo Marconi*; and Weightman, *Signor Marconi's Magic Box.* For more nuanced biographies, see: Masini, *Marconi,* and Hong, *Wireless.*
25 Jolly, *Marconi,* 251.
26 Marconi's lecture has been edited by Gamba and Schiera, *Fascismo e scienza.*
27 Maiocchi, "Fascist Autarky and the Italian Scientists."
28 Capristo, "The Exclusion of Jews from Italian Academies"; also Capristo, *L'espulsione degli ebrei.*
29 Richeri, "Italian Broadcasting and Fascism," 56; see also: Ghirardo, "Città Fascista."
30 Parravano, "Il fascismo e la scienza," as quoted in Russo, "Science and Industry in Italy," 294 and 296.
31 Severi, *Scritti di Guglielmo Marconi,* 418. For the close alliance between science and the regime see also: Maiocchi, *Gli scienziati del duce,* 219–27.
32 For the background, see: Birstein, *The Perversion of Knowledge.*
33 As quoted in Rossianov, "Editing Nature," 743.
34 Gould, *Hen's Teeth and Horse's Toe.*
35 Gestwa, *Die Stalinschen Grossbauten des Kommunismus,* 48–50.
36 Gerasimov, *Modernism and Public Reform,* 215–19; Gordin, "Was There Ever a 'Stalinist Science'?," 626.
37 Graham, *Science in Russia,* 133; Joravsky, *The Lysenko Affair,* 59, 34, and 131.
38 Joravsky, *The Lysenko Affair,* 43, 96, 110.
39 Roll-Hansen, "The Lysenko Effect," 146; Joravsky, *The Lysenko Affair,* 121.

40 Stanchevici, "Stalinist Genetics," 31–65; Mosterín, "Social Factors in the Development," 148 and 150; Sheehan, *Marxism and the Philosophy of Science*, 225.
41 Bailes, *Technology and Society*, and Siddiqi, "The Rockets' Red Glare."
42 Sheehan, *Marxism and the Philosophy of Science*, 226.
43 Schandevyl, "Soviet Biology," 105; Lyle, "Science bourgeoise"; Krementsov, "A 'Second Front' in Soviet Genetics," 249; Jong-Lambert, "Eugenics and Lysenkoism," and Köhler, "Lysenko Affair and Polish Botany."
44 Roll-Hansen, "The Lysenko Effect," 145.
45 Nye, "National Styles"; Metzler, *Internationale Wissenschaft*; Gerovich, *From Newspeak to Cyberspeak*.
46 Graham, *Science in Russia*, 131.
47 Mosterín, "Social Factors in the Development," 152.
48 Patel, *Fertile Ground for Europe*; see also: Cullather, *The Hungry World*.
49 Joravsky, *The Lysenko Affair*, 121; Groys and Hagemeister, *Die neue Menschheit*.
50 Fangerau, "From Mephistopheles to Isaiah."
51 See: Geyer, "The Militarization of Europe"; Edgerton, *Warfare State*; Reichherzer, *Alles ist Front*.
52 See the examples in Solomon, *Doing Medicine Together*. Generally, Fuchs, "Wissenschaftsinternationalismus in Kriegs- und Krisenzeiten."
53 Guse, "Volksgemeinschaft Engineers"; Schattenberg, "Stalinismus in den Köpfen."
54 Speer, *Inside the Third Reich*, 63.
55 Wieland, "Autarky and Lebensraum."
56 Aldcroft, *Europe's Third World*, 63–64.
57 Nord, *France's New Deal*.
58 Clarke, *France in the Age of Organization*, 14; see also: Clarke, "Engineering a New Order in the 1930's."
59 Petersen, *Bevölkerungsökonomie*, 234; see also: Unger, *Ostforschung in Westdeutschland*.

6 Experts in Exile

1 Karlsch, *Allein bezahlt*, 230–31, estimates that reparation costs in East Germany amounted to a total of 14 billion USD (at 1938 prices) between 1945 and 1953.
2 The term "intellectual reparations" was coined by John Gimbel in his seminal book: *Science, Technology and Reparations*.
3 Mick, *Forschen für Stalin*, 63–65.
4 Chertok, *Rockets and People*, Vol. 1, 365–66.
5 Most of the returning experts ended up in East Germany. The most prominent repatriate, after the Nobel laureate Gustav Hertz, was the physicist and inventor Manfred von Ardenne, who had twice received the Stalin Prize. Back in Dresden, von Ardenne succeeded in establishing a veritable scientific empire, the privately run "Forschungsinstitut Manfred von Ardenne," and engaged in party politics as a delegate to the "Volkskammer," the East German parliament, from 1963 to 1989.
6 Neufeld, "The Nazi Aerospace Exodus"; Mick, *Forschen für Stalin*; Nahum, "I Believe"; Glatt, "Reparations and the Transfer"; Ludmann-Obier, "Un aspect de la chasse aux cerveaux"; Jones, *The Employment of German Scientists*;

Stanley, *Rüstungsmodernisierung durch Wissenschaftsmigration?*; Presas i Puig, "Technoscientific Synergies"; Margolin, *Unauthorized Entry*"; Erichsen, "Die Emigration deutschsprachiger Naturwissenschaftler."

7 Neufeld, *Von Braun: Dreamer of Space*, which is by far the best biography of the multilayered personality of the baron, son of the conservative civil servant Magnus Freiherr von Braun, minister of agriculture during the Weimar Republic.

8 See: Johnston, *The Secret of Apollo*.

9 AEDC Staff and Office of Public Affairs, *Arnold Engineering Development Center*: *An Air Force Command Test Facility* (Arnold Air Force Base, TE, n.d.), as quoted in Ciesla and Trischler, "Legitimation through Use," 172.

10 Ciesla, "German High Velocity Aerodynamics," 100.

11 Apart from some hagiographies like Conradis, *Design for Flight*, there is no solid scholarly literature on Tank.

12 Stokes, "Forced Technology Transfer," 283.

13 Gimbel, *Science, Technology and Reparations*, 152.

14 Neebe, "Technologietransfer und Aussenhandel."

15 See: Oldenziel and Hård, *Consumers, Tinkerers, Rebels*, and Edgerton, *The Shock of the Old*.

16 Ludmann-Obier, "Un aspect de la chasse," 208.

17 Gimbel, *Science, Technology and Reparations*, 184; for the following argument, see: Stokes, "Forced Technology Transfer," 286–87, and Bähr et al., "The Politics of Ambiguity."

18 Beyerchen, "Anti-Intellectualism"; Krohn, "Deutsche Wissenschaftsemigration seit 1933."

19 Wolff, "Das Jahr 1933," and Ash and Söllner, "Introduction," 7. For an excellent evaluation of the effects resulting from the expulsion of scientists in various field of science, see: Waldinger, "Peer Effects in Science."

20 As quoted in Ash and Söllner, "Introduction," 3–4.

21 Moser et al., *German Jewish Émigrés and U.S. Invention*, 1.

22 See: Reisman, *Turkey's Modernization*; Strohmeier, "Der zeitgeschichtliche und politische Rahmen"; and Kinross, *Atatürk*.

23 Reisman, *Turkey's Modernization*, 279.

24 Erichsen, "Die Emigration deutschsprachiger Naturwissenschaftler," 77.

25 Bozdoğan, *Modernism and Nation Building*, 156–75, and Dogramaci, "Staatliche Repräsentation durch Emigranten," 65. For the following, see also: Dogramaci, *Kulturtransfer und nationale Identität*, 59–75.

26 Bozdoğan, *Modernism and Nation Building*, 108.

27 Durth, *Deutsche Architekten*, 93.

28 See: Keleş, "Der Beitrag Ernst Reuters."

29 Fred Forbat, Autobiography (manuscript), 206, Fred Forbat papers. Bauhaus archive. Berlin.

30 Dogramaci, "Ewig schönes Istanbul," 212.

31 Nicolai, "'Der goldene Käfig'," 240.

32 Flier, "Possibly the Greatest Task." See also: Bodenschatz et al., *Städtebau im Schatten Stalins*; Šarapov, "Deutsche Arbeiter in der Sowjetunion"; Sutton, *Western Technology*.

33 Kopp, "Foreign Architects in the Soviet Union," and Bliznakov, "The Realization of Utopia."

34 Bodenschatz et al., *Städtebau im Schatten Stalins*, 127; Quiring, "From Carp Pond."
35 Bodenschatz et al., *Städtebau im Schatten Stalins*, 35.
36 Rössler, "Die Wunderstadt, die niemals schläft."
37 Ash, "Scientific Changes in Germany"; see also: Ash, "Forced Migration and Scientific Change."
38 Sachar, *A History of the Jews in America*, 749.
39 Moser et al., *German Jewish Émigrés*, 2.
40 Marton, *The Great Escape*.
41 Siegmund-Schultze, "Hilda Geiringer-von Mises."
42 Ash and Söllner, "Introduction," 4.
43 Hoch and Platt, "Migration and the Denationalization of Science."

7 Geographies of Cooperation in Nuclear Europe

1 Blaizot et al., *Study of Potential Dangerous Events* and American Physical Society, Division of Particles & Fields, *Statement by the Executive Committee*.
2 Rossi, "Superconductivity," 15–16.
3 For the concept of technical failures and failed innovation, see the special issue of *Social Studies of Science* 22/2 (1992); for the inherent uncertainty of high-tech projects, see: Scranton, "The Challenge of Technological Uncertainty."
4 Aymar, "Basic Science in a Competitive World."
5 Krige and Pestre, "Some Thoughts on the History of CERN."
6 Most comprehensively by Herrmann et al., *History of CERN*.
7 Oreskes, "Science, Technology and Free Enterprise," in response to Krige, *American Hegemony*, and Krige, "Building the Arsenal of Knowledge."
8 Krige, *American Hegemony*, 57.
9 Carson, *Heisenberg in the Atomic Age*, 223.
10 Heisenberg, "European Cooperation for the Advancement of Nuclear Research," 293.
11 Gentner to Bothe, January 23, 1952, as quoted in Trischler, "Wolfgang Gentner," 106.
12 Carson, "Beyond Reconstruction." See also: Pestre, "The Difficult Decision."
13 Gentner, "Grossforschung als Problem moderner europäischer Zusammenarbeit."
14 Strasser, "The Coproduction of Neutral Science," 168. See also: Strasser and Joye, "Une science 'neutre' dans la guerre froide." For the co-construction of technology and Swiss identity, see: Schueler, *Materializing Identity*.
15 Strasser, "The Coproduction of Neutral Science," 173.
16 Kojevnikov, "The Great War."
17 See: Grahman, *Science in Russia*; Kojevnikov, *Stalin's Great Science*; and Kojevnikov, "The Phenomenon of Soviet Science."
18 Reisinger: *Energy and the Soviet Bloc*, 51.
19 Schmidt, "Nuclear Colonization," 125.
20 50 Years Dubna, "JINR 1956–2006."
21 Knorr Cetina, "How Superorganisms Change," and Knorr Cetina, "Culture in Global Knowledge Societies."
22 CMS Collaboration, "Transverse-Momentum and Pseudorapidity Distributions."
23 As quoted in Merali, "The Large Human Collider," 484.

24 This will likely change in the post-Higgs-boson era, as intense discussions in the U.S. in December 2012 to reallocate resources away from particle physics showed.
25 USSR International Service, press statement of Vladimir Veksler, Sept. 8, 1955, as quoted in Krige, "Atoms for Peace," 166 and 177.
26 Josephson, *Red Atom*, 175; see also: Hewlett and Holl, *Atoms for Peace and War*, 530–36; Weilemann, *Die Anfänge der Europäischen Atomgemeinschaft*; Guzzetti, *A Brief History of European Union Research Policy*, 7–33; Krige, "The Peaceful Atom as Political Weapon"; and Mallard, "The Atomic Confederacy."
27 Krige, "*Atoms for Peace*," 180.
28 Hewlett and Holl, *Atoms for Peace and War*, 324.
29 Memorandum of Understanding.
30 Hecht, *The Radiance* of *France*.
31 Buch and Vanderlinden, *L'uranium, la Belgique et les puissances*. See also: Moreau, "L'industrie nucléaire en Belgique," and Laes et al., *Kernenergie (on)besproken*. For a postcolonial perspective, see: Mollin, *Die USA und der Kolonialismus*.
32 "Où en est Euratom?," 5; numbers in Guzzetti, *A Brief History of European Union Research Policy*, 25–26, and Kramer, *Nuklearpolitik in Westeuropa*, 87. In addition to Ispra, the Euratom's Joint Research Centre (JCR) comprises establishments at Geel in Belgium, Petten in the Netherlands, and Karlsruhe in Germany.
33 Krige, "The Peaceful Atom as a Political Weapon," 44.
34 "Pour sortir Euratom de la crise," 1; Gérard, "La crise de l'Europe nucléaire."
35 Guzzetti, *A Brief History of European Union Research Policy*, 8.
36 Asbeek Brusse, "Euratom," 210.
37 Andreini, "EURATOM," 121. See also: Fischer, "Das Projekt einer trilateralen Nuklearkooperation"; Soutou, "Les accords de 1957–1958"; and Schrafstetter and Twigge, *Avoiding Armageddon*.
38 Presas i Puig, "Nota histórica"; Presas i Puig, "Science on the Periphery"; and Presas i Puig, "Technoscientific Synergies between Germany and Spain."
39 Cockroft, as quoted in Hanevik and Bendiksen, *The European Atomic Energy Society*, 12.
40 OECD, *NEA 50 Anniversary*, 13.
41 Ibid., 37 and 38.
42 Weilemann, *Anfänge der Europäischen Atomgemeinschaft*, 192.
43 Krige, "Peaceful Atom," 43; Guzzetti, *Brief History*, 31; Asbeek Brusse, "Euratom," 222; Weilemann, *Anfänge der Europäischen Atomgemeinschaft*, 157.
44 Mallard, "L'Europe puissance nucléaire," 142.
45 Even outright lobbying organizations like Foratom, the "European Atomic Forum," founded in 1960 to give nuclear industries a voice in European politics, and the European Nuclear Society (ENS), founded in 1975, have been presenting themselves successfully as expert networks. ENS claims on its website to pool "20,000 professionals from industry, the academic world, research centres and authorities: people who voluntary [sic] commit themselves to generate ideas and to take up responsibilities, who have the enthusiasm to get things done and the curiosity to learn from colleagues and from people outside the network."
46 "A white elephant" is a burdensome possession whose cost of upkeeping is out of proportion to its usefulness; the idiom is often used to describe the societal problems of large-scale technological projects, see: Laak, *Weisse Elefanten*.

47 Press release of the Federal Ministry of Research and Education 14/91, March 21, 1991, as quoted in Marth, *Geschichte von Bau und Betrieb*, 142.
48 Fjaestad, *Visionen om outtömlig energi*.
49 Hecht, "Technology, Politics, and National Identity in France." See also: Jasper, *Nuclear Politics*; Jasper compares the French style of nuclear politics with those of Sweden and the U.S., pinpointing divergences on the political level alongside convergences on the technical level.
50 Rotblat, *Nuclear Reactors*.
51 Keck, "The West German Fast Breeder Programme."
52 This question is asked by Fjaestad, *Why Did the Breed Reactor Fail?*; see also: Marth, *The Story of the European Fast Reactor Cooperation*.
53 "Euratom Unifies Reactor Research," *Chemical Engineering News* 44/3 (1966): 66.
54 The environmental scientist and energy expert Vaclav Smil likes to argue that fusion technology has always been 50 years in the future for the last 50 years or more; see: Smil, *Energy in World History*, and Smil, *Energy in Nature and Society*.
55 Euratom website.
56 McCray, "Globalisation with Hardware," 285. The following is based on this in-depth study and on Bromberg, *Fusion*, and Shaw, *Europe's Experiment in Fusion*.
57 McCray, "Globalisation with Hardware," 285.
58 Boenke, *Entstehung und Entwicklung*.
59 Shaw, "Joint European Torus," 166.
60 Quoted in McCray, "Globalization with Hardware," 289.
61 Willson, *A European Experiment*, ix and x.
62 Ibid., 126.
63 We here refer to Hans-Jörg Rheinberger's concept of local experimental systems and experimental cultures; see his most recent book: *On Historicizing Epistemology*.
64 Joint Soviet–United States Statement.

8 Contesting Europe in Space

1 Gerovitch, "Creating Memories," 228.
2 For a comprehensive list of all unmanned and manned Interkosmos-missions see: Wikipedia, "Intercosmos." There is no solid history of Interkosmos; instructive insights into the cosmonauts' training program for Interkosmos-missions are provided by Hall et al., *Russia's Cosmonauts*, 205–14; for the Soviet space program in general, see: Siddiqi, *Sputnik and the Soviet Space Challenge*; Siddiqi, *The Soviet Space Race with Apollo*; and Siddiqi, "Soviet Space Power during the Cold War."
3 Hein-Weingarten, *Das Institut für Kosmosforschung*.
4 See: Zickler, "Weltraumforschung in der DDR."
5 From the considerable hagiographical literature, see in particular: Hoffmann, *Sigmund Jähn*; see also the Deutsche Raumfahrtausstellung Museum's website. For space suits, see: Shaw, "Bodies out of This World."
6 See: Raj, *Reach for the Stars*, and Siddiqi, "Asia in Orbit."
7 Siddiqi, "Competing Technologies, National(ist) Narratives, and Universal Claims," 436; see also: Siddiqi, "National Aspirations on a Global Stage," and Ryan, "The Role of National Culture."
8 Abraham, *The Making of the Indian Atomic Bomb*, 9.

9   See chapter 7; for excellent studies in the inherent uncertainty of aerospace projects, see: Scranton, "Technology-Led Innovation," and Vaughan, *The Challenger Launch Decision*.
10 Johnson, "A Failure to Communicate," 17; see also: de Maria and Krige, "Early European Attempts." From the vast amount of literature, see in particular, Krige et al., *A History of the European Space Agency*.
11 Report of the Project Review Commission, May 19, 1972, as quoted in Trischler, "A Talkative Artefact," 16.
12 Schwarz, "European Policies on Space Science and Technology," 43.
13 The emphasis is here on the public. For the political and space administration, *Sputnik* did not come as so much of a shock; see: Divine, *The Sputnik Challenge*, and Launius et al., *Reconsidering Sputnik*. For a European perspective, see: Polianski and Schwartz, *Die Spur des Sputnik*.
14 Auger, "The Prehistory of ESRO." See also: Sebesta, *Alleati competitive*.
15 McDougall, "Space-Age Europe"; see also: McDougall, *…The Heavens and the Earth*.
16 For this crucial shift in European relations, see: Kaiser, *Using Europe, Abusing the Europeans*.
17 Sebesta, "Choosing its Own Way."
18 Trischler, *The "Triple Helix" of Space*. For the following, see: Trischler and Weinberger, "Engineering Europe."
19 Collett, *Making Sense of Space*.
20 Strasser, "The Coproduction of Neutral Science," 179; see also: Zellmeyer, *A Place in Space*.
21 Laureys, *La contribution de la Belgique*.
22 Stiernstedt, *Sverige i rymden*.
23 Wormbs, *Vem älskade Tele-X?*, and Wormbs, "A Nordic satellite project."
24 Krige, "Preface," v–vi.
25 Trischler, *Dokumente zur Geschichte der Luft- und Raumfahrtforschung*, 401.
26 The agreement is reproduced in Welck and Platzöder, *Weltraumrecht*, 157–62; for the following, see: Reinke, *The History of German Space Policy*, 113–17, and Krige et al., *A History of the European Space Agency*, Vol. 1.
27 For a detailed history of the complex political bargaining between the U.S. and Franco-German alliance, see: Krige et al., *A History of the European Space Agency*, Vol. 2, 437–49.
28 Interviews with Reimar Lüst, October 8, 2010, and Heinz Riesenhuber, May 5, 2010.
29 Reinke, *The History of German Space Policy*, 256.
30 The high hopes of Europe to gain access to the advanced American knowledge in launcher technology did not materialize. Neither did Europe have access to the U.S. Space Shuttle-technology in general nor the opportunity to use the know-how resulting from Spacelab experiments for direct economic applications. For space Europe in general, and Germany in particular, the sortie lab was neither a jump start into a bright future of manned space flight nor a big step into transatlantic cooperation on an equal footing. A slightly different interpretation is given by Krige et al., *A History of the European Space Agency*, Vol. 2, 62, for whom the project "reflected the very uneven balance of power between the partners with the odds stacked heavily in favour of the U.S. On the other hand, Europe achieved some of what it wanted: a cheap way into manned space

flight, a quantum leap in project management experience, and the laying of the foundations—political, industrial and personal—for a new kind of international collaborative venture with NASA and the US administration."

31 Nolte, *Der Wissenschaftsmacher*.
32 Shapin, *The Scientific Life*, 5.
33 Bayer, "Hermes: Learning from Our Mistakes."
34 Sheehan, *The International Politics of Space*, 84.
35 Chadeau, "Why Was Arianespace Established (1979)?," 281; the numbers are drawn from Wikipedia, "Arianespace."
36 Morton, *Fire Across the Desert*; Cawte, *Atomic Australia 1944–1990*; Launius and Johnston, *Smithsonian Atlas of Space Exploration*, 285; Redfield, *Space in the Tropics*.
37 See the European Space Agency website.
38 Madders and Thiebaut, "Two Europes in One Space," 128; see also: Madders, *A New Force at a New Frontier*, and interview with Reimar Lüst, September 8, 2010.
39 European Commission, *"Space: A New Frontier for an Expanding Union,"* 7.
40 European Commission, *Towards a Space Strategy*, 2 and 12.
41 Johnson, Secret of Apollo, 154–208.
42 Servan-Schreiber, *The American Challenge*; Strauss, *Challenge and Response*; see: Ritter et al., *Antworten auf die amerikanische Herausforderung*.
43 See: Harvey, *Russia in Space*, 291–3.
44 See: Graham, "Big Science in the Last Years."
45 Krige, "What is 'Military' Technology?".
46 Zabusky, *Launching Europe*, 103 and 123.
47 Reimar Lüst, *Die Zeit*, March 10, 1995, as quoted in Sheehan, *International Politics*, 181; interview with Reimar Lüst, September 8, 2010.
48 Mitrany, "The Functional Approach," 532; for a comprehensive layout of functionalism see his study: *The Functional Theory of Politics*.
49 Haas, *The Uniting of Europe*, which in 1997 the journal *Foreign Affairs* choose as one of the 50 most significant books in international relations in the twentieth century.
50 Suzuki, *Policy Logics*, 5.

9 Experts' Europe from a Bird's-Eye View

1 See: Brendecke, *Die Jahrhundertwenden*.
2 European Parliament, *Lisbon European Council*.
3 Castells, *The Rise of the Network Society*; Castells, *The Power of Identity*; Castells, *End of Millennium*.
4 For her conceptual ideas, see: Rodrigues, *European Policies for a Knowledge Economy*, and Rodrigues, *Europe, Globalization and the Lisbon Agenda*.
5 Commission of the European Communities, *Towards a European Research Area*, 8.—After many critical reports on failed expectations and limited effects, the Commission revised the Lisbon Agenda in March 2010: European Commission, *Europe 2020*.
6 Wirsching, *Der Preis der Freiheit*, 236.
7 Lyotard, *The Postmodern Condition*.
8 Historians, in contrast, have emphasized the many ruptures and deep-cutting transformations in the history of the European university. The scholastic

universities of the Middle Ages and the research universities of the post-Humboldtian periods did not have much in common, and they differed in many ways from the multi-purpose university of today; see, for example, Wittrock, "The Transformation of European Universities," and Kintzinger, "Universität."

9 Krige, *American Hegemony,* 208–25, and Gemelli, "Western Alliance and Scientific Diplomacy."
10 André, "L'espace européen de la recherché." Political scientists have quickly adopted the concept: see, for example, Edler et al., *Changing Governance of Research and Technology Policy,* and Frenken et al., *Towards a European Research Area.*
11 Commission of the European Communities, *Working Program in the Field of Research.*
12 André, "L'espace européen," 172 (English summary). Even the Commission frankly admitted that the concept of rebuilding Europe based on an integrated research area was "not a new idea." Commission of the European Communities, *Towards a European Research Area,* 8.
13 Servan-Schreiber, *Le défi américain.* The book sold 600,000 copies in France alone and was translated into fifteen languages.
14 For the importance of the subsidiarity principle in European affairs, see: Duff, *Subsidiarity within the European Community.*
15 Quoted from European Research Council, "Synergy Grants."
16 Quotes after the personal profile of Enserink, "Keeping Europe's Basic Research Agency on Track." For the vision of a "socially robust" mode of knowledge production, see: Nowotny et al., *Re-Thinking Science.*
17 Nordmann, "European Experiments," 278.
18 Nordmann, *Converging Technologies*; Roco and Bainbridge, *Converging Technologies for Improving Human Performance.*
19 Nordmann, "European Experiments," 291; from the vast literature on the two reports, see: Giorgi and Luce, *Converging Science and Technologies.*
20 European Commission, *Taking European Knowledge Society Seriously,* 79 and 81; the expert group also included numerous prominent scholars such as Michel Callon, Sheila Jasanoff, Dominique Pestre, and Arie Rip.
21 European Commission, *Building Europe Knowledge,* 15–16.
22 Caracostas and Muldur, *Society, the Endless Frontier.*
23 *Nature* 392, 641 (April 16, 1998).
24 Muldur et al., *New Deal for an Effective European Research Policy,* 93. For an excellent meta-critique of these impact studies, see: Luukkonen, "The Difficulties in Assessing the Impact"; see also the most recent meta-study of Arnold et al., *Understanding the Long Term Impact of the Framework Programme.*
25 Leclerc and Gagné, "International Scientific Cooperation," 279.
26 Braun and Glänzel, "International Collaboration," 251.
27 Leclerc and Gagné, "International Scientific Cooperation," 272; also Glänzel, "Science in Scandinavia."
28 Glänzel, et al., "A Bibliometric Analysis"; Glänzel and Schlemmer, "National Research Profiles"; Georghiou, "Global Cooperation in Research."
29 Waltman et al., *Globalisation of Science in Kilometres.*
30 Hoekman et al., "Research Collaboration at a Distance."

31 See: Hessler and Zimmermann, *Creative Urban Milieus*.
32 The programs Basic Research in Industrial Technology (BRITE) and European Research in Advanced Materials (EURAM) were launched in 1985 and 1986, respectively. They merged in 1989 under the Second Framework Programme, creating Brite-Euram. Etienne Davignon, Vice-President of the European Commission from 1977 to 1985, later concluded that ESPRIT and BRITE "helped to create a genuinely transnational relationship between companies, universities and government bodies, which was never the case before." See: Davignon (interview).
33 Grande, *The Erosion of State Capacity*, 9. ESPRIT, by far the largest EU program, involved over 9,000 organizations of various type throughout Europe while running from 1984 to 1998; see: Lichtenberg, *The European Strategic Program*.
34 Peterson, *High Technology and the Competition State*; Sandholtz, *High-Tech Europe*; Zysman and Borrus, "From Failure to Fortune?"; Peterson and Sharp, *Technology Policy in the European Union*.
35 European Commission, *Green Paper on Innovation*.
36 Glänzel et al., "A Bibliometric Analysis," 189.
37 Glänzel and Schlemmer, "National Research Profiles," 275.
38 Burke, *A Social History of Knowledge*; Dülmen and Rauschenbach, *Macht des Wissens*.
39 Leydesdorff, "Is the European Union," 274.
40 Geyer and Paulmann, *The Mechanics of Internationalism*. See also: Drori et al., *Science in the Modern World Polity*, 81–99, and Stichweh, *Die Weltgesellschaft*.
41 Grande and Peschke, "Organizing Science in Europe," 309; after unfolding the dense network of institutions, the authors come to the conclusion that European science and technology policy "still continues to suffer from serious deficiencies," remains fragmented, and struggles to find its proper role between the national and the transnational poles.

10 Conclusion

1 Saint-Simon, *De la réorganisation*. See also: Kaiser and Schot, *Writing the Rules for Europe*, and Oldenziel and Hård, *Consumers, Tinkerers, Rebels*.
2 Wolff, "Physicists in the 'Krieg der Geister'"; Ungern-Sternberg and Ungern-Sternberg, *Der Aufruf An die Kulturwelt*, 144–45.
3 Einstein, "Why Socialism?".
4 Chessa et al., "Is Europe Evolving," 651.
5 Misa and Schot, "Inventing Europe."
6 Wirsching, *Der Preis der Freiheit*, 253 and 265.

Bibliography

Abel, Richard, and Lyman W. Newlin, eds. *Scholarly Publishing: Books, Journals, Publishers, and Libraries in the Twentieth Century.* New York: Wiley, 2002.

Abraham, Itty. *The Making of the Indian Atomic Bomb: Science, Secrecy and the Postcolonial State.* London: Zed Books, 1998.

Acker, Wouter van. "Internationalist Utopias of Visual Education: The Graphic and Scenographic Transformation of the Universal Encyclopaedia in the Work of Paul Otlet, Patrick Geddes, and Otto Neurath." *Perspectives on Science* 19, no. 1 (2011): 32–80.

Agnew, Hugh LeCaine. "The Flyspecks on Palivec's Portrait: Francis Joseph, the Symbols of Monarchy, and Czech Popular Loyalty." In *The Limits of Loyalty: Imperial Symbolism, Popular Allegiances, and State Patriotism in the Late Habsburg Monarchy*, edited by Laurence Cole and Daniel L. Unowsky, 86–112. New York: Berghahn Books, 2007.

Aldcroft, Derek H. *Europe's Third World: The European Periphery in the Interwar Years.* Aldershot: Ashgate, 2006.

Alder, Ken. *Engineering the Revolution: Arms and Enlightenment in France, 1763–1815.* Princeton: Princeton University Press, 1997.

Alofsin, Anthony. *The Struggle for Modernism: Architecture, Landscape Architecture, and City Planning at Harvard.* New York: Norton, 2002.

Aly, Götz, and Susanne Heim. *Architects of Annihilation: Auschwitz and the Logic of Destruction.* Princeton: Princeton University Press, 2003.

American Physical Society, Division of Particles & Fields, *Statement by the Executive Committee of the DPF on the Safety of Collisions at the Large Hadron Collider,* 2008. Accessed March 23, 2011. http://www.aps.org/units/dpf/governance/reports/upload/lhc_safety_statement.pdf.

Anděl, Jaroslav, ed. *The New Vision for the New Architecture: Czechoslovakia 1918–1938.* Prague: Slovart, 2005.

Andersen, Hendrik C., and Ernest M. Hébrard. *Creation of a World Centre of Communication.* Paris : n.p., 1913.

André, Michel. "L'espace européen de la recherche: Histoire d'une idée." *Journal of European Integration History* 12, no. 2 (2006): 131–50.

Andreini, Ginevra. "EURATOM: An Instrument to Achieve a Nuclear Deterrent? French Nuclear Independence and European Integration during the Mollet Government (1956)." *Journal of European Integration History* 6, no. 2 (2000): 109–28.

Antoniou, Yiannis, Michalis Assimakopoulos, and Konstantinos Chatzis. "The National Identity of Inter-War Greek Engineers: Elitism, Rationalization, Technocracy, and Reactionary Modernism." *History and Technology* 23 (2007): 241–61.

Arnold, Eric et al. *Understanding the Long Term Impact of the Framework Programme: Final Report to the European Commission DG Research*, December 5, 2011. Accessed August 27, 2012. http://ec.europa.eu/research/evaluations/pdf/archive/other_reports_studies_and_documents/long_term_impact_of_the_fp.pdf.

Asbeek Brusse, Wendy. "Euratom." In *The Netherlands and the Integration of Europe, 1945–1957*, edited by Richard T. Griffiths, 209–24. Amsterdam: NEHA, 1990.

Ash, Mitchell G. "Forced Migration and Scientific Change: Steps towards a New Overview." In *Intellectual Migration and Cultural Transformation: Refugees from National Socialism in the English-Speaking World*, edited by Edward Timms and Jon Hughes, 241–63. Vienna: Springer, 2003.

———. "Scientific Changes in Germany 1933, 1945 and 1990: Towards a Comparison." *Minerva* 37 (1999): 329–54.

———. "Wissenschaft und Politik als Ressource füreinander." In *Wissenschaften und Wissenschaftspolitik: Bestandsaufnahmen zu Formationen, Brüchen und Kontinuitäten im Deutschland des 20. Jahrhunderts*, edited by Rüdiger vom Bruch and Brigitte Kaderas, 32–51. Stuttgart: Steiner, 2002.

——— and Alfons Söllner. "Introduction: Forced Migration and Scientific Change after 1933." In *Forced Migration and Scientific Change: Emigré German-Speaking Scientists and Scholars after 1933*, edited by Mitchell G. Ash and Alfon Söllner, 1–19. Cambridge: Cambridge University Press, 1966.

Auger, Pierre. "The Prehistory of ESRO – A Personal Memoir." In *Two Decades in Space: Recollections by Some of the Principal Pioneers*, edited by AAVV Europe. 12–15. Noordwijk: ESA Publications, 1984.

Aymar, Robert, "Basic Science in a Competitive World," first published in *Symmetry Magazine* (August 2006). Accessed March 31, 2011. http://public.web.cern.ch/public/en/About/BasicScience-en.html.

Bacon, Mardges. *Le Corbusier in America: Travels in the Land of the Timid*. Cambridge, MA and London: MIT Press, 2003.

Bähr, Johannes, Paul Erker, and Geoffrey G. Giles. "The Politics of Ambiguity: Reparations, Business Relations, Denazification and the Allied Transfer of Technology." In *Technology Transfer out of Germany after 1945*, edited by Matthias Judt and Burghard Ciesla, 131–44. Amsterdam: Harwood, 1996.

Bailes, Kendal E. *Technology and Society under Lenin and Stalin: Origins of the Soviet Technical Intelligentsia, 1917–1941*. Princeton: Princeton University Press, 1978.

Barr, Helen. *Neues Wohnen 1929/2009: Frankfurt und der 2. Congres International d'Architecture Moderne*. Berlin: Jovis, 2010.

"Bauhaus Man." *Time Magazine*, February 8, 1937.

Bayer, Martin. "Hermes: Learning from Our Mistakes." *Space Policy* 11 (1995): 171–80.

Belhoste, Bruno, and Konstantinos Chatzis. "From Technical Corps to Technocratic Power: French State Engineers and Their Professional and Cultural Universe in the First Half of the 19th Century." *History and Technology* 23 (2007): 209–25.

Benz, Wolfgang, Barbara Distel, and Angelika Königseder. "Der Ort des Terrors: Geschichte der nationalsozialistischen Konzentrationslager in neun Bänden." *Dachauer Hefte 25* (2009): 31–33.

Berger, Silvia. *Bakterien in Krieg und Frieden: Eine Geschichte der medizinischen Bakteriologie in Deutschland 1890–1933*. Göttingen: Wallstein, 2009.

Berner, Boel. "Educating Men: Women and the Swedish Royal Institute of Technology, 1880–1930." In *Crossing Boundaries, Building Bridges: Comparing the History of Women Engineers 1870s–1990s*, edited by Annie Canel, Ruth Oldenziel, and Karin Zachmann, 75–102. Amsterdam: Harwood, 2000.

Bernhardt, Maciej. "Szkoła Wawelberga i Politechnika Warszawska w Latach 1940–1944." *Zeszyty Historyczne* 118 (1996): 95–108.

Bertrams, Kevin. "Planning and the 'Techno-Corporatist Bargain' in Western Europe and the United States, 1914–44: Diffusion and Confusion of Economic Models." In *Expert Cultures in Central Eastern Europe: The Internationalization of*

Knowledge and the Transformation of Nation States since World War I, edited by Martin Kohlrausch, Katrin Steffen, and Stefan Wiederkehr, 43–61. Osnabrück: fibre, 2010.

Beyerchen, Alan. "Anti-Intellectualism and the Cultural Decapitation of Germany under the Nazis." In *The Muses Flee Hitler: Cultural Transfer and Adaption, 1930–1945*, edited by Jerold C. Jackman and Carla M. Borden, 29–44. Washington, DC: Smithsonian Institution Press, 1983.

———. "Rational Means and Irrational Ends: Thoughts on the Technology of Racism in the Third Reich." *Central European History* 30 (1997): 386–402.

Birstein, Vadim J. *The Perversion of Knowledge: The True Story of Soviet Science.* Cambridge, MA: Westview, 2004.

Blackbourn, David. *The Conquest of Nature: Water, Landscape, and the Making of Modern Germany.* New York: Norton, 2006.

Blaizot, J.P. et al. *Study of Potential Dangerous Events during Heavy Ion Collisions at the LHC: Report of the LHC Safety Study Group*, Geneva 2003, iii (CERN-2003–001). Accessed March 31, 2011. http://cdsweb.cern.ch/record/613175/files/CERN-2003–001.pdf.

Bliznakov, Milka. "The Realization of Utopia: Western Technology and Soviet Avant-Garde Architecture." In *Reshaping Russian Architecture: Western Technology, Utopian Dream*, edited by William C. Brumfield, 145–75. Cambridge: Cambridge University Press, 1990.

Bluche, Lorraine, Veronika Lipphardt, and Kiran K. Patel, eds. *Der Europäer—ein Konstrukt: Wissensbestände, Diskurse, Praktiken.* Göttingen: Wallstein, 2009.

Bocquet, Denis. "Engineers and the Nation in Italy (1750–1922): Local Traditions and Different Conceptions of Unity and Modernity." *History and Technology* 23 (2007): 227–42.

Bodenschatz, Harald, and Christiane Post, eds. *Städtebau im Schatten Stalins: Die internationale Suche nach der sozialistischen Stadt in der Sowjetunion 1929–35.* Berlin: Braun, 2003.

Boenke, Susan. *Entstehung und Entwicklung des Max-Planck-Instituts für Plasmaphysik, 1955–1971.* Frankfurt a.M. and New York: Campus, 1991.

Bogdanov, Alexander. *Essays in Tektology: The General Science of Organization.* Seaside, CA: Intersystems Publications, 1980.

Bolenz, Eckhard. "Baubeamte in Preussen, 1799–1931: Aufstieg und Niedergang einer technischen Elite." In *Ingenieure in Deutschland: 1770–1990*, edited by Peter Lundgreen, 117–40. Frankfurt a.M. and New York: Campus, 1994.

———. *Vom Baubeamten zum freiberuflichen Architekten: Technische Berufe im Bauwesen (Preussen/Deutschland, 1799–1931).* Frankfurt a.M. and New York: P. Lang, 1991.

Bonin, Hubert, Yannick Lung, and Steven Tolliday. *Ford, 1903–2003: The European History.* Paris: P.L.A.G.E., 2003.

Boudia, Soraya. *Marie Curie et son laboratoire: Sciences et industrie de la radioactivité en France.* Paris: Ed. des archives contemporaines, 2001.

Bozdoğan, Sibel. *Modernism and Nation Building: Turkish Architectural Culture in the Early Republic.* Seattle, WA: University of Washington Press, 2001.

Braun, T., and Wolfgang Glänzel. "International Collaboration: Will It Be Keeping Alive East European Research?" *Scientometrics* 36, no. 2 (1996): 247–54.

Brendecke, Arndt. *Die Jahrhundertwenden. Eine Geschichte ihrer Wahrnehmung und Wirkung.* Frankfurt a.M. and New York: Campus, 1999.

Brocke, Bernhard vom, and Hubert Laitko, eds. *Die Kaiser-Wilhelm-/Max-Planck-Gesellschaft und ihre Institute: Studien zu ihrer Geschichte.* Berlin: Walter de Gruyter, 1996.

Bromberg, Joan. *Fusion, Science, Politics and the Invention of a New Energy Source.* Cambridge, MA: MIT Press, 1982.

Brown, John K., Gary Lee Downey, and Maria Paula Diogo. "The Normativities of Engineers: Engineering Education and History of Technology." *Technology and Culture* 50 (2009): 737–52.

Bruisch, Katja, "'...that the Future Belongs to Us'. Agricultural Experts in Early 20th Century Russia." In *Performing Expertise: Scientists between State and Society, 1860–1960*, edited by Evert Peeters and Joris Vandendriessche, forthcoming.

Buch, Pierre, and Jacques Vanderlinden. *L'Uranium, la Belgique et les puissances: Marché de dupes ou chef-d'oeuvre diplomatique?* Brussels: De Boeck-Wesmael, 1995.

Buchanan, Robert A. *The Engineers: A History of the Engineering Profession in Britain, 1750–1914.* London: Kingsley, 1989.

Buckland, Michael. "On the Cultural and Intellectual Context of European Documentation in the Early Twentieth Century." In *European Modernism and the Information Society: Informing the Present, Understanding the Past*, edited by Warden B. Rayward, 44–57. Aldershot and Burlington, VT: Ashgate, 2008.

Bucur, Maria. *Eugenics and Modernization in Interwar Romania.* Pittsburgh, PA: University of Pittsburgh Press, 2002.

Burke, Peter. *A Social History of Knowledge: From Gutenberg to Diderot.* Cambridge: Polity Press, 2000.

Canales, Jimena. "Einstein, Bergson, and the Experiment that Failed: Intellectual Cooperation at the League of Nations." *MLN* 120 (2005): 1168–91.

Canel, Annie. "Maintaining the Walls: Women Engineers at the École Polytechnique Féminine and the Grandes Écoles in France." In *Crossing Boundaries, Building Bridges: Comparing the History of Women Engineers 1870s–1990s*, edited by Annie Canel, Ruth Oldenziel, and Karin Zachmann, 127–58. Amsterdam: Harwood, 2000.

Capristo, Annalisa. *L'espulsione degli ebrei dalle accademie Italiane.* Turino: Zamorani, 2002.

———. "The Exclusion of Jews from Italian Academies." In *Jews in Italy under Fascist and Nazi Rule, 1922–1945*, edited by Joshua D. Zimmerman, 81–95. Cambridge and New York: Cambridge University Press, 2005.

Caracostas, Paraskevas, and Ugur Muldur. *Society, the Endless Frontier: A European Vision of Research and Innovation Policies for the 21st Century.* Luxembourg: Office for Official Publications of the European Communities, 1998.

Carson, Cathryn. "Beyond Reconstruction: CERN's Second-Generation Accelerator Program as an Indicator of Shifts in West German Science." In *Physics and Politics: Research and Research Support in Twentieth Century Germany in International Perspective*, edited by Helmuth Trischler and Mark Walker, 107–30. Stuttgart: Steiner, 2010.

———. *Heisenberg in the Atomic Age: Science and the Public Sphere.* Cambridge: Cambridge University Press, 2010.

Castells, Manuel *The Rise of the Network Society. The Information Age: Economy, Society and Culture*, Vol. I. Oxford and Cambridge, MA: Blackwell, 1996, second edition, 2000.

———.*The Power of Identity. The Information Age: Economy, Society and Culture,* Vol. II. Oxford and Cambridge, MA: Blackwell, 1997, second edition, 2004.

———. *End of Millennium. The Information Age: Economy, Society and Culture,* Vol. III. Oxford and Cambridge, MA: Blackwell, 1998, second edition, 2000.

Cawte, Alice. *Atomic Australia 1944–1990.* Sidney: University of New South Wales Press, 1992.

Chadeau, Emmanuel. "Le contexte de la loi sur le titre d'ingénieur: Les tensions de l'économie." In *Les ingénieurs de la crise: Titre et profession entre les deux guerres,* edited by André Grelon, 49–59. Paris: Editions de l'Ecole des Hautes Études en Sciences Sociales, 1986.

———. "Why Was Arianespace Established (1979)?" In *History of European Scientific and Technological Cooperation,* edited by John Krige and Luca Guzzetti, 271–81. Luxembourg: Office for Official Publications of the European Communities, 1997.

Charles, Daniel. *Between Genius and Genocide: The Tragedy of Fritz Haber, Father of Chemical Warfare.* London: Jonathan Cape, 2005.

Chertok, Boris E. *Rockets and People,* Vol. 1. Washington, DC: NASA History Office, 2005.

Chessa, A., A. Morescalchi, F. Pammolli, O. Penner, A.M. Petersen, and M. Riccaboni. "Is Europe Evolving toward an Integrated Research Area?" *Science* 339 (February 8, 2013): 650–1.

Chionne, Roberta. "Blok e praesens: Dagli ideali del costruttivismo alla sperimentazione funzionle." In *Costruttivismo in Polonia,* edited by Silvia Parlagreco, 157–98. Torino: Bollati Boringhieri, 2005.

Chmielewski, Jan, and Szymon Syrkus. *Warszawa Funkcjonalna.* Warszawa: Towarzystwo Urbanistów Polskich, 1934.

Christy, Jim. *The Price of Power: A Biography of Charles Eugène Bedaux.* New York: Doubleday 1989.

Ciesla, Burghard. "German High Velocity Aerodynamics and Their Significance for the U.S. Air Force 1945–1952." In *Technology Transfer out of Germany after 1945,* edited by Matthias Judt and Burghard Ciesla, 93–106. Amsterdam: Harwood, 1996.

———, and Helmuth Trischler. "Legitimation through Use: Rocket and Aeronautics Research in the Third Reich and the USA." In *Science and Ideology: A Comparative History,* edited by Mark Walker, 156–85. London: Routledge, 2002.

Clark, Vincent. "A Struggle for Existence: The Professionalization of German Architects." In *German Professions, 1800–1950,* edited by Geoffrey Cocks and Konrad H. Jarausch, 143–60. New York: Oxford University Press, 1990.

Clarke, Jackie. "Engineering a New Order in the 1930's: The Case of Jean Coutrot." *French Historical Studies* 24 (2001): 63–86.

———. *France in the Age of Organization: Factory, Home and Nation from the 1920s to Vichy.* Oxford and New York: Berghahn Books, 2011.

Clavin, Patricia, and Jens-Wilhelm Wessels. "Transnationalism and the League of Nations: Understanding the Work of Its Economic and Financial Organisation." *Contemporary European History* 14 (2005): 465–92.

CMS Collaboration. "Transverse-Momentum and Pseudorapidity Distributions of Charged Hadrons in pp Collisions at Root s=0.9 and 2.36 TeV J. High Energy." *Journal of High Energy Physics* (2010), doi.org/10.1007/JHEP02(2010)041.

Cohen, Jean-Louis. "'Unser Kunde ist unser Herr': Le Corbusier trifft Baťa." In *Zlín: Modellstadt der Moderne*, edited by Winfried Nerdinger, 112–47. Berlin: Jovis, 2009.

Collett, John P., ed. *Making Sense of Space: The History of Norwegian Space Activities.* Oslo: Scandinavian University Press, 1995.

Collins, Christiane C. *Werner Hegemann and the Search for Universal Urbanism.* New York: Norton, 2005.

Commission of the European Communities. *Towards a European Research Area*, COM (2006) 6, Brussels, January 18, 2000. Accessed August 2, 2011. www.bologna-berlin2003.de/pdf/toward_Education_En.pdf.

———. *Working Program in the Field of Research, Science and Education*, SEC (1973) 2000, Brussels, May 23, 1973.

Conradis, Heinz. *Design for Flight: The Kurt Tank Story.* London: Macdonald, 1960.

Coopersmith, Jonathan. *The Electrification of Russia, 1880–1926.* Ithaca: Cornell University Press, 1992.

Coppens, Chris, Mark Derez, and Jan Roegiers. *Leuven University Library, 1425–2000.* Leuven: Leuven University Press, 2005.

Courtiau, Catherine. "Le création d'une ville internationale autonome selon Paul Otlet." *Transnational Associations* 55, nos. 1–2 (2003): 60–71.

Couturat, Louis. *Étude sur la dérivation dans la langue internationales.* Paris: Delagrave, 1910.

Crawford, Elisabeth. *Nationalism and Internationalism in Science, 1880–1939 – Four Studies of the Novel Population.* Cambridge: Cambridge University Press, 1992.

Crawford, Margaret. *Building the Workingman's Paradise: The Design of American Company Towns.* London and New York: Verso, 1995.

Cullather, Nick. *The Hungry World: America's Cold War Battle against Poverty in Asia.* Boston, MA: Harvard University Press, 2010.

Davignon, Etienne (interview with). Accessed February 15, 2013. http://ec.europa.eu/research/growth/competitive/en/1–2.html.

Debschitz, Uta von, and Thilo von Debschitz. *Fritz Kahn – Man Machine / Maschine Mensch.* Vienna: Springer 2009.

Deicer, Gregory. "Influence and Intercultural Exchange: Engineers, Engineering Schools and Engineering Works in the Nineteenth Century." *History and Technology* 12 (1995): 163–77.

deutsche raumfahrtausstellung museum website. Accessed March 3, 2011. http://www.deutsche-raumfahrtausstellung.de/index.html.

Devinat, Paul. "Working Conditions in a Rationalised Undertaking: The Bata System and Its Social Consequences." *International Labour Review* 21, no. 1 (1930): 45–69.

———."Working Conditions in a Rationalised Undertaking: The Bata System and its Social Consequences II." *International Labour Review* 21, no. 2 (1930): 163–86.

Dluhosch, Eric and Rostislav Švácha, eds. *Karel Teige. 1900–1951: L'Enfant Terrible of the Czech Modernist Avant-Garde.* Cambridge, MA: MIT Press, 1999.

Dienel, Hans-Liudger, ed. *Der Optimismus der Ingenieure: Triumph der Technik in der Krise der Moderne um 1900.* Stuttgart: Steiner, 1998.

Diogo, Maria Paula, and Dirk van Laak. *Europeans Globalizing: Mapping, Exploiting, Exchanging.* London: Palgrave Macmillan, forthcoming.

Disco, Nil. *Made in Delft: Professional Engineering in the Netherlands: 1880–1940.* Amsterdam: Universiteit van Amsterdam, 1990.

Dittrich, Klaus. "Experts Going Transnational: Education at World Exhibitions during the Second Half of the 19th Century." PhD diss., University of Portsmouth, 2010.

Divine, Robert A. *The Sputnik Challenge: Eisenhower's Response to the Soviet Satellite.* London and New York: Oxford University Press, 1993.

Dogramaci, Burcu. "'Ewig schönes Istanbul – Daima hasret ediyoruz': Ernst Reuter und Gustav Oelsner als Urbanisten im türkischen Exil." In *Ernst Reuter: Kommunalpolitiker und Gesellschaftsreformer, 1921–1953*, edited by Heinz Reif and Moritz Feichtinger, 203–37. Bonn: Dietz, 2009.

———. *Kulturtransfer und nationale Identität: Deutschsprachige Architekten, Stadtplaner und Bildhauer in der Türkei nach 1927.* Berlin: Gebr. Mann, 2008.

———. "Staatliche Repräsentation durch Emigranten. Der Anteil deutschsprachiger Architekten und Bildhauer an der Etablierung und Selbstdarstellung der Türkischen Republik nach 1933." In *Neue Staaten – neue Bilder? Visuelle Kultur im Dienst staatlicher Selbstdarstellung in Zentral- und Osteuropa seit 1918*, edited by Arnold Bartetzky and Thomas Fichtner, 61–74. Köln: Böhlau, 2005.

Domschke Jan P., and Peter Lewandrowski. *Wilhelm Ostwald.* Cologne: Pahl-Rugenstein, 1982.

Drori, Gili S., John Meyer, Francisco Ramirez, and Evan Schofer. *Science in the Modern World Polity: Institutionalization and Globalization.* Stanford, CA: Stanford University Press, 2003.

Drozdowski, Marian M. *Eugeniusz Kwiatkowski.* Rzeszów: Wysza Szkoła Informatyki i Zarządzania, 2005.

Ducheyne, Steffen. "Paul Otlet's Theory of Knowledge and Linguistic Objectivism." *Knowledge Organization* 32, no. 3 (2005): 110–16.

Dülmen, Richard van, and Sina Rauschenbach. *Macht des Wissens: Die Entstehung der modernen Wissensgesellschaft.* Cologne: Böhlau, 2004.

Duff, Andrew, ed. *Subsidiarity within the European Community.* London: Federal Trust, 1993.

Durth, Werner. *Deutsche Architekten: Biographische Verflechtungen. 1900–1970.* Stuttgart, Zurich: Krämer, 2001.

Dybczyńska-Bulyszko, Anna. "Pawilon Polski na Wystawie Światowej w Paryżu 1937 r." *Kwartalnik Architektury i Urbanistyki* 50 (2005):143–62.

Edgerton, David. *The Shock of the Old: Technology and Global History since 1900.* Oxford: Oxford University Press, 2006.

———. *Warfare State: Britain 1920–1970.* Cambridge: Cambridge University Press, 2006.

Edler, Jakob, Stefan Kuhlmann, and Maria Behrens, eds. *Changing Governance of Research and Technology Policy: The European Research Area.* Cheltenham: Edward Elgar, 2003.

Edwards, Steve. "The Accumulation of Knowledge or, William Whewell's Eye." In *The Great Exhibition of 1851: New Interdisciplinary Essays*, edited by Louise Purbrick, 26–52. Manchester: Manchester University Press, 2001.

Edwards, Trystan. "A World Centre of Communication." *Town Planning Review* 5 (1914): 14–30.

Eesteren, Cornelis van, and Vincent van Rossem, eds. *Het Idee van de functionele Stad: Een Lezing met Lichtbeelden 1928.* Rotterdam: NAi Publishers, 1997.

Efmertová, Marcela. "Les professeurs électrotechnicies Tchèques dans le monde: Formation et impact des travaux scientifiques dans les années 1918–1938." In

The Quest for a Professional Identity: Engineers between Training and Action, edited by Ana de Cardoso Matos, André Grelon, Irina Gouzévitch, and Maria Paula Diogo, 513–23. Lisbon: Colibri, 2009.

Einstein, Albert. "Why Socialism?" *Monthly Review* (May 1949), reprinted 61 (2009), Issue 1. Accessed April 10, 2013. http://monthlyreview.org/2009/05/01/why-socialism.

Enserink, Martin. "Keeping Europe's Basic Research Agency on Track." *Science* 331 (March 4, 2011): 1134–35.

Epstein, Stephan R., and Maarten Prak, eds. *Guilds, Innovation, and the European Economy*. Cambridge and New York: Cambridge University Press, 2008.

Erichsen, Regine. "Die Emigration deutschsprachiger Naturwissenschaftler von 1933 bis 1945 in die Türkei in ihrem sozial- und wissenschaftshistorischen Wirkungszusammenhang." In *Die Emigration der Wissenschaften nach 1933; Disziplingeschichtliche Studien*, edited by Herbert A. Strauss, Klaus Fischer, Christhard Hoffmann, and Alfons Söllner, 73–104. Munich: Saur, 1991.

Etzemüller, Thomas. "Social Engineering als Verhaltenslehre des kühlen Kopfes." In *Die Ordnung der Moderne: Social Engineering im 20. Jahrhundert*, edited by Thomas Etzemüller, 11–39. Bielefeld: transcript, 2009.

"Eugenics after 1945." Special issue of the *Journal of Modern European History* 10, no. 4 (2012).

"Euratom Unifies Reactor Research." *Chemical Engineering News* 44, no. 3 (1966): 66.

Euratom website. Accessed May 3, 2011. http://ec.eur opa.eu/research/energy/euratom/fusion/at-a-glance/history/index_En.htm.

European Commission. *Building Europe Knowledge: Towards the Seventh Framework Programme 2007–2013* (April 2005). Accessed February 15, 2013. http://www.eurosfaire.prd.fr/bibliotheque/pdf/FP7_Complete_presentation_April_2005.pdf.

———. *Europe 2020: A Strategy for Smart, Sustainable and Inclusive Growth*, COM (2010) 2020, Brussels, March 3, 2010. Accessed August 10, 2011. http://europa.eu/press_room/pdf/complet_En_barroso___007_-_Europe_2020_-_En_version.

———. *Green Paper on Innovation*, COM (95) 688 final (Brussels 1995). Accessed September 21, 2012. http://aei.pitt.edu/1218/1/innovation_gp_COM_95_688.pdf.

———. *Taking European Knowledge Society Seriously: Report of the Expert Group on Science and Governance to the Science, Economy and Society Directorate, Directorate-General for Research, European Commission*. Luxembourg: Office for Official Publications of the European Communities, 2007.

———. *Towards a Space Strategy for the European Union That Benefits Its Citizens*, Brussels, April 4, 2011, COM/2011/152.

———. White *Paper (COM/2002/672) "Space: A New Frontier for an Expanding Union."* Luxembourg: Office for Official Publications of the European Communities, 2003.

European Nuclear Society website. Accessed May 2, 2011. http://www.euronuclear.org/1-about/organisation.htm.

European Parliament. *Lisbon European Council 23 and 24 March 2000: Presidency Conclusions*. Accessed August 2, 2011. http://www.europarl.europa.eu/summits/lis1_En.htm.

European Research Council. "Synergy Grants." Accessed February 15, 2013. http://erc.europa.eu/funding-schemes/synergy-grants.

European Space Agency website. Accessed March 29, 2011. http://www.esa.int/SPECIALS/About_ESA/SEMP936LARE_0.html.

F.A., "Kaisertum und Zeitalter," *B.Z. am Mittag*, January 27, 1908.

Fabian, Jeanette, and Ulrich Winko, eds. *Architektur zwischen Kunst und Wissenschaft: Texte der tschechischen Architektur-Avantgarde 1918–1938*. Berlin: Gebr. Mann, 2010.

Faliasecca, Gabriele, and Barbara Valotti, eds. *Guglielmo Marconi: Genio, storia e modernità*. Milano: Editorale Giorgio Mondadori, 2003.

Fangerau, Heiner. "From Mephistopheles to Isaiah: Jacques Loeb, Technical Biology and War." *Social Studies of Science* 39 (2009): 229–56.

Fava, Valentina. "People's Cars and People's Technologies: Škoda and FIAT Experts Face the American Challenge." In *Expert Cultures in Central Eastern Europe: The Internationalization of Knowledge and the Transformation of Nation States since World War I*, edited by Martin Kohlrausch, Katrin Steffen, and Stefan Wiederkehr, 105–26. Osnabrück: fibre, 2010.

Feuerhahn, Wolf, and Pascale Rabault-Feuerhahn. *La fabrique internationale de la science: Les congrès scientifiques de 1865 à 1945*. Paris: CNRS Éditions 2010.

Fischer, Frank. *Technocracy and the Politics of Expertise*. New York: Sage Publications, 1990.

Fischer, Peter. "Das Projekt einer trilateralen Nuklearkooperation: Französisch-deutsch-italienische Geheimverhandlungen 1957/1958." *Historisches Jahrbuch* 112 (1992): 143–56.

Fjaestad, Maja. *Visionen om outtömlig energi: Bridreaktorn i svensk kärnkraftshistoria 1945–80*. Stockholm: Gidlunds förlag, 2010.

———. *Why Did the Breed Reactor Fail? Swedish and International Nuclear Reactor Development in a Cold War Context*. Stockholm: CESIS, 2009.

Fleury, Antoine. "The League of Nations: Towards a New Appreciation of Its History." In *The Treaty of Versailles: A Reassessment after 75 Years*, edited by Manfred F. Boemeke, 507–22. New York: Cambridge University Press, 1998.

Flier, Thomas. "'Possibly the Greatest Task an Architect Ever Faced': Ernst May in the Soviet Union (1930–1933)." In *Ernst May 1886–1970*, edited by Claudia Quiring, Wolfgang Voigt, Peter Cachola Schmal, and Eckhard Herrel, 157–95. Munich: Prestel, 2011.

Fox, Robert, and Anna Guagnini. "Introduction." In *Education, Technology and Industrial Performance in Europe: 1850–1939*, edited by Robert Fox and Anna Guagnini, 1–9. Cambridge: Cambridge University Press, 1993.

Frenken, Koen, Jarno Hoekman, and Frank van Oort. *Towards a European Research Area*. Rotterdam: NAi Publishers, 2007.

Friedel, Robert. *A Culture of Improvement: Technology and the Western Millennium*. Cambridge, MA and London: MIT Press, 2007.

Frohmann, Bernd. "The Role of Facts in Paul Otlet's Modernist Project of Documentation." In *European Modernism and the Information Society: Informing the Present, Understanding the Past*, edited by Warden B. Rayward, 75–88. Aldershot: Ashgate, 2008.

Frost, Robert L. "Fordism and the American Dream in France, 1919–1939." Accessed July 7, 2013. http://www.cddc.vt.edu/digitalfordism/fordism_materials/frost.htm.

Fuchs, Eckhardt. "The International Catalogue of Scientific Literature as a Mode of Intellectual Transfer: Promise and Pitfalls of International Scientific Co-operation before 1914." In *Transnational Intellectual Networks: Forms of Academic*

Knowledge and the Search for Cultural Identities, edited by Christophe Charle, Jürgen Schriewer, and Peter Wagner, 165–93. Frankfurt a.M. and New York: Campus, 2004.

———. "Wissenschaft, Kongressbewegung und Weltausstellungen: Zu den Anfängen der Wissenschaftsinternationale vor dem Ersten Weltkrieg." *Comparativ* 6 (1996): 156–77.

———. "Wissenschaftsinternationalismus in Kriegs- und Krisenzeiten: Zur Rolle der USA bei der Reorganisation der internationalen Scientific Community, 1914–1925." In *Wissenschaft und Nation in der europäischen Geschichte*, edited by Ralph Jessen and Jakob Vogel, 263–84. Frankfurt a.M and New York: Campus, 2002.

Füeg, Jean-François. "Ordo ab chaos: Classer est la plus haute opération de l'esprit." *Transnational Associations* 55, nos. 1–2 (2003): 29–35.

Gamba, Aldo, and Pierangelo Schiera, eds. *Fascismo e scienza: Le celebrazioni voltiane e il Congresso internazionale dei fisici del 1927*. Bologna: Il Mulino, 2005.

Garnier, Tony. *Une cité industrielle: Etude pour la construction des villes*. Princeton: Princeton Architectural Press, 1996 [first published 1918].

Gartman, David. *From Autos to Architecture: Fordism and Architectural Aesthetics in the Twentieth Century*. New York: Princeton Architectural Press, 2009.

———. "Why Modern Architecture Emerged in Europe, Not America: The New Class and the Aesthetics of Technocracy." *Theory, Culture & Society* 17, no. 5 (2000): 75–96.

Geier, Manfred. *Der Wiener Kreis: Mit Selbstzeugnissen und Bilddokumenten*. Reinbek bei Hamburg: Rowohlt, 1992.

Gemelli, Giuliana. "Western Alliance and Scientific Diplomacy in the Early 1960s." In *The American Century in Europe*, edited by Laurence Moore and Maurizio Vaudagna, 171–92. Ithaca: Cornell University Press, 2003.

Gentner, Wolfgang. "Grossforschung als Problem moderner europäischer Zusammenarbeit." In *Physik und Kosmologie: Stand und Zukunftsaspekte naturwissenschaftlicher Forschung in Deutschland*, edited by Ruprecht Kurzrock, 137–48. Berlin: Colloquium, 1971.

Georghiou, Luke. "Global Cooperation in Research." *Research Policy* 27 (1998): 611–26.

Geppert, Alexander C.T. "Space *Personae*: Cosmopolitan Networks of Peripheral Knowledge, 1927–1957." *Journal of Modern European History* 6 (2008): 262–86.

Gérard, Francis. "La crise de l'Europe nucléaire: Douze ans après sa creation va-t-on enterrer l'Euratom et soustraire la recherche et la technologie au domain de la coopération internationale?" *L'Europe en formation*, no. 106 (January 1969): 5–9.

Gerasimov, Ilya V. *Modernism and Public Reform in Late Imperial Russia: Rural Professions and Self-Organisation, 1905–1930*. Houndmills and New York: Palgrave Macmillan, 2009.

"Gerd Arntz." Accessed July 7, 2013. http://www.gerdarntz.org/content/gerd-arntz#isotype.

Gerovich, Slava. "Creating Memories: Myth, Identity, and Culture in the Russian Space Age." In *Remembering the Space Age*, edited by Steven J. Dick, 203–36. Washington, DC: NASA History Division, 2008.

———. *From Newspeak to Cyberspeak: A History of Soviet Cybernetics*. Cambridge, MA: MIT Press, 2002.

Gestwa, Klaus. *Die Stalinschen Grossbauten des Kommunismus: Sowjetische Technik- und Umweltgeschichte, 1948–1967*. Munich: Oldenbourg, 2010.

Geyer, Martin H. "One Language for the World: The Metric System, International Coinage, Gold Standard, and the Rise of Internationalism, 1850–1890." In *The Mechanics of Internationalism: Culture, Society, and Politics from the 1840s to the First World War*, edited by Martin H. Geyer and Johannes Paulmann, 55–92. Oxford: Oxford University Press, 2001.

———, and Johannes Paulmann. *The Mechanics of Internationalism: Culture, Society, and Politics from the 1840s to the First World War*. Oxford: Oxford University Press, 2001.

Geyer, Michael. "The Militarization of Europe, 1914–1945." In *The Militarization of the Western World*, edited by John G. Gillis, 65–102. New Brunswick: Rutgers, 1989.

Ghils, Paul. "Fonder le monde, fonder le savoir du monde, ou la double utopie de Paul Otlet." *Transnational Associations* 55, nos. 1–2 (2003): 36–48.

Ghirardo, Diane Y. "Città Fascista: Surveillance and Spectacle." *Journal of Contemporary History* 31 (1996): 347–72.

Giedion, Siegfried. *Mechanization Takes Command: A Contribution to Anonymous History*. Oxford: Oxford University Press, 1948.

Gimbel, John. *Science, Technology and Reparations: Exploitation and Plunder in Postwar Germany*. Stanford, CA: Stanford University Press, 1990.

Giorgi, Liane, and Jacquelyne Luce, eds. *Converging Science and Technologies: Research Trajectories and Institutional Settings*. Special issue of *Innovations: The European Journal of Social Science Research* 20, no. 4 (2007).

Gispen, Kees. *New Profession, Old Order: Engineers and German Society, 1815–1914*. Cambridge: Cambridge University Press, 1989.

———. "Engineers in Wilhelmian Germany: Professionalization, Deprofessionalization and the Development of Nonacademic Technical Education." In *German Professions, 1800–1950*, edited by Geoffrey Cocks and Konrad H. Jarausch, 104–22. New York: Oxford University Press, 1990.

Glänzel, Wolfgang. "Science in Scandinavia: A Bibliometric Approach." *Scientometrics* 48, no. 2 (2000): 121–50.

———, and Balázs Schlemmer, "National Research Profiles in a Changing Europe (1983–2003): An Exploratory Study of Sectoral Characteristics in the Triple Helix." *Scientometrics* 70, no. 2 (2007): 267–75.

———, A. Schubert, and H.-J. Czerwon. "A Bibliometric Analysis of International Scientific Cooperation of the European Union (1985–1995)." *Scientometrics* 45, no. 2 (1999): 185–202.

Glatt, Carl. "Reparations and the Transfer of Scientific and Industrial Technology from Germany: A Case Study of the Roots of British Industrial Policy and Aspects of British Occupation Policy in Germany between Post-World War II Reconstruction and the Korean War, 1943–1951." 2 Vol. PhD diss., European University Institute, 1994.

Gold, John R. "Creating the Charter of Athens: CIAM and the Functional City, 1933–43." *The Town Planning Review* 69 (1998): 225–47.

Goldberg, Ann. *Honor, Politics and the Law in Imperial Germany, 1871–1914*. Cambridge and New York: Cambridge University Press, 2010.

Gordin, Michael D. "Was There Ever a 'Stalinist Science'?" *Kritika: Explorations in Russian & Eurasian History* 9 (2008): 625–39.

Goschler, Constantin. *Rudolf Virchow: Mediziner – Anthropologe – Politiker*, 2nd ed. Cologne: Böhlau, 2009.

Gould, Stephen J. *Hen's Teeth and Horse's Toe: Further Reflections in Natural History.* New York and London: Norton, 1983.

Gouzevitch, Dmitri, and Irina Gouzevitch. "A Woman's Challenge: The Petersburg Polytechnic Institute for Women, 1905–1918." In *Crossing Boundaries, Building Bridges: Comparing the History of Women Engineers 1870s–1990s*, edited by Annie Canel, Ruth Oldenziel, and Karin Zachmann, 103–26. Amsterdam: Harwood, 2000.

Grazia, Victoria de. *Irresistible Empire: America's Advance through Twentieth-Century Europe.* Cambridge, MA: The Belknap Press of Harvard University Press, 2006.

Graham, Loren R. "Big Science in the Last Years of the Big Soviet Union." *Osiris* 7 (1992): 49–71.

———. *Science in Russia and the Soviet Union: A Short History.* Cambridge and London: Cambridge University Press, 1993.

Grande, Edgar. *The Erosion of State Capacity and the European Innovation Policy Dilemma.* Vienna: Institute for Advances Studies, 2000.

———, and Anke Peschke. "Organizing Science in Europe." *European Review* 7 (1999): 295–312.

Grelon, André. "Die deutschen Ingenieure aus französischer Sicht, 1770–1990." In *Ingenieure in Deutschland: 1770–1990*, edited by Peter Lundgreen, 369–89. Frankfurt a.M. and New York: Campus, 1994.

———. "French Engineers: Between Unity and Heterogeneity." *History of Technology* 27 (2006): 107–24.

Gresleri, Giuliano. "Convergences et divergences: de Le Corbusier à Otto Neurath." *Transnational Associations* 55, nos. 1–2 (2003): 72–81.

Gropius, Walter. *Internationale Architektur.* Mainz: Kupferberg, 1981.

Groys, Boris, and Michael Hagemeister, eds. *Die neue Menschheit: Biopolitische Utopien in Russland zu Beginn des 20. Jahrhunderts.* Frankfurt a.M.: Suhrkamp, 2005.

Gugerli, David, Patrick Kupper, and Daniel Speich. *Die Zukunftsmaschine: Konjunkturen der ETH Zürich 1855–2005.* Zurich: Chronos-Verlag, 2005.

Guillén, Mauro F. *The Taylorized Beauty of the Mechanical: Scientific Management and the Rise of Modernist Architecture.* Princeton: Princeton University Press, 2006.

Guse, John C. "Volksgemeinschaft Engineers: The Nazi 'Voyages of Technology'." *Central European History* 44 (2011): 447–77.

Guzetti, Luca. *A Brief History of European Union Research Policy.* Brussels: European Commission, 1995.

Györgi, Péteri, "Engineer Utopia: On the Position of Technostructure in Hungary's War Communism, 1919." *International Studies of Management & Organization* 19 (1989): 82–102.

Haas, Ernst B. *The Uniting of Europe.* Stanford, CA: Stanford University Press, 1958.

Hall, Rex D., David J. Shayler, and Bert Vis. *Russia's Cosmonauts: Inside the Yuri Gagarin Training Center.* Chichester: Springer, 2005.

Hanevik, A., and K. Bendiksen, *The European Atomic Energy Society 1954–2004.* Kjeller: IFA, 2005.

Hanisch, Tore J., and Even Lange. *Vitenskap for industrien. NTH – En høyskole i utvikling gjennom 75 År.* Oslo: Universitetsforlaget, 1985.

Hapke, Thomas. "Roots of Mediating Information: Aspects of the German Information Movement." In *European Modernism and the Information Society: Informing the Present, Understanding the Past*, edited by Warden B. Rayward, 307–27. Aldershot and Burlington, VT: Ashgate, 2008.

Hartmann, Frank. *Bildersprache: Otto Neurath Visualisierungen.* Vienna: Wiener Universitätsverlag, 2006.

Harvey, Brian. *Russia in Space: The Failed Frontier?* Chichester: Springer, 2001.

Hecht, Gabrielle. *The Radiance of France: Nuclear Power and National Identity after World War II.* Cambridge, MA and London: MIT Press, 1998.

———. "Technology, Politics, and National Identity in France." In *Technologies of Power: Essays in Honor of Thomas Parke Hughes and Agatha Chipley Hughes*, edited by Michael T. Allen and Gabrielle Hecht, 253–94. Cambridge, MA and London: MIT Press, 2001.

Heesen, Anke te. *Der Zeitungsausschnitt: Ein Papierobjekt der Moderne.* Frankfurt a.M.: Fischer, 2006.

Heinemann, Isabel. "Wissenschaft und Homogenisierungsplanungen für Osteuropa: Konrad Meyer, der 'Generalplan Ost' und die Deutsche Forschungsgemeinschaft." In *Wissenschaft – Planung – Vertreibung: Neuordnungskonzepte und Umsiedlungspolitik im 20. Jahrhundert*, edited by Isabel Heinemann and Patrick Wagner, 45–72. Stuttgart: Steiner, 2006.

Hein-Weingarten, Katharina. *Das Institut für Kosmosforschung der Akademie der Wissenschaften der DDR: Ein Beitrag zur Erforschung der Wissenschaftspolitik der DDR am Beispiel der Weltraumforschung von 1957 bis 1991.* Berlin: Duncker & Humblot, 2000.

Heisenberg, Werner. "European Cooperation for the Advancement of Nuclear Research." In *Werner Heisenberg: Collected Works*, Vol. c.v, edited by Walter Blum, Hans-Peter Dürr, and Helmut Rechenberg, 292–93. Munich and Zurich: Piper, 1989.

Herf, Jeffrey. *Reactionary Modernism: Technology, Culture, and Politics in Weimar and the Third Reich.* Cambridge and New York: Cambridge University Press, 1986.

Hermann, Armin, John Krige, Ulrike Mersits, and Dominique Pestre, eds. *History of CERN*, Vols. 1–3. Amsterdam: North Holland, 1987–96.

Herren, Madeleine. *Hintertüren zur Macht: Internationalismus und modernisierungsorientierte Aussenpolitik in Belgien, der Schweiz und den USA 1865–1914.* Munich: Oldenbourg, 2000.

Hessler, Martina. "Science in the City: European Traditions and American Models." In *Urban Machinery: Inside Modern European Cities*, edited by Mikael Hård and Thomas J. Misa, 211–29. Cambridge, MA: MIT Press, 2008.

———, and Clemens Zimmermann, eds. *Creative Urban Milieus: Historical Perspectives on Culture, Economy, and the City.* Frankfurt a.M. and New York: Campus, 2008.

Heuvel, Charles van den. "Building Society, Constructing Knowledge, Weaving the Web: Otlet's Visualizations of a Global Information Society and His Concept of a Universal Civilization." In *European Modernism and the Information Society: Informing the Present, Understanding the Past*, edited by Warden B. Rayward, 127–53. Aldershot and Burlington, VT: Ashgate, 2008.

———, Warden B. Rayward, and Pieter Uyttenhove. "L'architecture du savoir: Une recherche sur le Mundaneum et les précurseurs européens de l'internet." *Transnational Associations* 55, nos. 1–2 (2003): 16–28.

Hewlett, Richard G., and Jack M. Holl. *Atoms for Peace and War 1953–1961: Eisenhower and the Atomic Energy Commission.* Berkeley, CA: University of California Press, 1989.

Heymann, Matthias."Kunst" und Wissenschaft in der Technik des 20. Jahrhunderts: Zur Geschichte der Konstruktionswissenschaft.* Zurich: Chronos-Verlag, 2005.

Heywood, Anthony. *Engineer of Revolutionary Russia: Iurii V. Lomonosov (1876–1952) and the Railways*. Farnham and Burlington, VT: Ashgate, 2011.

Hilaire-Perez, Liliane, and Catherine Verna. "Dissemination of Technical Knowledge in the Middle Ages and the Early Modern Era." *Technology and Culture* 47 (2006): 536–65.

Hilpert, Thilo, ed. *Le Corbusiers 'Charta von Athen': Texte und Dokumente*. Braunschweig: Vieweg, 1988.

Hirose, Shin. "Two Classes of British Engineers: An Analysis of Their Education and Training, 1880s–1930s." *Technology and Culture* 51 (2010): 388–402.

Hirschi, Caspar. *The Origins of Nationalism: An Alternative History from Ancient Rome to Early Modern Germany*. Cambridge: Cambridge University Press, 2012.

Hitchcock, Henry-Russell, and Philip Johnson. *The International Style: Architecture Since 1922*, 3rd edition. New York: Norton, 1995.

Hobsbawm, Eric. *The Age of Extremes: The Short Twentieth Century, 1914–1991*. London: Michael Joseph, 1994.

Hoch, Paul, and Jennifer Platt. "Migration and the Denationalization of Science." In *Denationalizing Science: The Contexts of International Scientific Practice*, edited by Elisabeth Crawford, Terry Shinn, and Sverker Sörlin, 133–52. Dordrecht: Kluwer, 1993.

Hoekman, Jarno, Koen Frenken, and Robert J.W. Tijssen. "Research Collaboration at a Distance: Changing Spatial Patterns of Scientific Collaboration within Europe." *Research Policy* 39 (2010): 662–73.

Hoffmann, Horst. *Sigmund Jähn: Der fliegende Vogtländer*. Berlin: Das Neue Berlin, 1999.

Holden, Len. "Fording the Atlantic: Ford and Fordism in Europe." *Business History* 47 (2005): 122–27.

Hong, Sungook. *Wireless: From Marconi's Black-Box to the Audion*. Cambridge: Cambridge University Press, 2010.

Horňáková, Ladislava. "Baťa Satellite Towns around the World." In *A Utopia of Modernity: Zlín: Revisiting Baťa's Functional City*, edited by Katrin Klingan, 117–37. Berlin: Jovis, 2009.

———. "Der Aufbau Zlíns in der Zeit zwischen den Kriegen – Städtebau, Firmenbauwesen und Architektur." In *Zlín: Modellstadt der Moderne*, edited by Winfried Nerdinger, 40–71. Berlin: Jovis, 2009.

Hortleder, Gerd. *Das Gesellschaftsbild des Ingenieurs: Zum politischen Verhalten der Technischen Intelligenz in Deutschland*. Frankfurt a.M.: Suhrkamp, 1974.

Hughes, Thomas P., and Agatha C. Hughes, eds. *Lewis Mumford: Public Intellectual*. New York: Oxford University Press, 1990.

"Intellectual Cooperation and International Bureaus Section, 1919–1946," Accessed July 7, 2013. http://biblio-archive.unog.ch/Detail.aspx?ID=408.

Iriye, Akira. *Global Community: The Role of International Organizations in the Making of the Contemporary World*. Berkeley, CA: University of California Press, 2002.

Janatková, Alena. *Modernisierung und Metropole: Architektur und Repräsentation auf den Landesausstellungen in Prag 1891 und Brünn 1928*. Stuttgart: Steiner, 2008.

Jarausch, Konrad H. *The Unfree Professions: German Lawyers, Teachers, and Engineers, 1900–1950*. New York: Oxford University Press, 1990.

Jasper, James M. *Nuclear Politics: Energy and the State in the United States, Sweden, and France*. Princeton: Princeton University Press, 1990.

Jessen, Ralph, and Jakob Vogel, eds. *Wissenschaft und Nation in der europäischen Geschichte*. Frankfurt a.M. and New York: Campus, 2002.

Johnson, Stephen B. "A Failure to Communicate: The Demise of ELDO." *History of Technology* 22 (2000): 1–23.

———. *The Secret of Apollo: Systems Management in American and European Space Programs*. Baltimore, MD: Johns Hopkins University Press, 2002.

Jolly, W.P. *Marconi*. London: Constable, 1972.

Joint Soviet–United States Statement on the Summit Meeting in Geneva, November 21, 1985. Accessed May 15, 2011. http://www.reagan.utexas.edu/archives/speeches/1985/112185a.htm.

Jones, Evan. *The Employment of German Scientists in Australia after World War II* (The University of Sydney School of Economics and Political Science, Working Paper ECOP2002–1, October 2002). Accessed June 9, 2011. http://pandora.nla.gov.au/pan/20949/20021018–0000/oldweb.econ.usyd.edu.au/pe/publication/ecop2002–1.pdf.

Jong-Lambert, William de. "Eugenics and Lysenkoism in Interwar and Postwar Polish Biology." *East Central Europe* 38 (2011): 115–32.

Joravsky, David. *The Lysenko Affair*. Chicago, IL: University of Chicago Press, 1986.

Josephson, Paul R. *Red Atom: Russia's Nuclear Power Program from Stalin to Today*. New York: Freeman, 2000.

———. *Totalitarian Sience and Technology*. Amherst: Prometheus Books, 2005.

———, Yury Ranyuk, Ivan Tsekhmistro, and Karl Hall. "Science and the Periphery under Stalin: Physics in Ukraine." In *Physics and Politics: Research and Research Support in Twentieth Century Germany in International Perspective*, edited by Helmuth Trischler and Mark Walter, 197–225. Stuttgart: Steiner, 2010.

Kahlow, Andreas. "French Influence on the Development of Applied Mechanics in Germany in the Nineteenth Century." *History and Technology* 12 (1995): 179–89.

Kaiser, Wolfram. "Cultural Transfer of Free Trade at the World Exhibitions 1851–62." *Journal of Modern History* 77 (2005): 563–90.

———. *Using Europe, Abusing the Europeans: Britain and European Integration, 1945–63*. Houndsmills: Macmillan, 1996.

Kaiser, Wolfram, and Johan W. Schot. *Writing the Rules for Europe: Experts, Cartels, and International Organizations*. London: Palgrave Macmillan, forthcoming.

Kamiński, Zygmunt. "Inauguracja Wydziału Architektury." In *Fragmenty Stuletniej Historii. 1899–1999: Relacje, Wspomnienia, Refleksje w Stulecie Organizacji Warszawskich Architektów*, edited by Tadeusz Barucki, 27–48. Warsaw: Drukoba, 2000.

Kargon Robert H., and Arthur P. Molella. *Invented Edens: Techno-Cities of the Twentieth Century*. Cambridge, MA and London: MIT Press, 2008.

Karvar, Anousheh. "Model Reception in the Domain of Engineering Education: Mediation and Negotiation." *History and Technology* 12 (1995): 81–95.

Keck, Otto. "The West German Fast Breeder Programme: A Case Study in Governmental Decision Making." *Energy Policy* 8 (1980): 277–92.

Kegler, Karl R. and Alexa Stiller. "Konrad Meyer." In *Handbuch der völkischen Wissenschaften*, edited by Ingo Haar and Michael Fahlbusch, 415–22. Munich: Saur, 2008.

Keleş, Ruşen. "Der Beitrag Ernst Reuters zur Urbanisierung der Türkei." In *Ernst Reuter: Kommunalpolitiker und Gesellschaftsreformer. 1921–1953*, edited by Heinz Reif and Moritz Feichtinger, 185-202. Bonn: Dietz, 2009.

Kentgens-Craig, Margret. *The Bauhaus and America: First Contacts 1919–1936.* Cambridge, MA: MIT Press, 1999.

Kinross, Patrick. *Atatürk: The Rebirth of a Nation.* London: Phoenix Press, 2003.

Kintzinger, Martin. "Universität." In *Europäische Erinnerungsorte, Vol. 2. Das Haus Europa*, edited by Pim de Boer, Heinz Duchhardt, Georg Kreis, and Wolfgang Schmale, 307–12. Munich: Beck, 2012.

Kline, Ronald R. "From Progressivism to Engineering Studies: Edwin T. Layton's *The Revolt of the Engineers.*" *Technology and Culture* 49 (2008): 1018–24.

Knoedler, Janet, and Anne Mayhew. "Thorstein Veblen and the Engineers: A Reinterpretation." *History of Political Economy* 31 (1999): 255–72.

Knorr Cetina, Karin. "Culture in Global Knowledge Societies: Knowledge Cultures and Epistemic Cultures." *Interdisciplinary Science Review* 32 (2007): 361–75. Accessed July 7, 2013, http://dx.doi.org/10.1179/030801807X163571.

———. "How Superorganisms Change: Consensus Formation and the Social Ontology in High-Energy Physics Experiments." *Social Studies of Science* 25 (1995): 119–47.

Köhler, Piotr. "Lysenko Affair and Polish Botany." *Journal of the History of Biology* 44 (2011): 305–43.

Kohlrausch, Martin. "Szymon Syrkus: Die Stadt imaginieren im Angesicht der Katastrophe. Warschau 1930–1950." *Historische Anthropologie* 18 (2010): 404–22.

———, Katrin Steffen, and Stefan Wiederkehr, eds. *Expert Cultures in Central Eastern Europe: The Internationalization of Knowledge and the Transformation of Nation States since World War I.* Osnabrück: fibre, 2010.

Kojevnikov, Alexei. *Stalin's Great Science: The Times and Adventures of Soviet Physicists.* London: Imperial College Press, 2004.

———. "The Great War, the Russian Civil War, and the Invention of Big Science." *Science in Context* 15 (2002): 239–75.

———. "The Phenomenon of Soviet Science." *Osiris* 23 (2008): 115–35.

König, Wolfgang. *Künstler und Strichezieher: Konstruktions- und Technikkulturen im deutschen, britischen, amerikanischen und französischen Maschinenbau zwischen 1850 und 1930.* Frankfurt a.M.: Suhrkamp, 1999.

———. "Technical Education and Industrial Performance in Germany: A Triumph of Heterogeneity." In *Education, Technology and Industrial Performance in Europe: 1850–1939*, edited by Robert Fox and Anna Guagnini, 65–87. Cambridge, MA: Cambridge University Press, 1993.

———. *Wilhelm II. und die Moderne: Der Kaiser und die technisch-industrielle Welt.* Paderborn: Schöningh, 2009.

Kopp, Anatole. "Foreign Architects in the Soviet Union during the First Two Five-Year Plans." In *Reshaping Russian Architecture: Western Technology, Utopian Dream*, edited by William C. Brumfield, 176–214. Cambridge: Cambridge University Press, 1990.

Krajewski, Markus. "Paper Parasite: F.M. Feldhaus and the Historiography of Technology." In *European Modernism and the Information Society: Informing the Present, Understanding the Past*, edited by Warden B. Rayward, 296–306. Aldershot and Burlington, VT: Ashgate, 2008.

———. *Restlosigkeit: Weltprojekte um 1900.* Frankfurt a.M.: Fischer, 2006.

Kramer, Alan. *Dynamic of Destruction: Culture and Mass Killing in the First World War.* Oxford: Oxford University Press, 2009.

Kramer, Heinz. *Nuklearpolitik in Westeuropa und die Forschungspolitik der Euratom.* Cologne: Carl Heymanns, 1976.

Kranakis, Eda. *Constructing a Bridge: An Exploration of Engineering Culture, Design, and Research in Nineteenth-Century France and America.* Cambridge, MA: MIT Press, 1997.

Krementsov, Nikolai. "A 'Second Front' in Soviet Genetics: The International Dimension of the Lysenko Controversy, 1944–1947." *Journal of the History of Biology* 29 (1996): 229–50.

Krige, John. *American Hegemony and the Postwar Reconstruction of Science in Europe.* Cambridge, MA and London: MIT Press, 2006.

———. "Atoms for Peace, Scientific Internationalism, and Scientific Intelligence." *Osiris* 21 (2006): 161–81.

———. "Building the Arsenal of Knowledge." *Centaurus* 52 (2010): 280–96.

———. "Preface." In *Choosing Big Technologies*, edited by John Krige, v–vi. Chur: Harwood, 1993.

———. "The Peaceful Atom as Political Weapon: Euratom and American Foreign Policy in the Late 1950s." *Historical Studies in the Natural Sciences* 38, no. 1 (2008): 5–44.

———. "What is 'Military' Technology? Two Cases of US–European Scientific and Technological Collaboration in the 1950s." In *The United States and the Integration of Europe: Legacies of the Postwar Era*, edited by Francis Heller and John Gillingham, 307–38. New York: St. Martin's Press, 1996.

———, and Dominique Pestre. "Some Thoughts on the History of CERN in the 50s and 60s." In *Big Science: The Growth of Large Scale Research*, edited by Peter Galison and Bruce Hevly, 78–99. Stanford, CA: Stanford University Press, 1992.

———, Arturo Russo, and Lorenza Sebesta. *A History of the European Space Agency 1958–1987*, Vols. 1 and 2. Noordwijk: ESA, 2000.

Krohn, Claus-Dieter. "Deutsche Wissenschaftsemigration seit 1933 und ihre Remigrationsbarrieren nach 1945." In *Wissenschaften und Wissenschaftspolitik: Bestandsaufnahmen zu Formationen, Brüchen und Kontinuitäten im Deutschland des 20. Jahrhunderts,* edited by Rüdiger vom Bruch and Brigitte Kaderas, 437–45. Stuttgart: Steiner, 2002.

Kudláček, Martina. "Zlín Views, 2009." In *A Utopia of Modernity: Zlín: Revisiting Baťa's Functional City*, edited by Katrin Klingan, 23–39. Berlin: Jovis, 2009.

Kühl, Stefan. *Die Internationale der Rassisten: Aufstieg und Niedergang der internationalen Bewegung für Eugenik und Rassenhygiene im 20. Jahrhundert.* Frankfurt a.M. and New York: Campus, 1997.

Laak, Dirk van. *Imperiale Infrastruktur: Deutsche Planungen für eine Erschliessung Afrikas 1880 bis 1960.* Paderborn: Schöningh, 2004.

———. *Weisse Elefanten: Anspruch und Scheitern technischer Grossprojekte im 20. Jahrhundert.* Stuttgart: Deutsche Verlags-Anstalt, 1999.

Laes, Erik, Lakshmi Chayapathi, Gilbert Eggermont, and Gaston Meskens, *Kernenergie (on)besproken. Een geschiedenis van het maatschappelijke debat over kernenergie in België*. Leuven: Acco, 2007.

Langbehn, Volker, and Mohammad Salama, eds. *German Colonialism: Race, the Holocaust, and Postwar Germany.* New York: Columbia University Press, 2011.

Latour, Bruno. *The Pasteurization of France.* Cambridge, MA: Harvard University Press, 1988.

Launius, Roger D., and Andrew K. Johnston. *Smithsonian Atlas of Space Exploration*. New York: HarperCollins, 2009.

———, John M. Logsdon, and Robert W. Smith, eds. *Reconsidering Sputnik: Forty Years since the Soviet Satellite*. Amsterdam: Harwood, 2000.

Laureys, Dawinka. *La contribution de la Belgique à l'aventure spatiale européenne: Des origines à 1973*. Paris: Beauchesne, 2008.

Layton, Edwin T. *The Revolt of the Engineers: Social Responsibility and the American Engineering Profession*. Cleveland, OH: Press of Case Western Reserve University, 1971.

Le Corbusier. *La charte d'Athènes*. Paris: Éditions de Minuit, 1957.

———. "In Defense of Architecture." *Oppositions* 4 (1974): 93–106.

Leclerc, Michel, and Jean Gagné. "International Scientific Cooperation: The Continentalization of Science." *Scientometrics* 31, no. 3 (1994): 261–92.

Léger, Fernand, and Jane Heap, ed. *Machine-Age Exposition: Catalogue*. New York: n.p., 1927.

Lepenies, Wolf. *Auguste Comte: Die Macht der Zeichen*. Munich: Hanser, 2010.

Lerchenmüller, Joachim. "Das Ende der Reichsuniversität Strassburg in Tübingen." *Bausteine zur Tübinger Universitätsgeschichte* 10 (2005): 113–72.

Levin, Miriam R. *Urban Modernity: Cultural Innovation in the Second Industrial Revolution*. Cambridge, MA: MIT Press, 2010.

Leydesdorff, Loet. "Is the European Union Becoming a Single Publication System?" *Scientometrics* 47, no. 2 (2000): 265–80.

Lichtenberg, Frank. *The European Strategic Program for Research in Information Technology (ESPRIT): An Ex-Post Analysis*. New York: Working Papers from Columbia – Graduate School of Business, 1996.

Lindenberger, Thomas. *Strassenpolitik: Zur Sozialgeschichte der öffentlichen Ordnung in Berlin 1900–1914*. Bonn: Dietz, 1999.

Lindqvist, Svante. "An Olympic Stadium of Technology: Deutsches Museum and Sweden's Tekniska Museet." *History and Technology* 10 (1993): 37–54.

Lintsen, Harry. *Ingenieur van Beroep: Historie, Praktijk, Macht en Opvattingen van Ingenieurs in Nederland*. Den Haag: Ingenieurspers, 1985.

Liulevicius, Vejas G. *The German Myth of the East: 1800 to the Present*. Oxford: Oxford University Press, 2009.

Long, David. "Who Killed the International Studies Conference?" *Review of International Studies* 32 (2006): 604-607.

Loose, Ingo. "How to Run a State: The Question of Knowhow in Public Administration in the First Years after Poland's Rebirth in 1918." In *Expert Cultures in Central Eastern Europe: The Internationalization of Knowledge and the Transformation of Nation States Since World War I*, edited by Martin Kohlrausch, Katrin Steffen, and Stefan Wiederkehr, 145–59. Osnabrück: Fibre, 2010.

Ludmann-Obier, Marie-France. "Un aspect de la chasse aux cerveaux: les transferts de techniciens Allemands en France: 1945–1949." *Relations Internationales* 46 (1986): 195–208.

"Ludwig Mies van der Rohe†." *Der Spiegel*, August 25, 1969.

Luukkonen, Terttu. "The Difficulties in Assessing the Impact of EU Framework Programmes." *Research Policy* 27 (1998): 599–610.

Lyle, Louise. "Science bourgeoise et science prolétarienne: French Literary Responses to the Lysenko Affair." In *The Lost Decade? The 1950s in European History, Politics, Society and Culture*, edited by Heiko Feldner, Claire Gorrara, and Kevin Passmore, 213–29. Newcastle: Cambridge Scholars Press, 2011.

Lyotard, Jean-François. *The Postmodern Condition: A Report on Knowledge.* Minneapolis, MN: University of Minnesota Press, 1984.

Maasberg, Ute, and Regina Prinz, *Die Neuen kommen!: Weibliche Avantgarde in der Architektur der zwanziger Jahre.* Hamburg: Junius, 2005.

MacLeod, Christine. *Heroes of Invention: Technology, Liberalism and British Identity, 1750–1914.* Cambridge and New York: Cambridge University Press, 2007.

MacLeod, Roy M. *Government and Expertise: Specialists, Administrators and Professionals, 1860 –1919.* Cambridge and New York: Cambridge University Press, 2003.

Madders, Kevin. *A New Force at a New Frontier: Europe's Development in the Space Field in the Light of Its Main Actors, Policies, Law and Activities from Its Beginnings to the Present.* Cambridge and New York: Cambridge University Press, 1997.

——— and W. Thiebaut. "Two Europes in One Space: The Evolution of Relations between the European Space Agency and the European Community in Space Affairs." *Journal of Space Law* 20 (1992): 117–32.

Mai, Gunther. "Politische Krise und Rationalisierungsdiskurs in den zwanziger Jahren." *Technikgeschichte* 62 (1995): 317–32.

Maier, Charles S. "Between Taylorism and Technocracy: European Ideologies and the Vision of Industrial Productivity in the 1920's." *Journal of Contemporary History* 5 (1970): 27–61.

———. "Consigning the Twentieth Century to History: Alternative Narratives for the Modern Era." *American Historical Review* 105 (2000): 807–31.

Maiocchi, Roberto. "Fascist Autarky and the Italian Scientists," *HoST* 3 (2009). Accessed May 12, 2011. http://johost.eu/vol3_fall_2009/vol3_rm.htm.

———. *Gli scienziati del duce: il ruolo dei ricercatori e del CNR nella politica autarchica del fascism.* Rome: Carocci, 2003.

Mallard, Grégoire. "L'Europe puissance nucléaire, cet obscur objet du désir." *Critique internationale* 42 (2009): 141–63.

———. "The Atomic Confederacy: Europe's Quest for Nuclear Weapons and the Making of the New World Order." PhD diss., Princeton University, 2008.

Marconi, Maria C., and Elettra Marconi. *Marconi My Beloved.* Boston, MA: Dante University of America Press, 1999.

Margolin, Howard. *Unauthorized Entry: The Truth about Nazi War Criminials in Canada, 1946–1956.* Toronto: University of Toronto Press, 2000.

Maria, Michelangelo de, and John Krige. "Early European Attempts in Launcher Technology: Original Sins in ELDO's Sad Parable." *History and Technology* 9 (1992): 109–37.

Marklund, Carl, and Peter Stadius. "Merging Modernity with Nationalism in the Stockholm Exhibition in 1930." *Culture Unbound* 2 (2010): 609–34.

Marth, Willy. *Die Geschichte von Bau und Betrieb des deutschen Schnellbrüter-Kernkraftwerk KNK II.* Karlsruhe: KfK, 1993.

———. *The Story of the European Fast Reactor Cooperation.* Karlsruhe: KfK, 1993.

Martin, Thierry, and Michèle Virol. *Vauban, architecte de la modernité?* Besançon: Presses Université de Franche-Comté, 2008.

Marton, Mati. *The Great Escape: Nine Jews Who Fled Hitler and Changed the World.* New York: Simon & Schuster, 2006.

Masini, Giancarlo. *Marconi.* New York: Marsilio, 1995.

Max Planck Gesellschaft. "Die Gründung der KWG." Accessed July 7, 2013. http://www.mpg.de/178569/Kaiser-Wilhelm-Gesellschaft.

Mayer, Arno J. *The Persistence of the Old Regime: Europe to the Great War.* London: Verso, 2010.

Mazower, Mark. *Dark Continent: Europe's Twentieth Century.* New York: Knopf, 1999.

McCray, W. Patrick. "Globalisation with Hardware: ITER's Fusion of Technology, Policy, and Politics." *History and Technology* 26 (2010): 283–312.

McDougall, Walter. "Space-Age Europe: Gaullism, Euro-Gaullism, and the American Dilemma." *Technology and Culture* 26 (1985): 179–203.

———. *...The Heavens and the Earth: A Political History of the Space Age.* New York: Knopf, 1985.

McNeely, Ian F., and Lisa Wolverton. *Reinventing Knowledge: From Alexandria to the Internet.* New York: Norton, 2008.

Meer, Elisabeth van. "'The Nation is Technological': Technical Expertise and National Competition in the Bohemian Lands, 1880–1914." In *Expert Cultures in Central Eastern Europe: The Internationalization of Knowledge and the Transformation of Nation States since World War I*, edited by Martin Kohlrausch, Katrin Steffen, and Stefan Wiederkehr, 85–104. Osnabrück: fibre, 2010.

Memorandum of Understanding Regarding the Joint Nuclear Power Program Proposed Between Euratom and the USA, May 29, 1958. Accessed April 29, 2011. http://www.ena.lu/memorandum_from_action_committee_united_states_Europe_programme_three_wise_february_1957-2-12188.pdf.

Merali, Zeeya. "The Large Human Collider." *Nature* 464 (March 25, 2010): 482–4.

Meriggi, Maurizio. *Una città possibile: Architetture di Ivan Leonidov 1926–1934.* Florence: Electa, 2007.

Metzler, Gabriele. *Internationale Wissenschaft und nationale Kultur: Deutsche Physiker in der internationalen Community 1900–1960.* Göttingen: Vandenhoeck & Ruprecht, 2000.

Meyer, Konrad. "Agrarprobleme des neuen Europa." *Forschungsdienst* 14 (1942): 113–31.

———. *Die Landeserschliessung in den EWG Ländern.* Bremen-Horn: Dorn, 1959.

———. "Ländliche Fördergebiete und ihre Sanierung." In *Festschrift der Akademie für Raumforschung und Landesplanung in Hannover: Raumforschung. 25 Jahre Raumforschung in Deutschland*, 367–81. Bremen: Jänecke, 1960.

Mick, Christof. *Forschen für Stalin: Deutsche Fachleute in der sowjetischen Rüstungsindustrie 1945–1958.* Munich and Vienna: Oldenbourg, 2000.

Middell, Matthias, ed. *Imagining Europeans: The Construction of the Homo Europaeus.* Leipzig: Universitätsverlag, 2013.

———. "Kompatibilität oder Diversität europäischer Wissenschaftssysteme – ein Blick auf die Transferprozesse im 19. Jahrhundert." In *Immer im Forschen bleiben: Rüdiger vom Bruch zum 60. Geburtstag*, edited by Marc Schalenberg and Peter Walther, 199–212. Stuttgart: Steiner, 2004.

Miller, Donald L. *Lewis Mumford: A Life.* New York: Weidenfeld & Nicolson, 1989.

Minorski, Jan. *Polska nowatorska myśl architektoniczna w latach 1918–1939.* Warsaw: PWN, 1970.

Misa, Thomas J. "Appropriating the International Style: Modernism in East and West." In *Urban Machinery: Inside Modern European Cities*, edited by Mikael Hård and Thomas J. Misa, 71–98. Cambridge, MA: MIT Press, 2008.

———, and Johan Schot. "Inventing Europe: Technology and the Hidden Integration of Europe." *History and Technology* 21 (2005): 1–21.

Mitrany, David. "The Functional Approach in Historical Perspective." *International Affairs* 47 (1971): 532–43.

———. *The Functional Theory of Politics*. New York: St. Martin's Press, 1975.

Mokyr, Joel. *The Gifts of Athena: Historical Origins of the Knowledge Economy*. Princeton: Princeton University Press, 2002.

Molella, Arthur P. "Science Moderne. Sigfried Giedion's Space, Time and Architecture and Mechanization Takes Command." *Technology and Culture* 43 (2002): 374–89.

Mollin, Gerhard T. *Die USA und der Kolonialismus: Amerika als Partner und Nachfolger der belgischen Macht in Afrika 1939–1965*. Berlin: Akademie-Verlag, 1996.

Moravčíková, Henrietta. "Social and Architectural Phenomenon of the Bataism in Slovakia: The Example of the Community Šimonovany – Baťovany – Partizánske." *Sociológia: Slovak Sociological Review* 36, no. 6 (2004): 519–43.

Moreau, Jean-Louis. "L'Industrie nucléaire en Belgique de 1945 à la mise en veilleuse d'Euratom: L'action du groupe Union minière." In *L'Énergie nucléaire en Europe: Des origines à Euratom*, edited by Michel Dumoulin, Pierre Guillen, and Maurice Vaisse, 65–97. Bern: Peter Lang, 1994.

Morton, Peter. *Fire Across the Desert: Woomera and the Anglo-Australian Joint Project 1946–1980*. Canberra: Australian Government Publication Service, 1989.

Moser, Petra, Alessandra Voena, and Fabian Waldinger. *German Jewish Émigrés and U.S. Invention* (Working Paper, Department of Economics, University of Warwick, March 2012). Accessed August 8, 2013. http://www2.warwick.ac.uk/fac/soc/economics/staff/academic/waldinger/research/tacit120320d.pdf.

Mosterín, Jesús. "Social Factors in the Development of Genetics and the Lysenko Affair." In *Epistemology and the Social*, edited by E. Agazzi, J. Echeverría, and A. Gómez, 143–55. Amsterdam and New York: Rodopi, 2008.

Muldur, Ugur, Fabienne Corvers, Henri Delanghe, Jim Dratwa, Daniela Heimberger, Brian Sloan, and Sandrijn Vanslembrouck. *New Deal for an Effective European Research Policy: The Design and Impacts of the 7th Framework Programme*. Dordrecht: Springer, 2006.

Mumford, Eric. *The CIAM Discourse on Urbanism, 1928–1960*. Cambridge, MA: MIT Press, 2000.

Musso, Pierre. *Saint-Simon: L'industrialisme contre l'état*. La Tour d'Aigues: éditions de l'Aube, 2010.

Muzeum Narodowe w Warszawie, ed. *Architektura Plakatu: Graficy-Architekci z Kręgu Politechniki Warszawskiej w Latach 1915–1939*. Warsaw, 2005.

Myllyntaus, Timo. "Foreign Models and National Styles in Teaching Technology in the Nordic Countries." In *La Formation des ingénieurs en perspective: Modèles de référence et réseaux de médiation—XVIIIe–XXe siècles*, edited by Irina Gouzévitch, André Grelon, and Anousheh Karvar, 141–52. Rennes: Presses Universitaires de Rennes, 2004.

Nahum, Andrew. "'I Believe the Americans Have not yet Taken Them All!' The Exploitation of German Aeronautical Science in Postwar Britain." In *Tackling Transport*, edited by Helmuth Trischler and Stefan Zeilinger, 99–138. London: NMSI Trading, 2003.

Neebe, Richard. "Technologietransfer und Aussenhandel in den Anfangsjahren der Bundesrepublik Deutschland." *Vierteljahrschrift für Sozial- und Wirtschaftsgeschichte* 76 (1989): 49–75.

Nelson, Robert L. "From Manitoba to the Memel: Max Sering, Inner Colonization and the German East." *Social History* 35 (2010): 439–57.

———, ed. *Germans, Poland, and Colonial Expansion to the East: 1850 through the Present.* New York: Palgrave Macmillan, 2009.

Nerdinger, Winfried. "Der Traum von der Universalbibliothek." In *Architektur wie sie im Buche steht: Fiktive Bauten und Städte in der Literatur*, edited by Winfried Nerdinger, 237–60. Salzburg: Pustet, 2006.

———. "Zlín: Sozial gelackte Modernität – Architektur und Leben im Gleichschritt." In *Zlín: Modellstadt der Moderne*, edited by Winfried Nerdinger, 16–39. Berlin: Jovis, 2009.

Neufeld, Michael. "The Nazi Aerospace Exodus: Towards a Global, Transnational History." *History and Technology* 28 (2012): 49–67.

———. *Von Braun: Dreamer of Space, Engineer of War.* New York: Knopf, 2007.

Neurath, Otto. "Bildliche Darstellung sozialer Tatbestände. " *Die Quelle* 77, no. 1 (1927): 130–36.

Nicolai, Bernd. "'Der goldene Käfig': Ernst Reuter und Martin Wagner im türkischen Exil." In *Ernst Reuter: Kommunalpolitiker und Gesellschaftsreformer, 1921–1953*, edited by Heinz Reif and Moritz Feichtinger, 239–50. Bonn: Dietz, 2009.

Noakowski, Stanisław. *Pisma.* Warsaw: Wydawnictwo Budownictwo i Architektura, 1957.

———. "Powstanie Wydziału Architektury." In *Fragmenty Stuletniej Historii. 1899–1999: Relacje, Wspomnenia, Refleksje w Stulecie Organizacji Warszawskich Architektów*, edited by Tadeusz Barucki, 29–32. Warsaw: Drukoba, 2000.

Nolan, Mary. *Visions of Modernity: American Business and the Modernization of Germany.* New York: Oxford University Press, 1994.

Nolte, Paul. *Der Wissenschaftsmacher: Reimar Lüst im Gespräch mit Paul Nolte.* Munich: Beck, 2008.

Nord, Philip. *France's New Deal: From the Thirties to the Postwar Era.* Princeton: Princeton University Press, 2010.

Nordmann, Alfred. *Converging Technologies—Shaping the Future of European Societies (High Level Expert Group Report).* Luxembourg: Office for Official Publications of the European Communities, 2004.

———. "European Experiments." *Osiris* 24 (2009): 278–302.

Nottmeier, Christian. *Adolf von Harnack und die deutsche Politik 1890–1930: Eine biographische Studie zum Verhältnis von Protestantismus, Wissenschaft und Politik.* Tübingen: Mohr Siebeck, 2004.

Nowotny, Helga, Peter Scott, and Michael Gibbons. *Re-Thinking Science: Knowledge and the Public in an Age of Uncertainty.* Cambridge: Polity Press, 2001.

Nye, Mary Jo. "National Styles? French and English Chemistry in the Nineteenth and Early Twentieth Centuries." *Osiris* 8 (1993): 30–49.

Nye, Robert. "Honor Codes and Medical Ethics in Modern France." *Bulletin of the History of Medicin* 69 (1995): 91–111.

OECD. *NEA 50 Anniversary, Issy-les-Moulineaux.* Paris: OECD, 2008.

Oldenziel, Ruth, and Mikael Hård, *Consumers, Tinkerers, Rebels: The People Who Shaped Europe.* Basingstoke: Palgrave Macmillan, 2013.

Oldenziel, Ruth, Annie Canel, and Karin Zachmann. "Introduction." In *Crossing Boundaries, Building Bridges: Comparing the History of Women Engineers 1870s–1990s*, edited by Annie Canel, Ruth Oldenziel, and Karin Zachmann, 1–10. Amsterdam: Harwood, 2000.

Omilanowska, Małgorzata. *Architekt Stefan Szyller 1857–1933*. Warsaw: Liber pro Arte, 2008.

Oreskes, Naomi. "Science, Technology and Free Enterprise." *Centaurus* 52 (2010): 297–310.

Ostwald, Wilhelm. *Zur Geschichte der Wissenschaft: Vier Manuskripte aus dem Nachlass von Wilhelm Ostwald*. Edited by Regine Zott. Thun: Deutsch, 1985.

Otlet, Paul. *Traité de documentation: Le livre sur le livre, théorie et pratique*. Brussels: Editiones Mundaneum, 1934.

Otter, Chris. *The Victorian Eye: A Political History of Light and Vision in Britain, 1800–1910*. Chicago and London: University of Chicago Press, 2008.

"Où en est Euratom? Un des plus grands chantiers d'Europe: Ispra." *Communauté européenne* (October 1960, no. 10): 5.

Pągowski, Stefan. "Architekt Architektów—Stanisław Noakowski (1867–1928)." *Kwartalnik Architektury i Urbanistyki* 35, nos. 3–4 (1990): 123–29.

Pallas, Jean-Claude, ed. *Histoire et architecture du Palais des Nations (1924–2001): L'Art Déco au service des relations internationales*. Genève: Nations Unies, 2001.

Parravano, Nicola. "Il fascismo e la scienza." *Richerche Scientifiche* 1 (1936): 361–68.

Patel, Kiran K. *Fertile Ground for Europe: The History of European Integration and the Common Agricultural Policy since 1945*. Baden-Baden: Nomos, 2009.

Paulmann, Johannes. "Feindschaft und Verflechtung: Anmerkungen zu einem scheinbaren Paradox." In *Vom Gegner lernen: Feindschaften und Kulturtransfers im Europa des 19. und 20. Jahrhunderts*, edited by Martin Aust and Daniel Schönpflug, 341–56. Frankfurt a.M. and New York: Campus, 2007.

———. "Reformer, Experten und Diplomaten: Grundlagen des Internationalismus im 19. Jahrhundert." In *Akteure der Aussenbeziehungen: Netzwerke und Interkulturalität im historischen Wandel*, edited by Hillard von Thiessen and Christian Windler, 173–97. Cologne: Böhlau, 2010.

Pearlman, Jill E. *Inventing American Modernism: Joseph Hudnut, Walter Gropius, and the Bauhaus Legacy at Harvard*. Charlottesville, VA: University of Virginia Press, 2007.

Peaucelle, Jean-Louis. *Henri Fayol, inventeur des outils de gestion: Textes originaux et recherches actuelles*. Paris: Économica, 2003.

Pelc, Ortwin, and Susanne Grötz, eds. *Konstrukteur der modernen Stadt: William Lindley in Hamburg und Europa 1808–1900*. Munich: Dölling und Galitz, 2008.

Peltcjean, Patrick. "Scientific Development, Engineering Schools and the Building of a Modern State." *History and Technology* 12 (1995): 191–204.

Pestre, Dominique. "The Difficult Decision, Taken in the 1960s, to Construct a 3–400 GeV Proton Synchrotron in Europe." In *History of CERN*, Vol. 3, edited by John Krige and Armin Hermann, 65–96. Amsterdam: North-Holland, 1996.

Petersen, Hans-Christian. *Bevölkerungsökonomie—Ostforschung—Politik: Eine biographische Studie zu Peter-Heinz Seraphim (1902–1979)*. Osnabrück: fibre, 2007.

Peterson, John. *High Technology and the Competition State: An Analysis of the Eureka Programme*. London: Routledge, 1993.

———, and Margaret Sharp. *Technology Policy in the European Union*. London: Macmillan Press, 1999.

Pfammatter, Ulrich. *Die Erfindung des modernen Architekten: Ursprung und Entwicklung seiner wissenschaftlich-industriellen Ausbildung*. Basel: Birkhäuser, 1997.

Picon, Antoine. "French Engineers and Social Thought, 18–20th Centuries: An Archeology of Technocratic Ideals." *History and Technology* 23 (2007): 197–208.

———. *L'invention de l'ingénieur moderne: L'Ecole des ponts et chaussées, 1747–1851.* Paris: Presses de l'Ecole nationale des ponts et chaussées, 1992.

Piłatowicz, Józef. *Kadra Inżynierska w II Rzeczypospolitej.* Siedlce: Wydawnictwo Wyższej Szkoły Rolniczo-Pedagogicznej, 1994.

———. "Technicy Lwowa i Krakowa wobec Perspektywy Odzyskania przez Polskę Niepodległości."*Kwartalnik Historii Nauki i Techniki* 3–4 (1999): 89–108.

Platzer, Monika. "Zlín – Ein architektonischer Sonderfall." In *Zlín: Modellstadt der Moderne*, edited by Winfried Nerdinger, 94–111. Berlin: Jovis, 2009.

Polianski, Igor J., and Matthias Schwartz, eds. *Die Spur des Sputnik: Kulturhistorische Expeditionen ins kosmische Zeitalter.* Frankfurt a.M. and New York: Campus, 2009.

Politechnika: Zakład Architektury Polskiej. *Warszawska Szkoła Architektury, 1915–1965.* Warsaw: Państwowe Wydawnictwo Naukowe, 1967.

Pollak, Martha D. *The Education of the Architect: Historiography, Urbanism, and the Growth of Architectural Knowledge: Essays Presented to Stanford Anderson.* Cambridge, MA: MIT Press, 1997.

Popplow, Markus. "Europa wider Willen? Konkurrenz um technische Innovationen als integratives Element des frühneuzeitlichen Europa." In *Europe en movement: Mobilisierungen von Europa-Konzepten im Spiegel der Technik*, edited by Angela Oster, 19–39. Berlin: Lit Verlag, 2009.

"Pour sortir Euratom de la crise: Un livre blanc de la Commission européenne," *Communauté européenne* (November 1968, no. 124): 1.

Presas i Puig, Albert. "Nota histórica: una conferencia de José Maria Albareda ante les autoridades académicas alemanas." *Arbor* (1998): 343–57.

———. "Science on the Periphery—the Spanish Reception of Nuclear Energy: An Attempt at Modernity." *Minerva* 43 (2005): 197–219.

———. "Technoscientific Synergies between Germany and Spain in the Twentieth Century: Continuity amid Radical Change." *Technology and Culture* 51 (2010): 80–98.

Pursell, Carroll. "'Am I a Lady or an Engineer?': The Origins of the Women's Engineering Society in Britain, 1918–1940." In *Crossing Boundaries, Building Bridges: Comparing the History of Women Engineers 1870s–1990s*, edited by Annie Canel, Ruth Oldenziel, and Karin Zachmann, 51–73. Amsterdam: Harwood, 2000.

Pyta, Wolfram. "'Menschenökonomie': Das Ineinandergreifen von ländlicher Sozialraumgestaltung und rassenbiologischer Bevölkerungspolitik im NS-Staat." *Historische Zeitschrift* 273 (2001): 31–94.

Quiring, Claudia. "From 'Carp Pond' to 'Getting Accustomed to Caviar': Insights into the Lives of May's Employees in Silesia, Frankfurt and the Soviet Union." In *Ernst May 1886–1970*, edited by Claudia Quiring, Wolfgang Voigt, Peter Cachola Schmal, and Eckhard Herrel, 131–55. Munich: Prestel, 2011.

Rabinbach, Anson. *The Human Motor: Energy, Fatigue, and the Origins of Modernity.* Berkeley, CA: University of California Press, 1992.

Radkau, Joachim. *Technik in Deutschland: Vom 18. Jahrhundert bis heute.* Frankfurt a.M. and New York: Campus, 2008.

Raj, Gopal. *Reach for the Stars: The Evolution of India's Rocket Programme.* New Delhi and New York: Viking, 2000.

Raphael, Lutz. "Die Verwissenschaftlichung des Sozialen als methodische und konzeptionelle Herausforderung für eine Sozialgeschichte des 20. Jahrhunderts." *Geschichte und Gesellschaft* 22 (1996): 165–93.

———. "Radikales Ordnungsdenken und die Organisation totalitärer Herrschaft: Weltanschauungseliten und Humanwissenschaftler im NS-Regime." *Geschichte und Gesellschaft* 27 (2001): 5–40.

———. "Sozialexperten in Deutschland zwischen konservativem Ordnungsdenken und rassistischer Utopie (1918–1945)." In *Utopie und politische Herrschaft im Europa der Zwischenkriegszeit*, edited by Wolfgang Hardtwig, 327–46. Munich: Oldenbourg, 2003.

Rathenau, Walther. *Die neue Wirtschaft*. Berlin: S. Fischer, 1918.

Raymaekers, Pieter. "Between Capital and Labour: The Social Turn of the Belgian Engineering Profession in the Second Half of the Nineteenth Century." Conférence Internationale: Un ingénieur, des ingénieurs. Expansion ou fragmentation? École des hautes études en sciences sociales, Paris, 6–7 Octobre 2011.

Rayward, Warden B. *The Universe of Information: The Work of Paul Otlet for Documentation and International Organisation*. Moscow: All-Union Institute for Scientific and Technical Information (VINITI), 1975.

Redfield, Peter. *Space in the Tropics: From Convicts to Rockets in French Guiana*. Berkeley, CA: University of California Press, 2000.

Reichherzer, Frank. *"Alles ist Front!" Wehrwissenschaften in Deutschland und die Bellifizierung der Gesellschaft vom Ersten Weltkrieg bis zum Kalten Krieg*. Paderborn: Schöningh, 2012.

Reinhard, Wolfgang. *Geschichte der Staatsgewalt: Eine vergleichende Verfassungsgeschichte Europas von den Anfängen bis zur Gegenwart*. Munich: Beck, 2000.

Reinke, Niklas. *The History of German Space Policy: Ideas, Influences, and Interdependences 1923–2002*. Paris: Beauchesne, 2007.

Reisinger, William M. *Energy and the Soviet Bloc: Alliance Politics after Stalin*. Ithaca, NY: Cornell University Press, 1992.

Reisman, Arnold. *Turkey's Modernization: Refugees from Nazism and Atatürk's Vision*. Washington, DC: New Academia Publishing, 2006.

Rheinberger, Hans-Jörg. *On Historicizing Epistemology: An Essay*. Stanford, CA: Stanford University Press, 2010.

Richeri, Guiseppe. "Italian Broadcasting and Fascism, 1924–1937." *Media, Culture & Society* 2 (1980): 49–56.

Rieger, Stefan. "Arbeit an sich: Ikonische Diskursformationen in der Geschichte der Moderne." In *Anthropologie der Arbeit*, edited by Ulrich Bröckling and Eva Horn, 79–96. Tübingen: Narr, 2002.

Ringrose, Daniel. "Work and Social Presence: French Public Engineers in Nineteenth-Century Provincial Communities." *History and Technology* 14 (1998): 293–312.

Ritter, Gerhard A., Margit Szöllösi-Janze, and Helmuth Trischler, eds. *Antworten auf die amerikanische Herausforderung: Forschung in der Bundesrepublik und der DDR in den "langen" siebziger Jahren*. Frankfurt a.M. and New York: Campus, 1999.

Roco, Mihail C., and William S. Bainbridge, eds. *Converging Technologies for Improving Human Performance: Nanotechnology, Biotechnology, Information Technology, and Cognitive Science*. Berlin: Links, 1993.

Rodgers, Daniel T. *Atlantic Crossings: Social Politics in a Progressive Age.* Cambridge, MA: The Belknap Press of Harvard University Press, 1998.

Rodrigues, Maria João, ed. *Europe, Globalization and the Lisbon Agenda.* Cheltenham: Edward Elgar, 2009.

———. *European Policies for a Knowledge Economy.* Cheltenham: Edward Elgar, 2003.

Rodrigues, Paulo S. "The Debate between Engineers and Architects in the Second Half of the 19th Century: What is Architecture?" In *The Quest for a Professional Identity: Engineers between Training and Action*, edited by Ana de Cardoso Matos, André Grelon, Irina Gouzévitch, and Maria Paula Diogo, 343–56. Lisbon: Colibri, 2009.

Rohdewald, Stefan. "Mimicry in a Multiple Postcolonial Setting: Networks of Techocracy and Scientific Management in Piłsudski's Poland." In *Expert Cultures in Central Eastern Europe: The Internationalization of Knowledge and the Transformation of Nation States since World War I*, edited by Martin Kohlrausch, Katrin Steffen, and Stefan Wiederkehr, 63–84. Osnabrück: fibre, 2010.

Rohkrämer, Thomas. "Antimodernism, Reactionary Modernism and National Socialism: Technocratic Tendencies in Germany, 1890–1945." *Contemporary European History* 8 (2000): 29–50.

Rolf, Malte. "Der Zar an der Weichsel—Repräsentationen von Herrschaft und Imperium im fin de siècle." In *Imperiale Herrschaft in der Provinz: Repräsentationen politischer Macht im späten Zarenreich*, edited by Jörg Baberowski, 145–71. Frankfurt a.M. and New York: Campus, 2008.

———. "Imperiale Herrschaft im städtischen Raum: Zarische Beamte und urbane Öffentlichkeit in Warschau (1870–1914)." In *Russlands imperiale Macht. Integrationsstrategien und ihre Reichweite in transnationaler Perspektive*, edited by Bianka Piotrow-Ennker, 123–53. Vienna: Böhlau, 2012.

Roll-Hansen, Nils. "The Lysenko Effect: Undermining the Autonomy of Science." *Endeavour* 29 (2005): 143–47.

Rossi, Lucio. "Superconductivity: Its Role, its Success, and its Setbacks in the Large Hadron Collider of CERN." *Superconductor Science and Technology* 23 (2010): 1–17.

Rossianov, Kirill O. "Editing Nature: Joseph Stalin and the 'New' Soviet Biology." *Isis* 84 (1993): 728–46.

Rössler, Hans-Christian. "Die Wunderstadt, die niemals schläft," *Frankfurter Allgemeine Zeitung*, no. 85, April 11, 2009.

Rössler, Mechtild. "Konrad Meyer und der 'Generalplan Ost' in der Beurteilung der Nürnberger Prozesse." In *Der "Generalplan Ost": Hauptlinien der nationalsozialistischen Planungs- und Vernichtungspolitik*, edited by Mechtild Rössler and Cordula Tollmien, 356–68. Berlin: Akademie Verlag, 1993.

Rotblat, J., ed. *Nuclear Reactors: To Breed or not to Breed: A Pugwash Debate on Fast Breeder Reactors Held at the Royal Society, London, on 28 Sept. 1976 under the Chairmanship of Alec Merrison.* London: Taylor and Francis, 1977.

Roulet, Daniel. "Le Corbusier in Vichy." *Tec21* 132 (2006): 22–25.

Rudberg, Eva. *The Stockholm Exhibition 1930: Modernism's Breakthrough in Swedish Architecture.* Stockholm: Stockholmia, 1999.

Rüger, Jan. *The Great Naval Game: Britain and Germany in the Age of Empire.* Cambridge: Cambridge University Press, 2007.

Russo, Arturo. "Science and Industry in Italy between the Two World Wars." *Historical Studies in the Physical and Biological Sciences* 16 (1986): 281–320.

Ryan, Mike. "The Role of National Culture in the Space-Based Technology Transfer Process." *Comparative Technology Transfer and Society* 2 (2004): 31–66.

Sachar, Howard M. *A History of the Jews in America.* New York: Knopf, 1992.

Saint, Andrew. *Architect and Engineer: A Study in Sibling Rivalry.* New Haven, CT: Yale University Press, 2008.

Saint-Simon, Henry. *De la réorganisation de la société européenne ou De la nécessité et des moyens de rassembler les peuples de l'Europe en un seul corps politique, en conservant à chacun son indépendance nationale.* Paris: A. Égron, 1814.

Sandholtz, Wayne. *High-Tech Europe: The Politics of International Cooperation.* Berkeley, CA: University of California Press, 1991.

Saraiva, Tiago. "A City for a Technical School: Spanish Civil Engineers and the Expansion of Madrid in the XIXth Century." In *The Quest for a Professional Identity: Engineers between Training and Action*, edited by Ana de Cardoso Matos, André Grelon, Irina Gouzévitch, and Maria Paula Diogo, 311–28. Lisbon: Colibri, 2009.

———. "Inventing the Technological Nation: The Example of Portugal (1851–1898)." *History and Technology* 23 (2007): 263–74.

Šarapov, N.P. "Deutsche Arbeiter in der Sowjetunion: Über die Teilnahme deutscher Arbeiter und Spezialisten am sozialistischen Aufbau in der UdSSR 1930 bis 1933." *Zeitschrift für Geschichtswissenschaft* 24 (1976): 1110–30.

Saunier, Pierre-Yves. "Sketches from the Urban Internationale, 1910–50: Voluntary Associations, International Institutions and US Philanthropic Foundations." *International Journal of Urban and Regional Research* 25, no. 2 (2001): 380–403.

Schandevyl, Eva. "Soviet Biology, Scientific Ethos and Political Engagement: Belgian University Professors and the Lysenko Case." *Journal of Communist Studies and Transition Politics* 19 (2003): 93–107.

Schattenberg, Susanne. "Stalinismus in den Köpfen: Ingenieure konstruieren ihre Welt." *Geschichte und Gesellschaft* 30 (2004): 94–117.

Schenk Dieter. *Der Lemberger Professorenmord und der Holocaust in Ostgalizien.* Bonn: Dietz, 2007.

Schivelbusch, Wolfgang. *Die Bibliothek von Löwen: Eine Episode aus der Zeit der Weltkriege.* Munich: Hanser, 1988.

Schmaltz, Florian. "Luftfahrtforschung unter deutscher Besatzung: Die Aerodynamische Versuchsanstalt Göttingen und ihre Aussenstellen in Frankreich im Zweiten Weltkrieg." In *"Fremde" Wissenschaftler im Dritten Reich: Die Debye-Affäre im Kontext*, edited by Dieter Hoffmann and Mark Walker, 384–407. Göttingen: Wallstein, 2011.

Schmidt, Sonja D. "Nuclear Colonization?: Soviet Technopolitics in the Second World." In *Entangled Geographies: Empire and Technopolitics in the Global Cold War*, edited by Gabrielle Hecht, 125–54. Cambridge, MA and London: MIT Press, 2011.

Schot, Johan, and Vincent Lagendijk. "Technocratic Internationalism in the Interwar Years: Building Europe on Motorways and Electricity Networks." *Journal of Modern European History* 6 (2008): 196–217.

———, Arie Rip, and Harry Lintsen. *Technology and the Making of the Netherlands: The Age of Contested Modernization, 1890–1970.* Cambridge, MA: MIT Press, 2010.

Schrafstetter, Susanne, and Stephen Twigge. *Avoiding Armageddon: Europe, the United States and the Struggle for Nuclear Nonproliferation, 1945–1970.* Westport: Praeger, 2004.

Schueler, Judith. *Materializing Identity: The Co-construction of the Gotthard Railway and Swiss National Identity*. Amsterdam: Aksant, 2008.

Schulz, Günther. "The Promotion of Industry—A Comment." In *Industrieentwicklung: Ein deutsch-britischer Dialog*, edited by Franz Bosbach, John R. Davis, and Andreas Fahrmeir, 69–72. Berlin: Walter de Gruyter, 2009.

Schumpeter, Joseph A. *Imperialism and the Social Classes*. New York: Augustus M. Kelley, 1951.

Schwarz, Michiel. "European Policies on Space Science and Technology 1960–1978." *Research Policy* 8 (1979): 204–43.

Schweitzer, Sylvie. "Der Ingenieur." in *Der Mensch des 19. Jahrhunderts*, edited by Ute Frevert, and Heinz-Gerhard Haupt, 67–85. Essen: Magnus, 2004.

———. *Des engrenages à la chaîne: Les usines Citroën, 1915–1935*. Lyon: Presses Universitaires de Lyon, 1982.

Scott, James C. *Seeing like a State: How Certain Schemes to Improve the Human Condition Have Failed*. New Haven, CT: Yale University Press, 1998.

Scranton, Philip. "Technology-Led Innovation: The Non-Linearity of US Jet Propulsion Development." *History and Technology* 22 (2006): 337–67.

———. "The Challenge of Technological Uncertainty." *Technology and Culture* 50 (2009): 513–18.

Sebesta, Lorenza. *Alleati competitive: Origini et sviluppo della cooperazione spaziale fra Europa et Stati Uniti, 1957–1973*. Rome and Bari: Laterza, 2003.

———. "Choosing its Own Way: European Cooperation in Space Europe as a Third Way between Science's Universalism and US Hegemony." *Journal of European Integration History* 12, no. 2 (2006): 2–56.

Seipp, Adam R. *The Ordeal of Peace: Demobilization and the Urban Experience in Britain and Germany, 1917–1921*. Farnham: Ashgate, 2009.

Sering, Max. *Die innere Kolonisation im östlichen Deutschland*. Leipzig: Duncker & Humblodt, 1893.

Servan-Schreiber, Jean-Jacques. *Le défi américain*. Paris: Denoël, 1967.

———. *The American Challenge*. London: Hamish Hamilton, 1968.

Severi, Francesco, ed. *Scritti di Guglielmo Marconi*. Rome: Reale Accademia d'Italia, 1941.

Shapin, Stephen. *The Scientific Life: A Moral History of a Late Modern Vocation*. Chicago and London: University of Chicago Press, 2008.

Shaw, Debra B. "Bodies out of This World: The Space Suit as Cultural Icon." *Science as Culture* 13 (2004): 123–44.

Shaw, Edwin N. *Europe's Experiment in Fusion: The JET Joint Undertaking*. Amsterdam: North Holland, 1990.

———. "Joint European Torus." In *History of European Scientific and Technological Cooperation*, edited by John Krige and Luca Guzzetti, 165–78. Luxembourg: Office for Official Publications of the European Communities, 1997.

Sheehan, Helena. *Marxism and the Philosophy of Science: A Critical History*. Atlantic Highlands: Humanities Press International, 1993.

Sheehan, Michael. *The International Politics of Space*. London and New York: Routledge, 2007.

Shioni, Haruhito. *Fordism Transformed: The Development of Production Methods in the Automobile Industry*. New York: Oxford University Press, 1995.

Siddiqi, Asif S. "Asia in Orbit: Asian Cooperation in Space." *Georgetown Journal of International Affairs* 11 (2010): 131–39.

———. "Competing Technologies, National(ist) Narratives, and Universal Claims: Toward a Global History of Space Exploration." *Technology and Culture* 51 (2010): 425–43.

———. "National Aspirations on a Global Stage: Fifty Years of Spaceflight." In *Remembering the Space Age*, edited by Steven J. Dick, 17–35. Washington, DC: NASA History Division, 2009.

———. "Soviet Space Power during the Cold War." In *Harnessing the Heavens: National Defense through Space*, edited by Paul G. Gillespie and Grant T. Weller, 135–50. Colorado Springs: United States Air Force Academy, 2008.

———. *Sputnik and the Soviet Space Challenge*. Gainesville, FL: University of Florida Press, 2003.

———. *The Rockets' Red Glare: Spaceflight and the Soviet Imagination, 1857–1957*. Cambridge: Cambridge University Press, 2010.

———. "The Rockets' Red Glare: Technology, Conflict, and Terror in the Soviet Union." *Technology and Culture* 44 (2003): 470–501.

———. *The Soviet Space Race with Apollo*. Gainesville, FL: University of Florida Press, 2003.

Siegmund-Schultze, Reinhard. "Hilda Geiringer-von Mises, Charlier Series, Ideology, and the Human Side of the Emancipation of Applied Mathematics at the University of Berlin during the 1920s." *Historia Mathematica* 20 (1993): 364–81.

Siegrist, Hannes. "The Professions in Nineteenth-Century Europe." In *The European Way: European Societies in the 19th and 20th Centuries*, edited by Hartmut Kaelble, 68–88. New York and Oxford: Berghahn, 2004.

Sigrist, Natalia T. "Les étudiantes étrangères dans les universités occidentales, des discriminations à l'exil universitaire (1870–1914)." In *Etudiants de l'exil: Université, refuges et migrations étudiantes (XVIe–XXe siècles)*, edited by Caroline Barrera and Patrick Fert, 105–18. Toulouse: Presses universitaires du Mirail, 2009.

Sinclair, Bruce. "The Power of Ceremony: Creating an International Engineering Community." *History of Technology* 21 (1999): 203–11.

Šlapeta, Vladimír. "Kulturelles und soziales Leben in Zlín." In *Zlín: Modellstadt der Moderne*, edited by Winfried Nerdinger, 72–93. Berlin: Jovis, 2009.

———. "The Baťa Legacy: The Realization of a Utopia." In *A Utopia of Modernity: Zlín: Revisiting Baťa's Functional City*, edited by Katrin Klingan, 52–67. Berlin: Jovis, 2009.

Smil, Vaclav. *Energy in Nature and Society: General Energetics of Complex Systems*. Cambridge, MA: MIT Press, 2007.

———. *Energy in World History*. Boulder, CA: Westview Press, 1994.

Smith, Cecil O. "The Longest Run: Public Engineers and Planning in France." *American Historical Review* 95 (1990): 657–92.

Solomon, Susan G., ed. *Doing Medicine Together: Germany and Russia between the Wars*. Toronto: University of Toronto Press, 2006.

Somer, Kees. *The Functional City: CIAM and the Legacy of van Eesteren*. Rotterdam: NAI Publishers, 2007.

Sonne, Wolfgang. *Representing the State: Capital City Planning in the Early Twentieth Century*. Munich: Prestel, 2003.

Soutou, Georges-Henri. "Les accords de 1957–1958: vers une communauté stratégique et nucléaire entre la France, l'Allemagne et l'Italie." In *La France et l'Atome: Études histoire nucléaire*, edited by Maurice Vaisse, 123–63. Paris: Fayard, 1994.

Späth, Manfred. "Der Ingenieur als Bürger: Frankreich, Deutschland und Russland im Vergleich." In *Bürgerliche Berufe: Zur Sozialgeschichte der freien und akademischen Berufe im internationalen Vergleich: acht Beiträge*, edited by Hannes Siegrist, 84–105. Göttingen: Vandenhoeck & Ruprecht, 1988.

Speer, Albert. *Inside the Third Reich*. New York and Toronto: Macmillan, 1970.

Spenkuch, Hartwin. *Das Preussische Herrenhaus: Adel und Bürgertum in der Ersten Kammer des Landtages, 1854–1918*. Düsseldorf: Droste, 1998.

Stanchevici, Dmitri. "Stalinist Genetics: The Constitutional Rhetoric of T.D. Lysenko." PhD diss., Texas Tech University, 2007.

Stanley, Ruth. *Rüstungsmodernisierung durch Wissenschaftsmigration? Deutsche Rüstungsfachleute in Argentinien und Brasilien 1947–1963*. Frankfurt a.M.: Vervuet, 1999.

Staudenmaier, John M. *Technology's Storytellers: Reweaving the Human Fabric*. Cambridge, MA: MIT Press, 1989.

Steffen, Katrin. "Wissenschaftler in Bewegung: Der Materialforscher Jan Czochralski zwischen den Weltkriegen." *Journal of Modern European History* 6 (2008): 237–61.

Steiner, Zara. *The Lights that Failed: European International History 1919–1933*. Oxford: Oxford University Press, 2005.

Steinführer, Annett. "Uncharted Zlín: The Forgotten Lifeworlds of the Baťa City." In *A Utopia of Modernity: Zlín: revisiting Baťa's Functional City*, edited by Katrin Klingan, 105–16. Berlin: Jovis, 2009.

Steinmann, Martin, ed. *CIAM: Dokumente 1928–1939*. Basel: Birkhäuser, 1979.

———. "Political Standpoints in CIAM: 1923–1933." *Architectural Association Quarterly* 4 (1972): 49–55.

Stichweh, Rudolph. *Die Weltgesellschaft: Soziologische Analysen*. Frankfurt a.M.: Suhrkamp, 2000.

Stiernstedt, Jan. *Sverige i rymden: svensk rymdverksamhet 1959–1972*. Solna, 1997.

Stoehr, Irene. "Von Max Sering zu Konrad Meyer: Ein 'machtergreifender' Generationenwechsel in der Agrar- und Siedlungswissenschaft." In *Autarkie und Ostexpansion: Pflanzenzucht und Agrarforschung im Nationalsozialismus*, edited by Susanne Heim, 57–90. Göttingen: Wallstein, 2002.

Stokes, Raymond G. "Forced Technology Transfer and Western Integration." In *The United States and the Integration of Europe: Legacies of the Postwar Era*, edited by Francis H. Heller and John R. Gillingham, 281–88. New York: St. Martin's Press, 1995.

Störtkuhl, Beate. "Ausstellungsarchitektur als Mittel nationaler Selbstdarstellung. Die 'Ostdeutsche Ausstellung' 1911 und die 'Powszechna Wystawa Krajowa' 1929 in Posen/Poznań." In *Nation, Style, Modernism*, edited by Jacek Purchla and Wolf Tegethoff, 237–55. Krakow: International Cultural Centre, 2006.

Straalen, Mariette. "Empirische Stadtanalysen." *Daedalus* 69/70 (1998): 60–67.

Strasser, Bruno J. "The Coproduction of Neutral Science and Neutral State in Cold War Europe: Switzerland and International Scientific Cooperation, 1951–69." *Osiris* 24 (2009): 165–87.

———, and Frédéric Joye. "Une Science 'neutre' dans la guerre froide? La Suisse et la coopération scientifique Européenne (1951–1969)." *Revue Suisse d'Histoire* 55 (2005): 95–112.

Strauss, Franz J. *Challenge and Response: A Programme for Europe*. New York: Atheneum, 1970.

Strohmeier, Martin. "Der zeitgeschichtliche und politische Rahmen der türkischen Universitätsreform und die Rolle der deutschen Wissenschaftsmigranten." In *Deutsche Wissenschaftler im türkischen Exil: Die Wissenschaftsmigration in die Türkei 1933–1945*, edited by Christopher Kubaseck and Günter Seufert, 67–75. Würzburg: Ergon, 2008.

Surman, Jan. "The Local and the Global: Scientific Communication in and beyond the Habsburg Empire Prior to World War I." Paper presented at the 4th Tensions of Europe Conference, Sofia, June 17–20, 2010.

Sussman, Gerald. *Communication, Technology, and Politics in the Information Age*. Thousand Oaks, CA: Sage, 1997.

Sutton, Antony C. *Western Technology and Soviet Economic Development*. Stanford, CA: Hoover Institute Press, 1968.

Suzuki, Kazuto. *Policy Logics and Institutions of European Space Collaboration*. Aldershot: Ashgate, 2003.

Švácha, Rostislav. "Before and after the Mundaneum: Teige as Theoretician of the Architectural Avant-Garde." In *Karel Teige: 1900–1951: L'Enfant Terrible of the Czech Modernist Avant-Garde*, edited by Eric Dluhosch and Rostislav Švácha, 107–39. Cambridge, MA: MIT Press, 1999.

———. "Tomáš Baťa and the Destruction of Old Zlín." In *A Utopia of Modernity: Zlín: Revisiting Baťa's Functional City*, edited by Katrin Klingan, 75–89. Berlin: Jovis, 2009.

Szcepanik, Petr. "The Aesthetics of Rationalization: The Media Network in the Baťa Company." In *A Utopia of Modernity: Zlín: Revisiting Baťa's Functional City*, edited by Katrin Klingan, 203–15. Berlin: Jovis, 2009.

Szöllösi-Janze, Margit. *Fritz Haber, 1868–1934: Eine Biographie*. Munich: Beck, 1998.

Teige, Karel. "Antwort auf Le Corbusier." In *Architektur zwischen Kunst und Wissenschaft: Texte der tschechischen Architektur-Avantgarde 1918–1938*, edited by Jeanette Fabian and Ulrich Winko, 281–86. Berlin: Gebr. Mann, 2010.

———. "Mundaneum." *Oppositions* 4 (1974): 83–93. Translation by Ladislav and Elisabeth Holovsky, and Lubamir Dolezel.

———. *Vývoj sovětské Architektury*. Prague, 1936.

———, and Eric Dluhosch. *The Minimum Dwelling*. Chicago: MIT Press; Graham Foundation for Advanced Studies in the Fine Arts, 2002.

Todorov, Tzvetan. *The Limits of Art: Two Essays*. London and New York: Seagull Books, 2010.

Tomlow, Jos. "Introduction: Building Science as Reflected in Modern Movement Literature." In *Climate and Building Physics in the Modern Movement*, edited by Jos Tomlow, 6–26. Zittau: Hochschule Zittau/Görlitz, 2006.

Trischler, Helmuth. "A Talkative Artefact: Germany and the Development of a European Launcher in the 1960s." In *Showcasing Space*, edited by Martin Collins and Doug Millard, 7–28. London: Science Museum, 2005.

———. *Dokumente zur Geschichte der Luft- und Raumfahrtforschung in Deutschland 1900–1970*. Cologne: DFVLR, 1991.

———. "The Syndrome of Falling Behind—Resource Constellations and Epistemic Orientations in the Natural and Engineering Sciences." In *The German Research Foundation 1920–1970: Funding Poised between Science and Politics*, edited by Mark Walker, Karin Orth, Ulrich Herbert, and Rüdiger vom Bruch, 105–17. Stuttgart: Steiner, 2013.

———. *The "Triple Helix" of Space: German Space Activities in a European Perspective.* Paris: ESA, 2002.

———. "Wolfgang Gentner und die Grossforschung im bundesdeutschen und europäischen Raum." In *Wolfgang Gentner: Festschrift zum 100. Geburtstag,* edited by Dieter Hoffmann and Ulrich Schmidt-Rohr, 95–120. Berlin: Springer, 2006.

———, and Hans Weinberger. "Engineering Europe: Big Technologies and Military Systems in the Making of 20th Century Europe." *History and Technology* 21 (2005): 49–83.

Trommler, Frank. "The Avant-Garde and Technology." *Science in Context* 8 (1995): 397–416.

Turda, Marius. *Modernism and Eugenics.* London: Palgrave Macmillan, 2010.

———, and Paul J. Weindling, eds. *"Blood and Homeland": Eugenics and Racial Nationalism in Central and Southeast Europe, 1900–1940.* Budapest and New York: Central European University Press, 2007.

Unger, Corinna. *Ostforschung in Westdeutschland: Die Erforschung des europäischen Ostens und die Deutsche Forschungsgemeinschaft, 1945–1975.* Stuttgart: Steiner, 2007.

Ungern-Sternberg, Jürgen von, and Wolfgang von Ungern-Sternberg. *Der Aufruf "An die Kulturwelt!": Das Manifest der 93 und die Anfänge der Kriegspropaganda im Ersten Weltkrieg: Mit einer Dokumentation.* Stuttgart: Steiner, 1996.

Union of International Associations. "The Union of International Associations: A World Centre (1914)." In *The International Organization and Dissemination of Knowledge: Selected Essays of Paul Otlet,* edited by W. Boyd Rayward, 112–30. Amsterdam: Elsevier, 1990.

Uyttenhove, Pieter, and Sylvia van Pethegem. "Ferdinand van der Haeghen's Shadow on Otlet: European Resistance to the Americanized Modernism of the Office International de Bibliographie." In *European Modernism and the Information Society: Informing the Present, Understanding the Past,* edited by Warden B. Rayward, 89–104. Aldershot and Burlington, VT: Ashgate, 2008.

Vaughan, Diana. *The Challenger Launch Decision: Risky Technology, Culture, and Deviance at NASA.* Chicago: University of Chicago Press, 1996.

Völckers, Hortensia. "Foreword." In *A Utopia of Modernity: Zlín: Revisiting Bat'a's Functional City,* edited by Katrin Klingan, 6–7. Berlin: Jovis, 2009.

Volkov, Shulamit. *Walther Rathenau: The Life of Weimar's Fallen Statesman.* New Haven, CT: Yale University Press, 2012.

Vossoughian, Nader. "Mapping the Modern City: Otto Neurath, the International Congress of Modern Architecture (CIAM), and the Politics of Information Design." *Design Issues* 22, no. 3 (2006): 48–65.

———. *Otto Neurath: The Language of the Global Polis.* Rotterdam: NAI Publishers, 2008.

———. "The Modern Museum in the Age of its Mechanical Reproducibility: Otto Neurath and the Museum of Society and Economy in Vienna." In *European Modernism and the Information Society: Informing the Present, Understanding the Past,* edited by Warden B. Rayward, 241–45. Aldershot and Burlington, VT: Ashgate, 2008.

Waldinger, Fabian. "Peer Effects in Science—Evidence from the Dismissal of Scientists in Nazi Germany." *The Review of Economic Studies* 79 (2012): 838–61.

Waltman, Ludo, Robert J. W. Tijssen, and Nees J. van Eck, *Globalisation of Science in Kilometres*, 03/2011. Accessed September 10, 2012. http://arxiv.org/ftp/arxiv/papers/1103/1103.3648.pdf.

Wazeck, Milena. *Einsteins Gegner: Die öffentliche Kontroverse um die Relativitätstheorie in den 1920er Jahren*. Frankfurt a.M. and New York: Campus, 2009.

Weber, Eugen. *Peasants into Frenchmen: The Modernization of Rural France, 1870–1914*. Stanford, CA: Stanford University Press, 1976.

Weber, Nicholas F., and Le Corbusier, *Le Corbusier: A Life*. New York: Knopf, 2008.

Wegner, Phillip E. *Imaginary Communities: Utopia, the Nation, and the Spatial Histories of Modernity*. Berkeley, CA: University of California Press, 2002.

Weightman, Gavin. *Signor Marconi's Magic Box: The Most Remarkable Invention of the 19th Century & the Amateur Inventor Whose Genius Sparked a Revolution*. Cambridge: Da Capo Press, 2003.

Weilemann, Peter. *Die Anfänge der Europäischen Atomgemeinschaft: Zur Gründungsgeschichte von EURATOM 1955–1957*. Baden-Baden: Nomos, 1983.

Weindling, Paul, ed. *International Health Organisations and Movements: 1918–1939*. Cambridge: Cambridge University Press, 1995.

———. "Public Health and Political Stabilisation: The Rockefeller Foundation in Central and Eastern Europe between the Two World Wars." *Minerva* 31 (1993): 253–67.

———. *Epidemics and Genocide in Eastern Europe 1890–1945*. Oxford: Oxford University Press, 2000.

Weiss, John H. *The Making of Technological Man: The Social Origins of French Engineering Education*. Cambridge, MA: MIT Press, 1982.

Weitz, Eric D. "Weimar Germany and Its Histories." *Central European History* 43 (2010): 581–91.

Welck, Stephan von, and Renate Platzöder, eds. *Weltraumrecht—Law of Outer Space*. Baden-Baden: Nomos, 1987.

Welter, Volker. *Biopolis: Patrick Geddes and the City of Life*. Cambridge, MA: MIT Press, 2002.

Wieland, Thomas. "Autarky and Lebensraum: The Political Agenda of Academic Plant Breeding in Nazi Germany," *HoST* 3 (2009). Accessed May 12, 2011. http://johost.eu/vol3_fall_2009/vol3_tw.htm.

Wijdeveld, Hendricus T., Erich Mendelsohn, and Amédée Ozenfant, eds. *Académie Européene "Méditerranée."* Amsterdam: n.p., 1933.

Wikipedia. "Arianespace." Accessed March 30, 2011. http://en.wikipedia.org/wiki/Arianespace.

Wikipedia. "Intercosmos." Accessed March 22, 2011. http://en.wikipedia.org/wiki/Intercosmos.

Wilkins, Mira. *American Business Abroad: Ford on Six Continents*. Cambridge and New York: Cambridge University Press, 2011.

Willson, Denis. *A European Experiment: The Launching of the JET Project*. Bristol: Adam Hilger, 1981.

Wildt, Michael. *Generation des Unbedingten: Das Führungskorps des Reichssicherheitshauptamtes*. Hamburg: Hamburger Edition, 2002.

Wirsching, Andreas. *Der Preis der Freiheit: Geschichte Europas in unserer Zeit*. Munich: Beck, 2012.

Wittrock, Björn. "The Transformation of European Universities: Disciplines and Professions in England, Germany and Russia." *Contemporary European History* 13 (2004): 101–16.

Wolff, Stefan. "Das Jahr 1933: Vertreibung und Emigration in der Physik." *Physik in unserer Zeit* 24 (1993): 267–73.

———. "Physicists in the 'Krieg der Geister': Wilhelm Wien's 'Proclamation'." *Historical Studies in the Physical and Biological Sciences* 33, no. 2 (2003): 337–68.

Wormbs, Nina. *Vem älskade Tele-X? Konflikter om satelliter i Norden 1974–1989*. Hedemora: Gidlunds förlag, 2003.

———. "A Nordic Satellite Project Understood as a Trans-National Effort." *History and Technology* 22 (2006): 257–75.

Yanni, Carla. "The Crystal Palace: A Legacy in Science." In *Die Weltausstellung von 1851 und ihre Folgen: The Great Exhibition and Its Legacy*, edited by Franz Bosbach and John R. Davis, 119–26. Munich: Saur, 2002.

Zabusky, Stacia E. *Launching Europe: An Ethnography of European Cooperation in Space Science*. Princeton: Princeton University Press, 1995.

Żarnowski, Janusz. "Learned Professions in Poland 1918–1939." In *State, Society and Intelligentsia: Modern Poland and Its Regional Context*, edited by Janusz Żarnowski, 407–26. Aldershot: Ashgate, 2003.

Zellmeyer, Stephan. A *Place in Space: The History of Swiss Participation in European Space Programmes, 1960–1987*. Paris: Beauchesne, 2007.

Zickler, Achim. "Weltraumforschung in der DDR." In *Ein Jahrhundert im Flug: Luft- und Raumfahrtforschung in Deutschland 1907–2007*, edited by Helmuth Trischler and Kai-Uwe Schrogl, 460–78. Frankfurt a.M. and New York: Campus, 2007.

Zimmerer, Jürgen. "The Birth of the Ostland out of the Spirit of Colonialism: A Postcolonial Perspective on the Nazi Policy of Conquest and Extermination." *Patterns of Prejudice* 39 (2005): 197–219.

———. *Von Windhuk nach Auschwitz? Beiträge zum Verhältnis von Kolonialismus und Holocaust*. Münster: LIT, 2011.

Zysman, John, and Michael Borrus. "From Failure to Fortune? European Electronics in a Changing World Economy." *The Annals of the American Academy of Political and Social Science* 531 (1994): 141–67.

Illustration Credits

Permissions to reproduce the illustrations and photographs in this book have generously been granted by the institutions and collections named here, and are gratefully acknowledged by the editors, the authors, the Foundation for the History of Technology, and Palgrave Macmillan.

The Foundation for the History of Technology has carefully tried to locate all rights holders. Parties who despite this feel that they are entitled to certain rights are requested to contact the Foundation for the History of Technology (www.histech.nl).

Cover Jan Antonín Baťa, Vladimír Karfík, and František Lydie Gahura in Zlín, 1935. Courtesy of State District Archive Zlín (Czech Republic). Photo collection of the Bata-Svit funds.

0.1 Original caption: "All the world going to see the Great Exhibition of 1851," illustration by George Cruikshank (1792–1878) in Henry Mayhew (1812–1887), *The world's show, 1851, or, The adventures of Mr. and Mrs. Sandboys and family who came up to London to "enjoy themselves" and to see the Great Exhibition* (London: David Bogue, 1851). 2

0.2 Chromolithograph "The Great Exhibition, North Germany Exhibit," 1852, by Joseph Nash (1809–1878). Permission by Museum of London. Image no. 002576. 5

0.3 A view of the construction of the Petrin tower, 1891. Permission by the City of Prague, Department of the Archives of the City of Prague. Reference no. II 57. 7

0.4 Photo with Friedrich Tamms and representatives of the city of Düsseldorf with plans of the Neue Messe, 1969. Courtesy of Friedrich Wolters, Architekt BDA, Wolters & Partner, Essen. 11

1.1 Count Bogdan Hutten-Czapski is honored with a doctor honoris causa at Warsaw's University of Technology, 1931. Permission by the Narodowe Archiwum Cyfrowe, Warzawa. Ref. no. I-N3382; NNC. 23

1.2 Original caption: "Doppeldecker in der Luft," 1910. Photo by Otto Haeckel. Permission by Bundesarchiv. Bild 146–1972–026–35. 29

1.3 Original caption: "Centenaire de 'l Ecole Polytechique 1794–1894." Illustration by Heliog Dujardin, 1895. Courtesy Collections Ecole Polytechnique (Palaiseau). 32

1.4 The human cantilever, photograph 1887. Permission by Institution of Civil Engineers, London. Ref. no. 2578201. 35

1.5 Know-how for the Continent: Map drawn by Foundation for the History of Technology based on Ortwin Pelc and Susanne Grötz, *Konstrukteur der Modernen Stadt: William Lindley in Hamburg und Europa, 1808–1900* (Hamburg: Dölling und Galitz Verlag, 2008), 318–24 (Anhang "Verzeichnis der Ingenieurprojekte der Familie Lindley im 19. und 20. Jahrhundert"). Permission by Foundation for the History of Technology, Eindhoven. 41

1.6 Engineers Abundant, graph by Foundation for the History of Technology based on Mauro F. Guillén, *The Taylorized Beauty of the Mechanical: Scientific Management and the Rise of Modernist Architecture* (Princeton: Princeton

University Press, 2006), 122. Permission by Foundation for the History of Technology, Eindhoven. 43
1.7 Original caption: "Fräulein Dipl. Ing.," 1913. Photo by Stauch, *Berliner Illustrirte Zeitung*, August 3, 1913. Courtesy Landesarchiv Berlin. 49
1.8 "Instituto Superior Técnico, Lisboa Portugal," Fotografia sem data. Produzida durante a actividade do Estúdio Mário Novais: 1933–1983. CFT003 122950.ic. Courtesy of Biblioteca de Arte-Fundação Calouste Gulbenkian. Reference www.flickr.com/photos/biblarte/3022459398/ 52
2.1 Original caption: "View of the interior of the Industrial Palace at the Exhibition Works in the Jubilee Exhibition,"1891. Photo by Eckert Jindřich. Permission by the City of Prague, Department of the Archives of the City of Prague. Ref. no. VIII 1029. 56
2.2 *Exhibiting Progress* (1916) painting by Georg Waltenberger (1865–1961). In collection Deutsches Museum, Munich. Permission by Picture Archive Deutsches Museum. 61
2.3 Original caption: "Visitado do Presidente da República Óscar Camona," 1936. Photo by Laboratório Fotográfico. Courtesy of Bibliotheca Nacional de Portugal. 64
2.4 Main source of graph: Kees Gispen, *New Profession, Old Order: Engineers and German Society, 1815–1914* (Cambridge and New York: Cambridge University Press, 1989), 172. Permission by Foundation for the History of Technology, Eindhoven. 67
2.5 Original caption: "Festplatsen i kvällsbelysning. Ljusreklam. Stockholmsutställningen," 1930. Permission by Nordiska Museet, Stockholm. Topografiska arkivet, *Album Stockholmsutställningen* 1930. Ref.no. N1_01881.TIF. 68
2.6 Funeral procession of Cornelis Lely (1854–1929) on January 26, 1929 in The Hague. Courtesy of Foundation for the History of Technology, Eindhoven. 75
2.7 "XVIII bal- 5 luty 38 - Warsawa swej Politechnice," poster by S. Malewicz. Reproduced by courtesy of Library of Congress, Prints and Photographic Division. Reproduction number LC-USZC4-7114. 77
3.1 Original caption: "Musee Mondial,"plan by Le Corbusier. Collection of Fondation Le Corbusier, © FLC/DACS, 2013. Permission by Fondation Le Corbusier via Pictoright. Ref.no. Plan PLC24510. 80
3.2 Original caption: "Documentation et Télé-Communication," drawing by P. Otlet (1868–1944). Permission by Centre d'Archives Wallonie-Bruxelles & Espace déxpositions Tempoiraires, Mundaneum collection. Ref.no. ARC-MUNDA-EU MC132-600. 83
3.3 Original caption: "An international world centre. Perspective generale," 1912. Drawing by Hendrik Christian Andersen (1872–1940) and Ernest M. Hébrard (1875–1933). Permission by Centre d'Archives Wallonie-Bruxelles & Espace déxpositions Tempoiraires, Mundaneum collection. 86
3.4 German excavation in Leuven, 1917. Published in Chris Coppens, Mark Derez, and Jan Rogiers, *Leuven University Library, 1425–2000* (Leuven: Leuven University Press, 2005). Permission by Leuven University Library. 88
3.5 Staff of the International Institute of Bibliography. Permission by Centre d'Archives Wallonie-Bruxelles & Espace déxpositions Tempoiraires, Mundaneum collection. Ref.no. PV-RBU1900. 93

| Illustration Credits | 371

3.6	Original caption: "Isotype 1940 greetings card," 1939. Permission by Otto and Marie Neurath Isotype Collection, Department of Typography & Graphic Communication, University of Reading.	99
3.7	Original caption: Both man and machine use plant products containing carbon for fuel (4), burning them through the addition of air (3), utilizing the heat obtained in this way and exhaling the carbon dioxide produced (1). The heat is distributed by a fluid circulating in pipes (2) and is utilized to move levers and joints (6). The non-combustible residues are expelled as ash or feces (5)."; Fritz Kahn: *Der Mensch gesund und krank*, Vol. II (Rüschlikon-Zürich: Albert Müller, 1939), p. 2. In: Uta von Debschitz and Thilo von Debschitz, *Fritz Kahn* (Köln: Taschen, 2013), P. 122. Permission by Thilo von Debschitz.	106
3.8	Solvay Conference Brussels, 1911. ©Roger-Viollet. Permission by Hollandse Hoogte. Ref.no. RV-325949.	110
4.1	Original caption: "Participants à bord du Patris II lors du CIAM IV à Athènes," 1933. ©FLC/DACS, 2013. Permission by Fondation Le Corbusier via Pictoright. Ref.no. L4(7)20.	116
4.2	Global Expertise: Published in: Eric Mumford, *The CIAM Discourse on Urbanism, 1928–1960* (Cambridge, MA: MIT Press, 2000), 109: Credits Van Eesteren Archives, Nederlands Architectuur Instituut Rotterdam.	118
4.3	Original caption: "CIAM 'battle plan' drawn by Le Corbusier for La Sarraz meeting," 1928. Permission by ETH – Swiss Federal Institute of Technology, Zürich – Institut für Geschichte und Theorie der Architektur (GTA). Ref.no. 42_1_F_13a.	121
4.4	A+B: Letter + attachement from CIAM to Stalin, May 4, 1932 in archival collection C. van Eesteren. Permission by Nederlands Architectuurinstituut Rotterdam / Collectie Van Eesteren-Fluck en Van Lohuizen Stichting, Den Haag.	122–23
4.5	Tomášovici. Courtesy of State District Archive Zlín (Czech Republic). Photo collection of the Bata-Svit funds. Ref.no. Photoarchives of The District Authorities_7352.	128
4.6	Baťa memorial. Courtesy of State District Archive Zlín (Czech Republic). Photo collection of the Bata-Svit funds. Ref.no. Photoarchives of The District Authorities_5412.	131
4.7	Cover Rudolf Philipp, *Stiefel der Diktatur* (Zürich: Resoverlag, 1936). Courtesy of State District Archive Zlín (Czech Republic). Photo collection of the Bata-Svit funds.	133
4.8	Original caption: "Der amerikanische Automobilkönig Henry Ford bei den Passionsspielen in Oberammergau!," 1930. Photo by Georg Pahl. Permission by Bundesarchiv. Bild 102-10478.	135
5.1	Original caption: "Kreis Kutno. Bestehende und künftige Besiedlung," 1941. Found in exhibition "Wissenschaft, Planung, Vertreibung. Der Generalplan Ost der Nationalsozialisten," 2006. Originally published in Udo von Schauroth, "Raumordnungsskizzen und ländliche Planung," in: *Neues Bauerntum* 33 (1941), Nr. 3, S. 125. Published by Deutsche Landbuchhandlung Berlin. Courtesy Deutsche Forschungsgemeinschaft.	150

5.2 Original caption: "Prozess RSHA SS – Angeklagte zu Beginn des letzten Verhandlungstages," 1948. Permission by Ullstein Bild DPA. Ref.no. 00022718. 154

5.3 Original caption: "Inventor Guglielmo Marconi (r), wearing uniform & decorations, standing next to Italian dictator Benito Mussolini (2nd from l)," 1937. Photo by Time Life Pictures. Permission by Time & Life Pictures / Getty Images. Ref.no. 50711164. 155

5.4 Original caption: "Guglielmo Marconi, radio inventor,"1914. Photo published by Alphonse Berget (1860–1934) in 1914. Permission by Science Photo Library. Ref.no. C008/3508. 156

5.5 Original caption: "In his element," 1920. Photo by Topical Press Agency. Permission by Getty Images. Ref.no. 3359375. 158

5.6 Original caption: "Lysenko measures wheat," photo taken in Odessa (Ukraine). ©Hulton-Deutsch Collection/CORBIS. Permission by Hollandse Hoogte. Ref.no. CB-HU044190. 162

5.7 Original caption: "Composite of three photos in Russia, T.D. Lysenko holding up onions in a meeting; crowd of people at meeting in Moscow, and group of people inspecting grain in field," 1949. Photo by Photovistavka Photo. Reproduced by courtesy of Library of Congress, Prints and Photographic Division. Reproduction number LC-USZC62-98686. 165

5.8 Orginal caption: "Entwurf für die Ordensburg Vogelsang in der Eifel, Innenhof des Hauptkomplexes," 1949. Photograph by Arthur Grimm. Permission by BPK – Bildagentur für Kunst, Kultur und Geschichte. Ref.no. 30024078. 174

6.1 Original caption: "Leipziger Frühjahrsmesse 1958, Luftfahrtexperten der sozialistischen Ländern trafen sich in Leipzig," 1958. Photo by Ulmer/Schmidt. Permission by Bundesarchiv. Ref.no. 183-53500-185. 180

6.2 Original caption: "Werner von Braun Rocket Team at Fort Bliss," 1946. Courtesy of NASA Media Services. 183

6.3 A: Original caption: "Die einzige Frau Deutschlands mit dem E.K. besucht ihre Heimatstadt," 1941. Photographer Ernst Schwahn. Permission by Bundesarchiv. Ref.no. 183-B02092. 184

6.3 B: Original caption: "Reitsch Hanna (29-3-1912 24-8-1979) Pilotin, Fliegerin mit dem indischen Ministerpraesidenten Nehru vor einem Rundflug im Segelflugzeug," 1959. Permission by Ullstein Bild. Ref.no. 00162233. 185

6.4 Original caption: "Franckfeier 1923, die Bonzen," 1923. Photo from estate of Friedrich Hund in collection of Gerard Hund. Courtesy Gerard Hund, available in Wikimedia Commons (http://commons.wikimedia.org/wiki/File:Franckfeier_1923_Die_Bonzen.jpg). 189

6.5 Original caption: "American engineer Hugh Cooper (left) advising a Soviet engineer building the Dnieper hydro-electric power plant," 1934. Permission by RIA Novosti. Ref.no.34826. 197

6.6 Cover *Das Neue Frankfurt*, September 1930. Courtesy of Nederlands Architectuurinstituut Rotterdam. 199

6.7 Original caption: "Ernst und Hanna Reuter bei der Rückreise nach Deutschland," 1946. Permission by Landesarchiv Berlin, Nachlass Reuter. 201

7.1 Original caption: "The LHC (large hadron collider) in its tunnel at CERN (the European particle physics laboratory) near Geneva,

	Switzerland." Photo by David Parker. Permission by Science Photo Library. Ref.no. A105/0235.	207
7.2	Installation of Nikolay Polissky, displayed in the Grand Hall of Mudam Luxembourg – Musée d'Art Moderne, 2009. By kind permission of Nikolay Polissky.	208
7.3	Original caption: "People checking out the US display at the Atoms for Peace Conference," 1955. Photo by Carl Mydans. Permission by Time & Life Pictures/Getty Images.	219
7.4	Original caption: "Igor V. Kurchatov (in the middle, with beard) during his visit at AERE Harwell, 25th April 1956. On his right is Nikita S. Khrushchev, to his left is Nikolai A. Bulganin. Opposite is Sir John D. Cockcroft, Director of AERE Harwell," 1956. Permission by EDFA.	220
7.5	Original caption: "The first thing that strikes the eye of a visitor to the 1958 World's Fair is the 320 foot high Atomium," 1958. ©Bettmann/Corbis. Permission by Hollandse Hoogte. Ref.no. BE072146.	222
7.6	Original caption: Atomreaktoren JEEP I. Tre fysikere jobber med en reaktorblokk, 1949–1951. Permission by Norsk Teknisk Museum, Oslo. Ref.no. NTM 2012/5–1.	226
7.7	Original caption: "Blick auf das Gelände des in der Bauphase abgebrochenen KKW/AKW Kalkar in Nordrhein-Westfalen," 2001. Photo AKG / euroluftbild. Permission by AKG.	231
7.8	Original caption: "Chairman of the JET Council Jacques Teillac (centre) shows Her Majesty the Queen and the French President, Francois Mitterand around Jet at its official opening on 9th April 1984." 1984. Permission by EDFA. Ref.no. CP84-681-04.	236
7.9	Original caption: "President Ronald Reagan and Soviet General Secretary Gorbachev meet in the boathouse during the Geneva Summit in Switzerland," 1985. Courtesy Ronald Reagan Library.	241
8.1	Carl-Zeiss MKF-6M Camera, 1978. In collection Deutsches Museum, Munich. Permission by Deutsches Museum. Ref. no. DMA LRD 06834	246
8.2	Original caption: "Dominating the approach to the main exhibition hall at the Farnborough Air Show is the Blue Streak," 1961. Permission by Mirropix. Ref.no. 00050957.	251
8.3	Europa 1 Rakete, 1971. In collection Deutsches Museum, Munich. Permission by Deutsches Museum. Ref.no. DMA, CD 272.	255
8.4	Stamp Deutsche Bundespost "Nachrichtensatellit," 1975. Designed by Beat Knoblauch. Courtesy Archiv für Philatelie. Museumsstiftung Post und Telekommunikation, Bonn.	258
8.5	Original caption: "The European Astronaut Centre in Cologne Germany." ESA – D. Baumbach, 2010. Permission by ESA.	262
8.6	Hotol impression. Courtesy BAE Systems.	269
8.7	Original caption: "View of the first Ariane 4 rocket during roll-out," 1988. Photo by David Parker, ESA/Science Photo Library. Permission by Science Photo Library. Ref.no. S220/0020.	270
8.8	Original caption: "Rosetta probe test," 2003/2004. Photo European Space Agency/Science Photo Library. Permission by Science Photo Library. Ref.no. R250/0156.	274
9.1	Orginal caption: "European Higher Education Area," 2011 by Inc Ru. Permission by Wikimedia Commons; reference http://commons.wikimedia.org/wiki/File%3ABolonga_zone.png.	279

9.2	Cover *L'Express* November 9–15, 2006, photo by Christian Taillandier /*L'Express*. ©*L'Express* no 2888/Novembre 2006. Permission through Virginie Le Trionnaire, Droits de Reproduction, Paris.	282
9.3	Original caption: "EU research: changing priorities," April 2005. Presentation European Commission, Research DG *Building Europe Knowledge. Towards the Seventh Framework Programme 2007–2013*. Page 58 of presentation. http://www.eurosfaire.prd.fr/bibliotheque/pdf/FP7_Complete_presentation_April_2005.pdf. Redrawn graph, permission by Foundation for the History of Technology, Eindhoven.	285
9.4	Redrawn graph based on Wolfgang Glänzel, A. Schubert, and H.J. Czerwon, "A Bibliometric Analysis of International Scientific Cooperation of the European Union (1985–1995),"*Scientometrics*, 45, no. 2 (1999): 185–202; here 189.	295
9.5	Logo Academia Europea. Permission by Academia Europea (www.acadeuro.org).	296
10.1	Prince Philip opening Department of Engineering of the University of Cambridge, 1952. Painting by Terence Cuneo (1907–1996). Permission by Bridgeman Art Library Ltd – Estate of Terence Cuneo.	309

Making Europe: Series Acknowledgements

Making Europe is the result/product of an unusual collaboration among a host of individuals and organizations. The *Making Europe* authors and series editors feel extremely fortunate to be working with them. We list here individuals and organizations who contributed to the entire series. Each volume in the series also has its own separate acknowledgements.

Making Europe was initiated by:

- Foundation for the History of Technology (www.histech.nl)

Making Europe is sponsored by:

- Eindhoven University of Technology (www.tue.nl)
- SNS REAAL Fonds (www.snsreaalfonds.nl)
- Next Generations Infrastructures (www.nextgenerationsinfrastructures.eu)
- Foundation for the History of Technology Corporate Program (www.histech.nl) that includes:
  - DSM (www.dsm.com)
  - EBN (www.ebn.nl)
  - FrieslandCampina(www.frieslandcampina.com)
  - Philips (www.philips.nl)
  - SIDN (www.sidn.nl)
  - TNO (www.tno.nl)

Making Europe has been made possible thanks to:

- Tensions of Europe Network (www.tensionsofeurope.eu)
- European Science Foundation EUROCORES Programme Inventing Europe – Technology and the Making of Europe, 1850 to the Present (www.esf.org)
- Research Theme Group Grant of the Netherlands Institute of Advanced Studies (NIAS) in 2010–11 (www.nias.nl)
- European University Institute in Florence, Italy for providing the support in developing the series and for the founding workshop (3–6 July 2008) (www.eui.eu)

Making Europe benefited from the feedback of a community of scholars who have been involved in the series from the start:

Håkon With Andersen, Alec Badenoch, Robert Bud, David Burigana, Cornelis Disco, Paul Edwards, Valentina Fava, Karen Johnson Freeze, Andrea Guintini, Gabrielle Hecht, Rüdiger Klein, Eda Kranakis, John Krige, Leonard Laborie, Vincent Lagendijk, Suzanne Lommers, Slawomir Lotysz, Dagmara Jaješniak-Quast, Karl-Erik Michelsen, Matthias Middell, Thomas J. Misa, Dobrinka Parusheva, Kiran Patel, Pierre-Yves Saunier, Emanuela Scarpellini, Frank Schipper, Michael Strang, Ivan Tchalakov, Frank Trentmann, Aristotle Tympas, Hans Weinberger

Making Europe relied on the unflagging support of:

Picture editors
- Katherine Kay-Mouat
- Giel van Hooff
- Jan Korsten (Management)
- Camiel Lintsen – Kade 05 Eindhoven (graphs and maps)

Text editors
- Lisa Friedman
- James Morrison

PalgraveMacmillan
- Jenny McCall (Publisher)
- Holly Tyler (Editorial Assistant)
- Philip Hillyer (Copy-Editor and Editorial Services Consultant)
- Susan Boobis (Indexer)

Office Foundation for the History of Technology:
- Sonja Beekers (Secretarial Support)
- Jan Korsten (Business Director)
- Loek Stoks (Bookkeeping)
- Henk Treur (Volunteer)

Board Foundation for the History of Technology:
- Hans de Wit (chair)
- Jacques Joosten (treasurer)
- Saskia Blom
- Dirk van Delft
- Herman de Boon
- Eric Fischer
- Frans Greidanus
- Emmo Meijer
- Wim van Gelder
- Michiel Westermann
- Harry Lintsen (advisor)
- Martin Schuurmans (advisor)

Index

academia 25, 47, 103, 161, 170, 188, 190, 278, 292, 293, 295, 296 Fig. 9.5
    see also higher education
Academia Europaea 295, 296 Fig. 9.5
academic associations 147, 295
academic titles 11, 40, 45, 48, 61
actors 5, 6–7, 8, 12, 13, 14, 17, 89, 166, 209, 216, 221, 228–29, 259–61, 271, 273, 283, 289, 302
Adenauer, Konrad 211, 225, 256
administration 22, 24, 28, 29, 34, 43 Fig. 1.6, 48, 71–73, 145, 147, 164, 177, 193, 240, 259, 283, 302
"Age of Extremes" 169–73, 174 Fig. 5.8, 175
Agnelli, Giovanni 134
Agol, Israel 167
agriculture 152–53, 161–69
agrobiology 166
airplanes 29 Fig. 12, 127, 128, 131 Fig. 4.6, 179, 180–84, 301
airships 63
Alsos Mission 181
Amaldi, Edoardo 250, 271
Americanization 137, 182, 183 Fig. 6.2, 281
ancien régime 30, 59
Andersen, Hendrik Christian 93, 95
André, Michel 281–83
annihilation 147–54
anti-science movement 147
anti-Semitism 109, 135, 158–59, 188
applied learning 42
appropriation of technology 10, 187, 229
architects 13, 47, 170, 191–94, 196
    of annihilation 15
    modernist 13, 40, 47, 116 Fig. 4.1, 120, 121 Fig. 4.3, 128–29, 137, 192, 194, 305
    see also engineers
architecture of ideas 92–93
aristocracy 30, 55, 62, 299
    see also nobility
Arkan, Seyfi 194
Armstrong, Neil 244
Arndt, Fritz 190

Arntz, Gerd 98
artifacts 2, 7, 8, 66, 81, 105, 206, 208 Fig. 7.2, 232, 274 Fig. 8.8, 306
arts 10, 90, 94
associations 10, 80, 84, 89, 101, 109, 146, 151, 226, 248
    academic 147, 295
    engineers 42, 43, 57, 60, 73
    professional 43
Atatürk, Mustafa Kemal 188–95
Atlas de la civilisation universelle 101
Auger, Pierre 210, 250
Austria 221
    museums 100
authoritarianism 15, 16, 26, 60, 64, 96, 111, 118, 120, 125, 139, 141, 170, 172–74, 174 Fig. 5.8, 194, 304
automobile industry 28, 132–34, 135 Fig. 4.8, 136–39
autonomy 34, 52, 253, 260, 273
    of experts 70, 119, 200, 201
    institutional 145
    political 55
    societal 174

Bade, Bruno 180 Fig. 6.1
Baker, Benjamin 35 Fig. 1.4
Bakewell, Frederick C. 2
Bannier, J.H. 254
Barroso, José Manuel 280–81
Bartók, Béla 89
Bašus, Albín 73
Baťa, Jan A. 128–29, 131 Fig. 4.6, 136, 306–7
Baťa, Tomáš 125–29, 130, 133 Fig. 4.7, 306–7
Bataism 128 Fig. 4.5, 129–30, 131 Fig. 4.6, 132, 135, 136–37
    and Nazi Germany 146, 304–5
Bauhaus 82, 136–37, 196
Bedaux, Charles 138–39, 172–73
Behrens, Peter 65–66, 105
Belgium 62, 87–88
    Office International de Bibliographie 8, 91
    Orbis Institute 101
    Palais de Cinquantenaire 145–46

377

Belgium – *continued*
　Research Centre for the Applications of Nuclear Energy 221, 222 Fig. 7.5
Belling, Rudolf 194
Bergson, Henri 89
Berlage, Heinrich Pertus 90
Berners-Lee, Tim 209
Besse, Georges 68
Best, Werner 145
Bethe, Hans 198
Beyen, Johan Willem 224
Big Science 64, 209, 214, 216
Bismark, Otto von 151
Bloch, Felix 198
Bogdanov, Alexander 138
Bohemia 55–57
　see also Czechoslovakia; Prague
Bolshevik revolution 144, 162
Bonapartism 31
Bonatz, Paul 192, 194
Born, Max 188, 189 Fig. 6.4
Borsig, August 40
botany 10
Bothe, Walther 212
bottom-up processes 288, 297, 307
Bourgois, Victor 96
brain-drain 22
Brauer, Alfred T. 198
Braun, Wernher von 170, 182, 186–87, 201
Brazil 308
Britain 1, 28, 31, 39, 42, 43, 45, 49–50, 60, 62, 67, 68–69, 134, 137, 138, 157, 159, 181, 184, 193, 200, 210, 220, 224, 225, 230, 232, 234–37, 241, 250, 252, 255 Fig. 8.3, 257, 261, 263, 265, 269 Fig. 8.6
　Association of Civil Engineers 43
　British Atomic Energy Research Establishment 226
　Great Exhibition 1–3, 2 Fig. 0.1, 3, 9, 300
　　museums 3
　　technical education 33–34, 42
　University of Cambridge, 309 Fig. 10.1
Brno 76, 125
Brown, Gordon 310
Bührer, Karl Wilhelm 105
Bulganin, Nikolai 220 Fig. 7.4, 238
Bureau International des Poids et Mesures 4
Busquin, Philippe 282

canals 64 Fig. 2.3, 66, 69
Caracostas, Paraskevas 287–88
Carnap, Rudolf 98
Castells, Manuel 278
catalogues 91–92
Central Office of International Associations 89
Central Place Theory 150 Fig. 5.1
CERN 7, 16, 205–6, 207 Fig. 7.1, 208 Fig. 7.2, 209–16, 234–35, 239, 250, 310
Charta of Athens 119
chemistry 13, 69–70, 102–7
Chernobyl 226, 227
Chertok, Boris E. 179
Chmielewski, Jan Olaf 124
Chrétien, Jean-Loup 244
CIAM 115, 116 Fig. 4.1, 117, 118 Fig. 4.2, 120, 121 Fig. 4.3, 122 Fig. 4.4, 123 Fig. 4.4, 124, 141, 146, 303–4
CIAM-Ost 125
circulation
　educational models 32–39
　of expert knowledge 15, 175, 181, 184 Fig. 6.3, 187, 191, 198, 301
　transnational xi, 17, 33
　see also knowledge; technology transfer
Citroën, André 134
civil society 24, 193, 229
civilian technology 169, 183, 211, 221, 223, 229
Clarke, Kenneth 264
class *see* social class
Cockroft, John 217, 220 Fig. 7.4, 225
Cold War 11, 13, 14, 177, 180 Fig. 6.1, 182–83, 184, 215, 240, 241 Fig. 7.9, 243, 271, 290, 306
collaboration
　bilateral 215, 224, 239, 240, 256, 257, 270
　cross-border 12, 301
　French with Nazi Germany 173
　global 217
　institutional 260, 265
　international 223–34, 294, 307
　political 250
　regional 306–7
　technoscientific 108–16, 301–2
　transatlantic 269
　transnational 16, 211, 213, 224–25, 239–41, 249

see also cooperation; and specific
   projects
colonialism 215, 270 Fig. 8.7
colonies 4, 39, 61, 240, 257
Colt, Samuel 3
*Commission préparatoire européenne
   de recherches spatiales*
   (COPERS) 250
communication technologies
   computers 248, 291, 292, 307
   media 12, 66, 81, 92, 99, 103, 109,
      132, 159, 160, 206, 249
   telegraph 3, 66
   telephone 132
   wireless 28, 63, 69
Communism 141, 170, 172, 305
   see also political movements
competition 16, 24, 39, 51, 80 Fig. 3.1,
      87, 94, 121, 196, 206, 215, 224,
      238, 239, 259, 281, 282, 301, 307,
      309, 310
computers 248, 291, 292, 307
Comte, Auguste 76, 90
*Conférence Générale des Poids et
   Mesures* 4
conflict *see* controversy; war
*Congrès internationaux d'architecture
   moderne see* CIAM
congresses 3, 69, 91, 102, 103,
      116 Fig. 4.1, 117, 119–21,
      122–23 Fig. 4.4, 125, 146, 157,
      165, 167, 300
*Conseil Européenne pour la recherche
   nucléaire see* CERN
Conservatives 28, 90, 194
Constations 119
construction 29
   airplane 29 Fig. 12, 127, 128,
      131 Fig. 4.6., 179, 184, 301
   building 2, 22, 23 Fig. 1.1, 34, 60,
      81–83
   naval 30, 38, 68–69
consumers 138
continentalization 289, 307
controversy 17, 63, 86, 129, 206, 237,
      260, 264, 273
Cook, Walter S. 188
Cooper, Hugh 197 Fig. 6.5
cooperation *see* collaboration
Cooperation in Science and
   Technology (COST) 283
Courant, Richard 198

courts of honor 45
Coutrot, Jean 172
Couturat, Louis 102
Cresson, Edith 287–88
Crimean War 28
crisis 70–77, 139, 163, 191, 228, 240
   economic 17, 46, 85, 125, 127, 140, 152,
      169, 195, 199 Fig. 6.6, 209, 254, 308
   of identity 134
   political 85, 98, 132–33, 136, 138
Crystal Palace *see* Great Exhibition
cultural decapitation 188
Curie, Marie Sklodowska 69, 89, 105
Curien, Hubert 262
Czechoslovakia 76, 125–26
   Bataism 125–30, 128 Fig. 4.5, 129–30,
      131 Fig. 4.6, 132, 135–36
   see also Bohemia; Prague

Dahrendorf, Ralf 281–82
d'Alembert, Jean-Baptiste le Rond 10
dams 6, 63, 67
Dautry, Raoul 209, 210
Davy, Humphrey 58
de Gaulle, Charles 220, 225, 256
degrees, academic 40, 44–46, 51, 62, 74
Delors, Jacques 267–68
Delville, Raphaël 96
demobilization 76
democratization 5
denationalization 200
Denmark 290
depression 136
deprofessionalization 44
design and designers 8, 15–17, 34, 40,
      47, 80 Fig. 3.1, 101, 107, 170, 177–84,
      191, 196, 219–20, 237, 245, 248–50
   see also Mundaneum; space race
Diderot, Denis 10
Die Brücke 105, 109
diplomas 30, 40, 45, 60
disasters
   political 260
   technical 157, 226, 248, 260
Dnieper Dam 6
documentation 9, 84, 90–92, 97, 102,
   105, 108, 109
Dornberger, Walter 184
Dosi, Giovanni 278
Dubreuil, Hyacinthe 130
Duke of Edinburgh (Prince Philip), 309
   Fig. 10.1

Durkheim, Émile 90
Düsseldorf, *Neue Messe* 11 Fig. 0.4
dynasty/dynastic 22, 34, 57, 58, 63, 88

East Germany 215, 245
Eastern Europe 290, 293
　emigration from 177–201
　see also individual countries
economic
　advantages 3, 147
　change 120
　competitiveness 271, 284
　constraints 9
　crisis 17, 46, 85, 100, 125, 127, 133, 135 Fig. 4.8, 138, 140, 169, 195, 199 Fig. 6.6, 307, 308
　development 192
　exchange 28, 103, 301
　goals 91
　growth 277, 284, 287, 291, 292, 309
　innovation 1
　integration 187, 280, 283
　organizations 172
　planning 98, 172
　power 60
　privilege 201
　progress 52, 219
　reconstruction 16
　recovery 186
　uncertainties 233
economy, knowledge-based 10, 17, 26, 85, 96–97, 130, 277–78, 308
Eden, Anthony 238
education
　higher 279 Fig. 9.1
　humanistic 60
　models of 32–39
　scientific 21–54
　technical 28–32, 36–38, 51
　visual 97–102
　see also individual institutions; educational systems; training
educational systems 47–48, 52, 189
efficiency 140
Egli, Ernst 192
Eiffel, Gustave 68
Eiffel Tower 3, 7 Fig. 0.3, 68
Einstein, Albert 86, 89, 109, 146, 188, 233–34, 306
Eisenhower, Dwight D. 218–19, 238
electricity 28, 155, 220, 229

electrification 73
electronic data storage 92
elites/elitism 9, 23, 26–27, 46, 59–65, 170
　see also technical elite
Elsässer, Martin 192
emigration *see* migration; exile
émigré experts 177–201
　as agents of modernity 187–95
　as endangered species 195–97
Emilianov, Vasilii Semenovich 218
energetics 102, 108
Engelhard, Paul 29 Fig. 1.2
engineer consciousness 72
engineers 13, 28–29, 43 Fig. 1.6
　degree-holding 44–5
　female 47–48, 49 Fig. 1.7, 50
　as generalists 29, 71
　as "heralds of the nation" 65–70
　honours 62
　military 40
　naval 30, 38, 68–69, 232
　overproduction 42–43
　professional associations 42, 44, 57, 60, 73
　state 31
　status 39–50, 61–62
　training 22, 26, 27, 28–40, 42, 45, 47, 50, 60, 71, 73, 74, 132, 212, 215, 227
Enlightenment 10
entropy 102
environmentalism 277, 283, 304
　Green Movement 158, 212
Escola Politécnica 32–33
Esperanto 102
Estonia 293
eugenics 141, 170
Euratom 218–20, 222–25, 228, 233–34
Europe/European 6–7
　experience 301
　expert communities 287–97
　geographies 238–41
　history 9–10
　identity 10, 17, 134, 209, 278, 286, 301, 309
　integration 6, 13, 264, 265 Table 8.1, 266–75
　knowledge circulation 33, 38–39, 185, 187, 198, 308–10
　market 104, 134
　nationalism 58–59

Index   381

nation(s) 170, 188, 192, 251–52, 257, 262, 293, 302
nuclear technology 205–41
perspective 75
technoscientific collaboration 208–29
totalitarianism 143–75
transnationalism 282 Fig. 9.2
see also individual countries
European Atomic Energy Community *see* Euratom
European Atomic Energy Society (EAES) 225–26
European Commission 223, 284, 308
  Framework Programmes 284, 285 Fig. 9.3, 286, 287–89
European Company for the Chemical Processing of Irradiated Fuels (Eurochemic) 227
European Economic Community (EEC) 153, 218
European Fast Reactor 230–31
European Nuclear Energy Agency (ENEA) 226–27
European Parliament 267
European Research Area 278, 280–87
European Research Council (ERC) 284
European Space Agency (ESA) 260–61, 262 Fig. 8.5, 263–34
European Space Foundation 283
European Space Research Organization (ESRO) 250
European Space Research and Technology Centre (ESTEC) 274 Fig. 8.8
European Space Vehicle Launcher Development Organization (ELDO) 247–50, 251 Fig. 8.2, 252–54, 255 Fig. 8.3, 256–57, 258 Fig. 8.4, 259–60
European Strategic Program on Research in Information Technology (ESPRIT) 291
European Union (EU) x–xi, 17, 267–68, 277–80, 292, 293, 307–8, 309
Europeanization x, 6, 26, 229, 288–90, 301, 307
exhibitions 57, 75–76
  Czechoslovakian National Exhibition 76

General Land Centennial Exhibition 55, 56 Fig. 2.1, 57
Great Exhibition 1, 2 Fig. 0.1, 3, 9
Stockholm Exhibition 68 Fig. 2.5, 75–76
exile 177–201
see also emigration; migration
existential minimum 140
experts 5, 8, 9, 10–11, 14, 50–53, 55–77, 147–48, 159–75, 302
  communities 287–97
  culture(s) 10
  education 21–54
  émigrés *see* émigré experts
  global circulation of 15–16, 181, 184 Fig. 6.3, 187
  network(s) 226–27, 289–90
  self-empowerment 15, 139–41
  technical *see* technical experts
  see also professionals; specialists
expertise 12, 16, 115–41, 302, 310
  investing in 67 Fig. 2.4
  neutral 305
  politics of 12, 71
  technoscientific 11, 123–24, 292
expertise with a cause 138, 140
Exposition Universelle 3, 7 Fig. 0.3

factories 125–32
fair return (*juste retour*) 224, 249
Fascism 141, 159–60, 170, 172, 305
fast breeder reactors 230–34
Faure, Edgar 238
Faustian bargain 143–75
Fayol, Henri 136
Fayolism 136
Feldhaus, Franz 108
FIAT reports 180–81, 183, 186
Finland 25–26, 221, 293
Finniston, Harold Montague (Monty) 237
firearms 2–3
First World War 11–12, 21, 27, 46, 71–73, 85, 111, 139, 140, 169, 294, 300, 302, 303
Fontaine, Henri La 87, 88, 94
Forbat, Fred 117, 193
forced migration 197–201, 305
Fordism 132–34, 135 Fig. 4.8, 136–39
France 3, 7 Fig. 0.3, 290
  École Centrale 40–41

France – *continued*
  École des Mines 30
  École des Ponts et Chaussées 30
  École Polytechnique 24, 30–31, 40, 302
  École Supérieure d'Electricité 41
  École Supérieure d'Electrotechnique 42
  Grandes Écoles 33–34, 47, 48
  Institute Aérotechnique de Saint-Cyr 145
  military technology 29–32
  polytechniciens 31, 32 Fig. 1.3, 33, 41, 44, 51
  Revolution 30
  Société des Ingénieurs Civils 44
  technical education 24–32
  Vichy regime 173
Franck, James 188, 189 Fig. 6.4, 198
Freeman, Christopher 278
Freud, Sigmund 89
Functional City 115–16, 123, 136
  Detroit 132–34, 136–39
  Zlín 125–30
functionalism 135–36, 257, 273, 294, 304
future 3, 18, 22, 35, 46, 51, 55, 57–58, 61, 65, 69–71, 73, 76, 97, 106, 119, 124–25, 130, 161, 209, 222, 232, 261, 263–64, 277, 278–80, 307

Gahura, František 129
Galip, Resit 189–90
Gargarin, Yuri 244
Garnier, Tony 94
Geddes, Patrick 96–97, 100
Geiringer, Hilda 199
gender equality 9, 48, 308
General Land Centennial Exhibition 55, 56 Fig. 2.1, 57, 58
generalists 29, 71
genetics 161–69
genocide 148, 152, 160
  see also Holocaust
Gentner, Wolfgang 212–13
Gérard, Francis 223
German-Turkish miracle 190
Germanization 35, 149
Germany 67–68, 290
  collaboration with Spain 225
  Deutsches Museum von Meisterwerken der Naturwissenschaft und Technik 57–58
  elites 60
  engineers 44–45
  female 48
  industrialization 45, 60
  Jahrhundertausstellung 57
  Kaiser-Wilhelm-Gesellschaft 63
  Kriegswirtschaftsmuseum 97–98
  Mittelstand 42
  museums 97–98, 100, 137, 245
  Nazi *see* Nazi Germany
  Promotionsrecht 61
  Realschulen 60
  technical education 24, 34, 42
  Technisch-Wissenschaftliche Lehrmittelzentrale (TWL) 107
  Technische Hochschulen 24, 34, 46, 60–61
  totalitarianism 141
  unification 28
  Verband Deutscher Diplom-Ingenieure (VDDI) 45–46
  Verein Deutscher Ingenieure (VDI) 42, 60
  see also National Socialism; Nazi Germany
Giedion, Siegfried 110, 115, 117, 119–21
globalization x, 229, 289–90, 301, 307
  see also internationalization
Gödel, Kurt 198
Goethert, Bernhardt 182, 185
Gorbachev, Mikhail 238, 241 Fig. 7.9
Göring, Hermann 146
Gould, Stephen Jay 161
graduates 23, 25, 30, 31, 32 Fig. 1.3, 38, 42, 44–45, 60, 74, 138
Gramsci, Antonio 132
Grandes Écoles 33–64, 47, 48
Great Exhibition 1, 2 Fig. 0.1, 3, 9, 300
Great War *see* First World War
Greece 75
Green Movement 158, 212
Gropius, Walter 40, 115, 137, 194
Gröttrup, Helmut 178
Gumbel, Emil J. 198
Guterres, António 278

Haas, Ernst B. 273
Haber, Fritz 65
Hahn, Otto 181
Harnack, Adolf von 63–65

Hébrard, Ernest 93
Hegemann, Werner 136
hegemony 153, 214–15, 239, 243, 245
    American hegemony in Europe 211, 218, 220, 240, 252, 269
Heisenberg, Werner 181, 211–12, 225, 235
Hertz, Heinrich 156 Fig. 5.4
Hevesi, Gyula 72
Heydrich, Reinhard 146
Heymans, Maurice 96
hidden integration 17, 272, 307
higher education 279 Fig. 9.1, 293, 309
    see also academia
Himmler, Heinrich 148, 149, 150–51
Hitler, Adolf 15, 134
Hobsbawn, Eric ix
Holocaust 15, 148, 152
    see also genocide
Holzhäuser, Clemens 192
honours 62
Hoover, Herbert 86
Hoste, Huib 96
Hotol project 269 Fig. 8.6
housing 137, 140
Hulthén, Lamek 254
Hungary 72–73, 290, 293
hygiene 24, 69, 73, 79, 99, 100, 140, 141

idealism 300, 303
identity 216
    civic 35
    corporate 128, 131–32
    European 10, 17, 134, 209, 278, 286, 301, 309
    national 58, 213, 220, 244, 290
Ido (universal language) 102
immigration *see* migration
imperialism 57, 215
individuality 64, 70, 100
industrial revolution 28, 74
industrialization 5, 10, 28, 39, 44, 60, 107, 134, 300
information sharing 102, 209, 291–92
infrastructure 4, 17, 26, 28, 29, 34, 39, 61, 64 Fig. 2.3, 66–67, 130, 172, 177, 184, 210, 268, 289, 308
innovation 63, 130, 200, 209, 245, 278–79, 284, 290
    culture 181, 200, 291, 307
    economic 1
    incremental 186
    language 102–4
    military 16, 179
    social 107
    technical 5 Fig. 0.2, 9, 11, 13, 27, 38, 42, 60, 84, 140, 148, 156, 286, 288, 291, 293, 302
institutional autonomy 145
institutional collaboration 260, 265
institutions 7, 21–24, 51
    collaboration 260, 265
    see also individual institutions
Instytut Politechniczny im. Cara Mikołaja 22
integration
    economic 280, 283
    European 264, 265 Table 8.1, 266–75
    hidden 17, 272, 307
    political 16, 187, 267, 301, 306, 308, 310
    societal 284
intellectual data 93
intellectual reparations 177, 179, 180 Fig. 6.1, 181 Table. 6.1, 182, 183 Fig. 6.2, 184 Fig. 6.3, 186–87
Intelligenzaktion 144
International Atomic Energy Agency (IAEA) 237
International Committee of Intellectual Cooperation (ICIC) 89
International Conference of International Associations (1907) 88
International Conference on the Peaceful Uses of Atomic Energy 217, 219 Fig. 7.3
International Federation for Documentation (FID) 91
International Institute of Intellectual Cooperation (IICI) 89
International Labour Organisation (ILO) 82, 130
International Thermonuclear Experimental Reactor (ITER) 234, 238
internationalism 4, 218, 294
    cultural 90
    technocratic 228
    see also universalism

internationalization 104, 136, 257
  see also globalization
Internet 209
Intze, Otto 67–68
Ireland 293
Iron Curtain 218, 227, 290
isotype 14, 97–98, 99 Fig. 3.6, 100–2
Italy 69, 221, 290, 294

Jähn, Sigmund 245
Jansen, Hermann 192
Jassinski, Stanislas 96
Jenner, Edward 69
Jewish experts 12–13, 15, 145, 158–59, 188, 198
Johnson, Alvin 188
Joint Establishment for Nuclear Energy Research (Jener) 224, 226 Fig. 7.6, 239–40
Joint European Torus (JET) 236 Fig. 7.8, 237, 241
juste retour see fair return

Kahn, Fritz 100, 106 Fig. 3.7
Kalkar breeder reactor 229–30, 231 Fig. 7.7
Kamiński, Zygmunt 21, 25
Kharkiv Institute 144–45
Khrushchev, Nikita 217, 220 Fig. 7.4
Kisch, Egon Erwin 128, 130
know-how 45
knowledge 45
  architectures of 79–111
  creation 17, 33, 279, 280, 290, 307, 309
  deliberate destruction of 86
  European 277–97
  exchange 5 Fig. 0.2, 9, 10–11
  flow 247
  forced migration 197–201
  global circulation 15, 175, 181, 181 Table 6.1, 185 Fig. 6.3, 187, 191, 198, 301–2
  informal 39, 40
  orders of 87–97, 197–201
  organization 92, 93 Fig. 3.3
  production 292
  sharing x, 15–16
  standardization 102–7
knowledge societies 10–11, 14, 175, 181, 187, 216, 237, 254, 271, 280–87, 291, 307

knowledge tourism 126
knowledge-based economy 10, 17, 26, 85, 97, 130, 277–78, 308
Koch, Robert 69
Kourou spaceport 16, 247, 262, 267, 270 Fig. 8.7
Kriegswirtschaftsmuseum 97
Kurchatov, Igor V. 217, 220 Fig. 7.4
Kwiatkowski, Eugeniusz 74

laboratories 7, 13, 50, 52, 76, 140
Lamarck, Jean-Baptiste de 161
Landau, Lew 144
language 102–4, 109, 293, 295 Fig. 9.4
  universal 33, 103, 293
Large Hadron Collider 205–6, 207 Fig. 7.1, 208 Fig. 7.2
large technical systems 302
  see also infrastructures
Latour, Bruno 13, 289
Lavoisier, Antoine Laurent de 10
law 159, 187
Le Corbusier 40, 79–93, 80 Fig. 3.1, 81–84, 87, 92, 94, 96, 115, 117, 118–22, 124, 126, 130, 146, 174, 303, 309
League of Nations 80–81, 84, 89
Leibniz, Gottfried Wilhelm 90, 103
leisure 116, 127, 229, 230, 231 Fig. 7.7
Lely, Cornelis 71, 75 Fig. 2.6
Leonidov, Ivan 97
Lesseps, Ferdinand de 69
Levin, Max 167
Levit, Solomon 167
Leydesdorff, Loet 294
libraries
  Universal Decimal Classification 91
  University of Leuven/Louvain 86, 88 Fig. 3.4
Lindley, William 39–40, 41 Fig. 1.6
Linnaeus, Carl 10
Lippisch, Alexander 181
Lisbon Agenda 17–18, 277–78, 279 Fig. 9.1, 280, 281, 283, 286–87, 307–8
Lloyd Wright, Frank 90
Loewi, Otto 198
logical positivism 98
Lomonosov, Iurii Vladimirovich 69

"Long Twentieth Century, The" x, 6, 8, 9, 11, 12, 18, 268, 281, 305–6, 308, 309, 310
Lundvall, Bengt-Åke 278
Lüst, Reimar 254, 260–61, 264, 267–68, 273
Luxembourg 307, 310
Lyotard, Jean-François 280
Lysenkoism 161, 162 Fig. 5.6, 163–64, 165 Fig. 5.7, 166–69, 170, 171, 173, 305

Maastricht Treaty 283
machine age 137
machines 2, 96, 137, 213, 234
Madariaga, Salvador de 89
Malche, Albert 189
Manhattan Project 198, 210, 220, 221
Manifesto of the Ninety-Three 301–2
Mann, Thomas 89
Marconi, Guglielmo 6, 69, 154, 155 Fig. 5.3, 156 Fig. 5.4, 157, 158 Fig. 5.5, 159–60, 171, 305
master narrative 216, 280–81, 283, 286
master plan 25, 115, 130, 131, 148, 150 Fig. 5.1
May, Ernst 196
media 12, 66, 81, 92, 99, 103, 109, 131–32, 159, 160, 206, 249
Mendel, Gregor 161
Mendelsohn, Erich 89
Merbold, Ulf 263
metric system 3
Meyer, Hannes 82
Meyer, Konrad 147–53, 154 Fig. 5.2, 170, 171, 173, 305
Meyerhof, Otto 198
Michurin, Ivan 164
Mies van der Rhoe, Ludwig 40
migration 44, 100, 187, 195
    forced 197–201, 305
    transatlantic 182
    see also émigré experts; exile
milieu 291
military technology 16, 29–30, 179, 210–11, 231–32
Miller, Oskar von 58, 61 Fig. 2.2
mining 30
Mises, Richard von 198, 199
Mitrany, David 273
mobile museums 100–101

models of technical education 10, 26, 27, 33, 38
modernism 47, 68 Fig. 2.5, 76, 87, 117, 131 Fig. 4.6, 137, 192, 304
    architectural 13, 40, 47, 116 Fig. 4.1, 120, 121 Fig. 4.3, 128–29, 137, 192, 194, 305
    reactionary 139
modernity 22, 56 Fig. 2.1, 66, 123, 162, 187–95, 246, 294
    as "iron cage" 59
    nuclear 219 Fig. 7.3, 229
modernization 22, 30, 34, 39, 45, 69, 127, 134–35, 141, 189, 191–92, 194
Moholy-Nagy, László 117
Mollet, Guy 224
Monnet, Jean 218, 223, 228
Mościcki, Ignacy 69–70
Muldur, Ugur 287–78
Multhopp, Hans 181
Mumford, Lewis 137
Mundaneum 79, 80 Fig. 3.1, 81–82, 83 Fig. 3.2, 84–97, 108, 303, 309
museums
    Austria 100
    Britain 3
    Germany 97–98, 100, 137, 245
    mobile 100–101
    Mundaneum see Mundaneum
    science and technology 57–58, 61 Fig. 2.2
Mussolini, Benito 15, 154–60, 171
"myth of the engineer" 69

Napoleonic Wars 28, 300
nation-building 16, 24, 221
nation states 5, 16, 63, 221, 302–3
National Socialism 141, 147, 153, 170, 173, 182, 201, 305
    see also Nazi Germany; Nazification
nationalism 4, 5, 6, 12, 39, 56, 57, 58, 65, 66, 77, 229, 302, 303
Nazi Germany 13, 111, 141, 143–54, 172–73, 304–5
    Generalplan Ost 148–49, 150 Fig. 5.1, 151, 152
    Generalsiedlungsplan 151
    Gesetz zur Widerherstellung des Berufsbeamtentums 187–88
    Nuremberg Trials 154 Fig. 5.2
    Ordensburg Vogelsang 174 Fig. 5.8

Nazi Germany – *continued*
  Volksdeutsche
    Forschungsgemeinschaften 151
  see also Second World War
Nazification 144
Nelson, Richard 278
Netherlands 29, 46–47, 71, 75 Fig. 2.6, 221, 290
  Nationaal Luchtvaartlaboratorium 145
  Peace Palace 94
  Technische Universiteit Delft 145
networks 226–27, 289–90
  regional 290–91
Neumann, John von 198
Neumark, Fritz 193
Neurath, Otto 14, 97–102, 115, 117, 119, 124, 145–46, 303
neutral fields 214
New Economy (concept) 71–72
nobility 22, 46, 59, 61, 299
  see also aristocracy
Noether, Emmy 198
Nordmann, Alfred 285–86
Norway 26, 221, 227, 290
  Institute of Atomic Energy 224
  space program 255–56
Nowotny, Helga 284–85
nuclear fusion 233–35, 236 Fig. 7.8, 237, 241
nuclear technology 7, 16, 198–99, 205–41

Oelsner, Gustav 192
Office International de Bibliographie 8, 91
Operation Osoaviakhim 177–79
Operation Overcast 181
Organisation for European Economic Cooperation (OEEC) 227
organizations 5–7, 16, 89, 145, 217, 278, 287, 295–97
  economic 172
  international 3–4, 10–11, 17, 81, 103, 226, 228, 267, 295
  leisure 127
  professional 42, 48, 52, 71, 73
  self-help 24
  see also *individual organizations*
Orientalism 189–95, 201 Fig. 6.7
Ostwald, Wilhelm 90, 97, 102–7
Otlet, Paul 8, 14, 79–97, 101, 145, 303, 309
Ottoman Empire 30, 191

Oud, Jacobus P. 40, 137
Ozenfant, Amédée 89

Palumbo, Donato 235
Pabst von Ohain, Hans 181
Paris 3, 7 Fig. 0.3
Parravano, Nicola 160
Peenemünde 8, 182, 183 Fig. 6.2, 185, 186, 187, 201
Perez, Carlota 278
periphery 26, 144, 189, 195, 200, 289, 303
Perret, August 90
physics 7, 69, 90, 110 Fig. 3.8, 144, 155, 157, 168, 181, 188, 189 Fig. 6.4, 198, 244, 249
  nuclear 205–41
Picot, Alfred 213
planning 50, 72, 74, 97–98, 125–26, 129–30, 141, 147, 172, 232, 238, 261
  spatial 148, 149, 150 Fig. 5.1, 151–53, 154 Fig. 5.2
  urban 24, 41 Fig. 1.5, 82, 96, 123, 193
Pohl, Robert W. 189 Fig. 6.4
Poincaré, Henri 105
Poland 21–25, 23 Fig. 1.1, 73–75, 290
  Instytut Politechniczny im. Cara Mikołaja see Warsaw University of Technology
  Intelligenzaktion 144
  Politechnika Warszawska see Warsaw University of Technology
  Sonderaktion Krakau 143–44
  urbanism 124–25
  Warsaw University of Technology 21–25, 23 Fig. 1.1, 26–28, 42, 74–75, 143
Poelzig, Hans 192
Polish Polytechnic Society 44
political autonomy 55
political collaboration 250
political crisis 85, 98, 132–33, 136, 138
political integration 16, 187, 267, 301, 306, 308, 310
political movements
  Communism 141, 170, 172, 305
  Greens 158, 212
  National Socialism 141, 147, 153, 170, 173, 182, 201, 305
  Socialism 72, 166
politics of comparison 1, 4, 7 Fig. 0.3, 38, 50, 124, 304, 309, 310

politics of expertise 12, 16, 71, 287
Polonization 152
polytechniciens 31, 32 Fig. 1.3, 33, 41, 44, 51
Portugal 66, 293
    Canal do Tejo 64 Fig. 2.3
    Escola Politécnica 32–33
    Fascism 172
    School of Engineering (*Instituto Superior Técnico*) 52 Fig. 1.8
positivism 22, 94, 109
    logical 98
poverty 1, 100
practice 4, 9, 11, 15, 25, 38, 39, 48, 71, 85, 124, 148, 151, 164, 171, 188, 216, 268, 273, 286, 310
Praetorius, Ernst 193
Prague 30, 55–57, 58
professional associations 44
professional organizations 42, 48, 52, 71, 73
professionalization 40, 44, 45, 48
professionals 3, 8, 14, 18, 28, 131
    see also experts; specialists
progress 1, 7 Fig. 0.3, 15, 31, 41 Fig. 1.5, 42, 52, 61 Fig. 2.2, 80, 84, 136, 165, 275
    belief in 125
    social 48, 99, 108, 137
    technological 10, 11, 27, 43 Fig. 1.6, 56, 58, 59, 64, 85, 93 Fig. 3.5, 94–95, 99, 106 Fig. 3.7, 120, 138, 139, 286, 303, 310
progressivism 48, 58, 67, 89, 117, 122, 174–75
protectionism 5 Fig. 0.2
protest 145, 164, 189 Fig. 6.4, 199, 232
    antinuclear 233, 240–41
pseudoscience 166
    see also Lysenkoism
public health 71, 74, 76, 141
public sphere 70, 108
purges, Stalinist 13, 144, 168, 196, 214

Rabi, Isidor I. 210
racism 141, 147, 151, 170
    see also anti-Semitism
railroads 1, 3, 4, 10, 28, 47, 66, 68, 69
Randers, Gunnar 225
Rathenau, Walther 71–72
rationalization 52, 106, 129, 172

Rayleigh, Lord 105
reactionary modernism 139
Reagan, Ronald 238, 241 Fig. 7.9, 243
regenerationism 66
Reich, Max 189 Fig. 6.4
Reichsuniversität Strassburg 145
Reinhardt, Max 90
Reitsch, Hanna 184 Fig. 6.3
religion 189, 190
respect 39–50
Reuter, Ernst 192–94, 201 Fig. 6.7
Richet, Charles 90
Riesenhuber, Heinz 260–61
Rietveld, Gerrit 40
road construction 32, 46
Rodrigues, Maria João 278
Röntgen, Wilhelm 105
Roosevelt, Franklin 136
Rossi, Lucio 207
Ruberti, Antonio 282
Russel, John Scott 38
Russia see Soviet Union
Russian Revolution (1905) 24, 47
Russification 22
Rutherford, Ernest 105

Saager, Adolf 105
Saint-Simon, Henri de 51, 76, 300
satellite-based telecommunication 257
Schleyer, Martin 103
Schroedinger, Erwin 188
Schumpeter, Joseph 59, 72
Schütte, Wilhelm 192
Schütte-Lihotzky, Margarete 192
Schwartz, Phillip 190
science 8, 13, 63, 64, 70, 74, 84, 85, 89, 96, 98, 103–5, 108, 122, 140, 152, 160, 161, 171, 188, 198, 200, 209, 211–12
    anti-science movement 147
    nuclear 7, 16, 198–99, 205–41
    see also pseudoscience; technoscience
Science Cities 310
scientific methods 122
Second World War 11, 13, 143–54, 209, 211, 300, 306
Seidel, Heinrich 70
Seraphim, Hans-Jürgen 173
Sering, Max 152
Servan-Schreiber, Jean-Jacques 282 Fig. 9.2, 283

sewer systems 40
shipbuilding 30, 38, 68–69
shipping 63
shop culture 42
"Short Twentieth Century, The" ix
Siemens, Werner 40
skills 15, 27, 31, 33, 38, 40, 42, 46, 61, 73, 97, 178, 179, 182, 191, 211, 245, 252
Skowron, Stanisław 168
Slaby, Adolf 62–63
Slovakia 307
Slovenia 293
social change 90, 96, 108, 139
social class 45
social engineering 71, 139–41
social mission 61, 120
social mobility 46
social reform 76, 100
social unrest 1
social welfare 15, 136
Socialism 72, 166
    see also political movements
socialist realism 110, 120
societal autonomy 174
societal integration 284
society 9–13, 15, 18, 22, 24, 27, 34, 48, 50–51, 59, 63, 66, 71–73, 85, 90, 93–94, 97–100, 117–18, 120, 123, 138, 140–41, 147, 172–73, 189, 193, 196, 216, 229, 246, 293
Soete, Luc 278
Solvay conferences 110 Fig. 3.8
Solvay, Ernest 105
Sonderaktion Krakau 143–44
Soviet Union 69, 161–69
    Bolshevik revolution 144, 162
    émigré experts 195–96, 197 Fig. 6.5, 199 Fig. 6.6
    female engineers 47
    Interkosmos program 243–45, 246 Fig. 8.1, 247
    Joint Institute for Nuclear Research 214–16
    Lenin All-Union Academy of Agricultural Sciences (VASKhNIL) 161
    Lysenkoism 161, 162 Fig. 5.6, 163–64, 165 Fig. 5.7, 166–69, 170, 171, 173, 305
    nuclear technology 214–16, 217–20

Operation Osoaviakhim 177–79
St. Petersburg State Polytechnical Institute 32–33
    see also Stalinism
Sovietization 144
Spaak, Paul-Henri 221
space race 243–75
Spain 66, 221, 294
    City of Culture 310
    collaboration with Germany 225
    Fascism 172
    technical education 33
spatial planning 148, 149, 150 Fig. 5.1, 151–53, 154 Fig. 5.2
specialists 3, 8, 13, 27, 65, 73, 119, 144, 146, 178–79, 182, 195, 201, 215, 218, 222, 228
    see also experts; professionals
specialization 30, 42, 44, 47
Speer, Albert 170
Spengler, Oswald 98
sport see leisure
Stalinism 13, 111, 120–21, 141, 144, 164, 165 Fig. 5.7, 166–67, 179
standardization 3–4, 92, 99, 102, 103, 105, 107, 108, 131, 133
state engineers 31
state and society 11, 12, 27, 50, 51, 173, 189, 193
statistics 117, 124
steam power 158 Fig. 5.5
Stern, Otto 198
Stevenson, Robert 68
Stockholm Exhibition 68 Fig. 2.5, 75–76
Strauss, Franz Josef 269
Stravinsky, Igor 90
Strughold, Hubertus 186
students 24, 25, 30, 31, 33, 38, 74, 145, 172, 278, 308
    female 47–48, 49 Fig. 1.7
superpowers 16, 212, 218, 235, 239, 243, 247, 249, 250, 252, 271
Sweden 221, 290
    Kungliga Tekniska högskolan 48
    space program 255–56
    Stockholm Exhibition 68 Fig. 2.5, 75–76
Switzerland 290
    CERN see CERN

École Centrale des Arts et Manufactures 50
Eidgenössische Technische Hochschule (ETH) 34–35, 38
Exposition nationale 57
Gotthard Tunnel 66–67
Landesausstellung 57
Symphonie satellite program 257, 258 Fig. 8.4, 259
Syrkus, Szymon 124
Szilárd, Leó 188, 198

Tank, Kurt 185–86
Taut, Bruno 192, 201
Taylorism 134, 136, 138
technical drawing 47
technical education 10, 14, 22, 23 Fig. 1.1, 24, 26, 28–32, 33, 34, 38 Table 1.1, 39, 40, 42, 44, 47–48, 51, 57, 60, 62, 70, 73, 107, 145, 192
   by country 36–38
   investment in 67
   models of 10, 26, 27, 33, 38
   see also individual institutions
technical elite 25, 27, 45, 51, 59–65, 74
technical experts 12, 27, 52–53, 71, 73, 77 Fig. 2.7, 108, 170
technical intelligentsia 13
technicians 15, 31, 71, 74, 119, 120, 164, 187, 195, 222, 239
Technische Hochschulen 24, 34, 46, 60–61
techno-celebrities 69
techno-intellectuals 15, 85, 110
technocracy and technocrats 31, 46, 60, 70–5, 96, 125, 130, 137, 138, 151, 152, 172–73, 228, 235, 237
technological heroes 65–66, 70
technology 35 Fig. 1.4
   appropriation of 10, 187, 229
   civilian 169, 183, 211, 221, 223, 229
   diffusion 28, 96
   military 29–30, 210–11, 231–32
   of racism 140, 151
   and skill 15, 27, 31, 33, 38, 40, 42, 47, 61, 73, 97, 178, 179, 182, 191, 211, 245, 252
   social promise of 107–11
technology transfer 179
   see also knowledge
technopolitics 221, 238

technoscience/technoscientific 4–5, 8, 12, 16, 51, 59, 172
   collaboration 108–16, 208–29, 248, 301–2
   expertise 11, 123, 292
   internationalism 218
tectology 138
Teige, Karel 79–85, 110–11, 115, 119, 139–40, 303
Teknillinen Korkeakoulu 25–26
telegraph 3, 66
telephone 131
Teller, Edward 198
Thatcher, Margaret 264
theory 45, 60, 71, 136, 156 Fig. 5.4, 165, 167, 273
Three Mile Island 227
Todorov, Tzvetan 107, 139
Tokamak fusion reactor 235, 237, 241
Tomášovici 127–28 Fig. 4.5
Toqueville, Alexis de 31
totalitarianism 12, 15, 119, 141, 143–75, 305–6
tradition 60, 62–63
training
   academic 30, 45, 60, 279 Fig. 9.1
   astronauts 262 Fig 8.5, 271,
   engineers 22, 26, 27, 28–40, 42, 44–45, 48, 60, 71, 73, 74, 132, 212, 215, 227, 279
   on-the-job 48
   technical 28–32, 36–38, 51
   see also education
transatlantic
   cooperation 240–41, 269
   migration 182
transnational circulation xi, 17, 33
transnationalism 6, 229, 282 Fig. 9.2, 290
transportation 9, 28, 81, 126, 129, 131, 136, 193, 263, 267, 283
   see also specific modes of transportation
Traube, Klaus 233
tunnels 66–67
Turkey, migration to 189–95, 201 Fig. 6.7

Ukraine, Kharkiv Institute 144–45
unification 28
Union of International Associations (UIA) 88–89

United Kingdom *see* Britain
Universal Decimal Classification (UDC) 91
universal language 33, 103, 293
  see also internationalism
universalism 12, 86, 92, 125, 145, 252, 301–3
  utopian 90
universities *see* academia; education; and individual universities
University of Leuven/Louvain, book burning 86, 88 Fig. 3.4
urban planning 24, 41 Fig. 1.5, 82, 96, 123–24, 193
urbanism 47, 84, 94, 95, 115, 124, 136, 191, 193
urbanization 5, 47, 71, 192
USA
  Americanization 137, 182, 183 Fig. 6.2, 281
  hegemony in Europe 211, 218, 220, 240, 252, 269
  see also space race
users 17
utopian universalism 90

Vago, Joseph 130
Valéry, Paul 89, 90
van der Velde, Henry 90
van Doesburg, Theo 40
van Eesteren, Cornelis 124
Vauban, Sébastien Le Prestre de 29
Vavilov, Nikolai 165–66, 167
Veblen, Thorstein 71
Veksler, Vladimir 214, 217
Verne, Jules 70
Vienna Circle 98
Vienna method 100, 117
visual education 97–102
Volapük (universal language) 103

Wagner, Martin 192, 193, 194
Wang Ganchang 214
war ix, 27, 39, 70–77
  see also First World War; Second World War
war socialism 72
Warszawa funkcjonalna 124–25
Watanabe, Kaichi 35 Fig. 1.4
Watt, James 68
Wattendorf, Frank L. 183
Weber, Max 59
weights and measures 3–4
Weisskopf, Victor 198
Wells, H.G. 90–91
Westermann, Franz 134
Weyl, Hermann 198
Wigner, Eugene 198
Wijdeveld, Theodorus 89
Wilson, Denis 237
wireless 28, 63, 69
Wirtz, Carl 225
women
  engineers 47–48, 49 Fig. 1.7
  right to vote 9
workshops 76, 135, 237, 302
World Center of Communication 86 Fig. 3.3, 93–95
  see also Mundaneum Affair
world fairs *see* exhibitions
World Federation of Education Associations (WEFA) 101
World Organization of Chemists 104
world peace 80, 89, 94
World Wars *see* First World War; Second World War
World Wide Web *see* Internet

Zamenhof, Lazar Ludwig 102–3
Zeppelin, Ferdinand Count 69
Zlín, role in Bataism 125–30

Printed and bound by CPI Group (UK) Ltd, Croydon, CR0 4YY